Engineering Materials

This series provides topical information on innovative, structural and functional materials and composites with applications in optical, electrical, mechanical, civil, aeronautical, medical, bio- and nano-engineering. The individual volumes are complete, comprehensive monographs covering the structure, properties, manufacturing process and applications of these materials. This multidisciplinary series is devoted to professionals, students and all those interested in the latest developments in the Materials Science field, that look for a carefully selected collection of high quality review articles on their respective field of expertise.

More information about this series at http://www.springer.com/series/4288

Shaoxian Song · Bowen Li
Editors

Adsorption at Natural Minerals/Water Interfaces

Springer

Editors
Shaoxian Song
School of Resources and Environmental Engineering
Wuhan University of Technology
Wuhan, China

Bowen Li
Department Materials Science and Engineering
Michigan Technological University
Houghton, MI, USA

ISSN 1612-1317 ISSN 1868-1212 (electronic)
Engineering Materials
ISBN 978-3-030-54450-8 ISBN 978-3-030-54451-5 (eBook)
https://doi.org/10.1007/978-3-030-54451-5

This Springer imprint is published by the registered company Springer Nature Switzerland AG
The registered company address is: Gewerbestrasse 11, 6330 Cham, Switzerland

Preface

Adsorption at natural mineral/water interfaces is referred to adsorbates adsorbing on natural minerals in the water. It is not only an important process occurred in the elemental geochemistry, but also widely used as the best method for wastewater treatment to remove contaminants from water. Compared with artificial adsorbents, natural minerals have shown great potentials to be used as adsorbents because of the versatility, low costs and easy operation. As the development of nanosheet minerals technology, natural minerals as effective adsorbents in water will attract increasing attention in many fields.

This book summarized the recent advances on natural minerals as adsorbents in water, including characterization and modification of mineral adsorbents, sorption processes, applications for removing inorganic, organic and microorganism pollutants from contaminated water, as well as recycling of the adsorbents. It is composed of eight well-designed and aligned chapters and is authored collectively by 19 experts from 13 institutions.

There are thousands of kinds of natural minerals on the earth. Only a part of them have unique mineralogical characteristics such as porous structure, high specific surface area and high ion exchange capacity, leading to be potential adsorbents, for example, zeolite and sepiolite. Some layered minerals such as montmorillonite and molybdenite could be exfoliated into two-dimensional nanosheets and then be used as adsorbents for wastewater treatment. In chapter "Mineral Adsorbents and Characteristics", Y. Zhao et al. describe various kinds of minerals used as adsorbents, including the crystalline structure, physical and chemical characteristics, microporous and layered structure, etc. Also, the techniques for characterizing the mineral adsorbents are briefly given.

Surface chemistry is the fundamentals for the adsorption at natural minerals/water interface. In chapter "Surface Chemistry of Mineral Adsorbents", myself and co-authors provide the comprehensive theories related to adsorption at solid/liquid interfaces, including internal and external surface structures, surface charges, surface reactions, adsorption thermodynamics and kinetics and the characterization of adsorptions. The general interactions and mechanisms of the adsorption at mineral/water interface were discussed, including ion exchange, electrostatic interactions, hydrophobic/hydrophilic interactions, ligand exchange, cation bridge, water bridge, etc.

In order to improve the adsorption capacity, natural minerals have to be often treated to modify the surface properties and microstructures. Chapter “Modification of Mineral Surfaces and Microstructures” by S. Bao focuses on the modification of a natural mineral before used as an adsorbent, including adsorption, intercalation, acidic activity, thermal modification, grafting and coating, etc. The commonly used modification agents have also been comprehensively summarized.

As for the perspective adsorbates, pollutants in contaminated water could be classified as anions, cations and organic matters. According to this category, the adsorption at natural minerals/water interface is discussed, respectively, in the adsorption of anionic pollutants, cationic pollutants and organic pollutants, which appear, respectively, in chapters “Adsorption of Anions on Minerals”, “Adsorption of Cations on Minerals” and “Adsorption of Organic Compounds on Minerals”. In chapter “Adsorption of Anions on Minerals”, F. Jia, et al. describe the adsorption of minerals to main anionic pollutants such as arsenic, phosphorus, fluorine and nitrogen in the water. The factors influencing the adsorption of the pollutants on various kinds of minerals and the mechanisms are deeply discussed. Chapter “Adsorption of Cations on Minerals” by F. Rao et al. discusses the adsorption of minerals to heavy metals of lead, copper, cadmium, mercury and chromium and highlights the gaps between fundamental researches and engineering applications for the removal of cationic pollutants from water. Chapter “Adsorption of Organic Compounds on Minerals” by J. Su et al. describes the adsorption of dyes, phenolic compounds, antibiotics and perfluorooctane sulphonate at various kinds of minerals/water interface. The related physical and chemical principles as well as adsorption mechanisms are discussed in detail.

Besides, microorganism pollutants adsorbed on minerals frequently occur in natural environment, while minerals are often used for removing harmful microorganisms from water. Chapter “Adsorption of Microorganisms to Minerals” by L. Xia et al. addresses the phenomenon of bacteria, fungi, algae and biomacromolecules adsorbing onto natural minerals. Also, the applications of microorganism–mineral interaction in environmental engineering such as eliminating heavy metals and contaminated organic matters are given.

The last but not the least, dewatering of used mineral adsorbents for recycling and reuse of the adsorbents is important for the adsorption of the mineral/water interface. Chapter “Dewatering of Mineral Adsorbents” by Y. Li describes the dewatering methods and processes as well as the mechanisms in water/solid separation.

This book will be a good reference for scientists, engineers and graduate students who are working on research and development in the fields of mineral processing, water treatment and environmental-related industries and also for anyone who are interested in the adsorption at natural mineral/water interface.

Finally, we would like to appreciate all the authors for their great contributions and the reviewers for their constructive comments and suggestions. The thanks would be to our former graduate students for their research results that are presented in this book. Also, our gratitude would go to Dr. Mayra Castro and Jialin Yan of the Springer Publishing for their superior handling of this project.

Wuhan, China
Houghton, USA

Shaoxian Song, Ph.D.
Bowen Li, Ph.D.

Contents

Mineral Adsorbents and Characteristics

Yunliang Zhao, Wei Wang, and Hao Yi

Abstract Removal of heavy metals from aqueous solutions has gotten considerable attention over the years by using adsorbents derived from low-cost materials, especially natural minerals. Treating the pollution of wastewater with the natural minerals, has advantages of wide sources of materials, cheap price, low energy consumption, the higher removal efficiency. This chapter introduces the latest research on the natural mineral adsorbents including montmorillonite, kaolinite, sepiolite, diatomite, vermiculite, perlite, zeolite, etc., which are classified as layered minerals, porous minerals and others according to the structural characteristics. The mineralogical characteristics of the mineral adsorbents including pore structure, specific surface area, morphology and ion-exchange property are presented and the effect of the mineral structure on the adsorption performance of the mineral adsorbents is discussed. The surface characterization technologies for determining surface chemical component, surface charge, surface areas, surface morphology and surface interaction are also presented.

Keywords Nature mineral adsorbents · Structural characteristics · Layered minerals · Porous minerals · Characterization technology

1 Introduction

There has been growing necessity to develop technologies for pollution control and environmental remediation, in consequence of the damage of pollution on the health and longevity of both human beings and the Earth's fragile ecosystems.

Y. Zhao (✉) · W. Wang · H. Yi
Hubei Key Laboratory of Mineral Resources Processing and Environment, Wuhan University of Technology, Luoshi Road 122, Wuhan 430070, Hubei, China
e-mail: yunliang286@163.com

Y. Zhao · W. Wang · H. Yi
School of Resources and Environmental Engineering, Wuhan University of Technology, Luoshi Road 122, Wuhan 430070, Hubei, China

S. Song and B. Li (eds.), *Adsorption at Natural Minerals/Water Interfaces*, Engineering Materials,
https://doi.org/10.1007/978-3-030-54451-5_1

Treatment of wastewaters containing polluting species especially the heavy metal ions and organic pollutants become indispensable. Several methods such as chemical precipitation, coagulation–flocculation, flotation, ion-exchange, have been proposed for removal of contaminants from wastewaters. However, most of them become less effective and more expensive in situations involving high volumes and low concentrations. Moreover, these methods can also create sludge disposal problem.

Adsorption is often considered as the best method to treat wastewater due to its universal character, inexpensiveness and easy to use. A good sorbent must be eco-friendly, cost-effective and industrially viable for a wide range of wastewater. But, traditional adsorbents such as activated carbon and ion exchange resins are high cost, preventing their use in industrial applications. The adsorption process using low cost natural minerals is an attractive option because of their efficient removal rate for contaminants at even trace levels. Removal of heavy metals from aqueous solutions has gotten considerable attention over the years by using adsorbents derived from low-cost materials, especially natural minerals.

Natural minerals having unique structure, properties and optimal environmental harmony, are extensively used in various fields of environment pollution treatments. The presence of micropores and canals in the structure, high specific surface and the elongated nature give some natural minerals high sorption capacity for different pollution species. Moreover, treating wastewater using the natural minerals have advantages of wide source, cheap price, low energy consumption, higher removal efficiency. This has already attracted great interest from domestic and international environmental protection scholars at present.

The classification of natural minerals has been proposed according to structural characteristic of natural minerals as follows:

(1) *Layered minerals*: Layered materials, with their structure consisting of stacked sheets, represent a promising material for treating wastewater. Due to the versatility in composition, morphology, and architecture, as well as their unique structural properties (intercalation, topological transformation, and self-assembly with other functional materials), layered materials display great potential in the design and fabrication of adsorbents applied in environmental treatment processes.
(2) *Porous minerals*: Nature porous minerals have been emerging as prospective hybrid materials to prevent and control pollution. Nature porous minerals possess accessible pores, high specific surface area, and good chemical, thermal, and mechanical stabilities. These properties usually provide high permeability and significant selectivity towards different pollution.
(3) *Other minerals*: Other minerals often have stronger surface activity as adsorbents. The surface and chemical property of the minerals determine the ability and type of the reactions happened between mineral surface and pollution.

This chapter is intended to introduce the latest research on natural mineral adsorbents, including montmorillonite, kaolinite, sepiolite, diatomite, vermiculite, perlite, and zeolite, etc. The mineralogical characteristics of the mineral adsorbents including pore structure, specific surface area, morphology and ion-exchange property are presented and the effects of the mineral structure on the adsorption performance of the mineral adsorbents are discussed. The surface characterization technologies for determining surface chemical component, surface charge, surface areas, surface morphology and surface interaction are also presented.

2 Mineral Adsorbents Classification

2.1 Layered Minerals

2.1.1 Montmorillonite

Montmorillonite, an important 2:1 layered silicate mineral, has been used in many industries and has received many interests in terms of practical application due to its swelling, colloidal, rheological, and electrical properties. It is known that a unit of this clay (a primary particle) consists of a quite thin platelet with tetrahedral layers of silicon oxide and octahedral layer formed by aluminum, magnesium, or iron oxide sandwiched (Fig. 1). Its general formula is $(Na)_{0.7}(Al_{3.3}Mg_{0.7})Si_8O_{20}(OH)_4 \cdot nH_2O$. Due to the Al^{3+} substitutes the Si^{4+} in tetrahedral sites and Mg^{2+} substitutes Al^{3+} in octahedral sites, the montmorillonite particles have permanent negative charges, which is balanced by cationic counterions occupying interlayer space [1]. The counterions can be exchanged by other organic/inorganic cations, making clay minerals as efficient adsorbents for cationic contaminants. Besides, the broken bonds located at the edges of the platelet (alumina sheet) have a capacity to adsorb H^+ or OH^-, depending on pH value. This determines the adsorption properties of the clays with respect to various ions.

Original montmorillonite have been used widely as adsorbents in the removal of various contaminants, particularly heavy metal cations, cationic dyes and radionuclides [2–7]. The cation exchange process has been considered as the

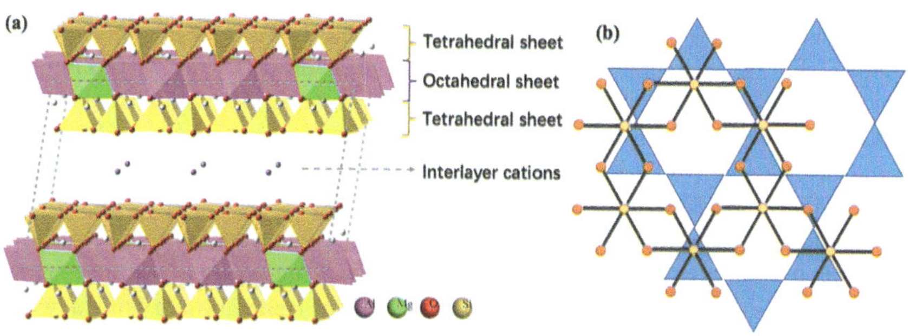

Fig. 1 Schematic MMT platelet structure

primary mechanism for the uptake of cationic contaminants. The adsorption affinity of heavy metal cations follows the order of $Pb^{2+} > Cd^{2+} > Zn^{2+} > Cu^{2+}$ [7]. But the adsorption capacities of heavy metal cations on montmorillonite are varied, depending on the montmorillonite from different sources. Bhattacharyya et al. [8] studied the adsorption of Rhodamine B from aqueous solution using montmorillonite. The adsorption experiments were conducted at a pH range from 2 to 12. Rhodamine B in aqueous solution therefore could be used successfully to separate the dye from water. For the adsorption of organic cations, the hydrophobic moieties of pre-adsorbed organic cations on montmorillonite can have hydrophobic interaction with the counter part of organic cations in bulk water, which can further contribute to the uptake of organic cations from water. As such, the adsorption capacity of many organic cations can be larger than the CEC of montmorillonite. With the relatively high adsorption capacity and expandable interlayer spaces, montmorillonite has also been shown to be efficient adsorbents for radionuclides, including Cs, Sr, Eu, Th, U, Co, Pu, etc. [9].

The ability of montmorillonite to attract and hold contaminants can be enhanced by suitable modifications. The modification methods for montmorillonite mainly includes acidification, thermal treatment, organic/inorganic intercalation [10–16]. Acid-treated montmorillonite, washing by HCl or H_2SO_4, can eliminate mineral impurities and replace exchangeable cations with H^+, which then increases the specific surface area of montmorillonite and enhances its adsorption capacity toward many contaminants. For example, Banat et al. found that the acid-treated montmorillonite possess the better adsorption capacity toward methylene blue dye from aqueous solutions compared with the untreated montmorillonite [17]. Thermal treatment of montmorillonite can change their physicochemical properties through the dehydration and dehydroxylation, which can also be accompanied by the movements of octahedral cations within the octahedral sheet. In addition, thermal treatment may also change the textural properties and influence the dispersibility of montmorillonite in water. These changes can enhance the adsorption capacity of the resulting sample toward cationic contaminants and hydrophobic organic compounds. El Mouzdahir et al. [18] compared the adsorption capacity of methylene blue on montmorillonite at 100–800 °C. In the case of thermally modified clay minerals, the adsorption capacities were found to vary depending on the treating temperature. Modification of montmorillonite by interlayering and pillaring with organic/inorganic show numerous promising applications in contamination prevention and environmental remediation. Fu et al. carried out the adsorption and desorption experiment for the industrial sewage containing chromium using organic modified montmorillonite. The results showed that the hexavalent chromium ion could be effectively adsorbed by the organic modified montmorillonite, compared with sodium-modified and acidification.

Although, various methods have been developed to promote the adsorption capacity of montmorillonite, the limited interlayer spacing do not allow the entrance of macromolecules, resulting in a poor performance. Hence, it is a tendency that turning montmorillonite into nanosheets. Bai et al. [19] have successfully exfoliated the montmorillonite into the nanosheets with the average thickness of 2 nm via

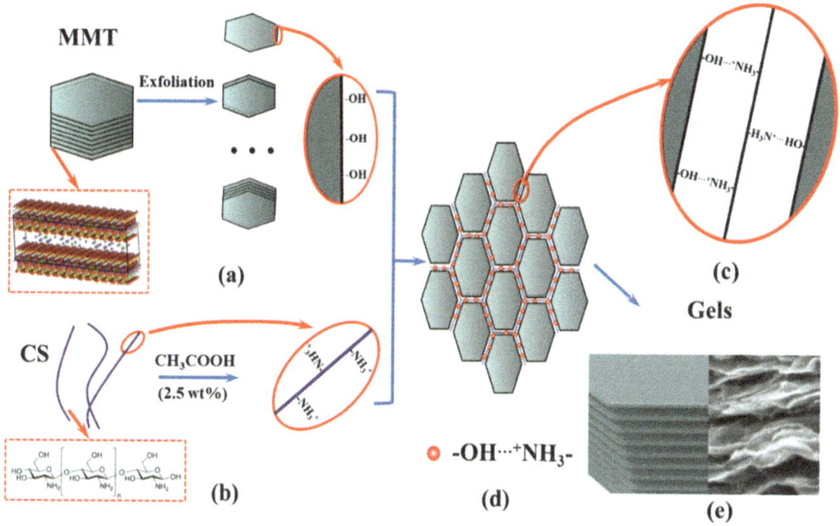

Fig. 2 Self-assembly mechanism schematic of MMTNS/CS hydrogels [21]

ultrasonic treatment. The hydration swelling occurs once montmorillonite was dispersed into water, which weaken the interlayer force [20]. Hence, the layers are easy separated with the appropriate energy input. The montmorillonite nanosheets possess high surface area and abundant adsorption sites, facilitating the removal of macromolecules pollutants that hard to be separated from water. To maintain the excellent properties of montmorillonite nanosheets and further develop the aggregate advantages such as high porosity, excellent mechanical properties, high separability and reusability are current research hotspot. Wang et al. [21] successfully prepared the self-assembled montmorillonite nanosheets/chitosan hydrogel as shown in Fig. 2. The exfoliated montmorillonite nanosheets carried –OH on the edge (Fig. 2a). Chitosan showed electropositive with $-NH_3^+$ in the acid solution (Fig. 2b). Hydrogen bonds (–OH ··· $^+NH_3$–) forming montmorillonite nanosheets and chitosan were mixed together (Fig. 2c) and produced huge plane layer by layer (Fig. 2d, e). The prepared montmorillonite nanosheets/chitosan hydrogel possessed three-dimensional porous structure, facilitating free entrance and combination with the reactive sites existing in montmorillonite.

2.1.2 Kaolinite

As one of the most important clay minerals, kaolinite has diverse industrial applications due to its excellent properties. Kaolinite has 1:1 layer structure with the basic unit consisting of a tetrahedral sheet and an octahedral sheet (Fig. 3). The basic structural element is an asymmetric layer with a chemical composition of $Al_2Si_2O_5(OH)_4$. There is no substitution of Si^{4+} with Al^{3+} in the tetrahedral layer

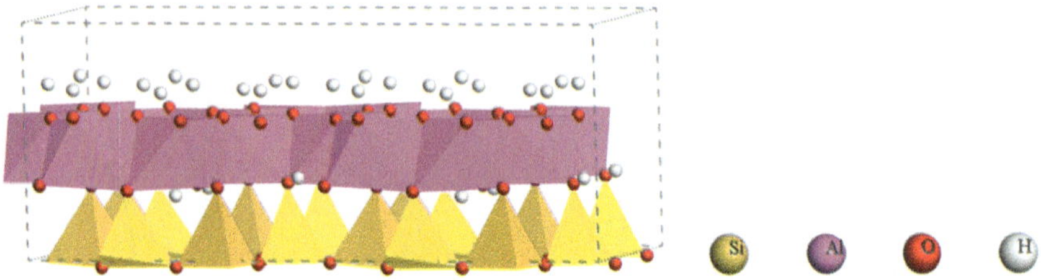

Fig. 3 Schematic kaolinite platelet structure

and no substitution of Al^{3+} with other ions (e.g., Mg^{2+}, Zn^{2+}, Fe^{2+}, Ca^{2+}, Na^{+} or K^{+}) in the octahedral layer. Kaolinite has different surface structures between base planes and edge planes. The charge on the edges is due to the protonation/deprotonation of hydroxyl groups and depends on the pH of the solution. The hydroxyl groups located at the edge planes are considered as the major reactive sites of kaolinite surfaces.

Though kaolinite is less reactive clay, its protonation/deprotonation nature react with different kinds of metal ions according to the solution pH. The metal ion adsorption is usually accompanied by the release of hydrogen (H^{+}) ions from the edge sites of the mineral. The flat exposed planes of the silica and the alumina sheets can also react with metal ions. Properties changes indicate the creation of empty spaces in the clay structure. Similar spaces could also be produced if the adsorption of metal ions produces a reduction in the van der Waals forces between the elements in kaolinite, but it is not certain exactly how these forces are affected.

Kaolinite has been studied extensively as adsorbents in the removal of various contaminants, particularly heavy metal cations, cationic dyes, proteins, etc. [22–27]. Srivastava et al. investigated the adsorption of Cd(II), Cu(II), Pb(II), and Zn(II) onto kaolinite in single- and multi-element systems as a function of pH and concentration [24]. The selective sequence of the adsorption edges of these metals was Cu > Pb > Zn > Cd in single-element system, while it was Pb > Cu > Zn > Cd in the multi-element system. The pretreatment of dye-effluents is indispensable prior to their discharge into the receiving water bodies. Khan et al. found that the kaolinite can efficiently remove the cationic dyes (Rhodamine B dye) from wastewater due to its short adsorption time, high adsorption capacity, natural pH [26]. In addition, kaolinite is also a very suitable adsorbent for the retention of proteins contained on dairy industry by-products. Duarte-Silva et al. found that kaolinite has high retention capacity and different selectivity for three proteins, along with its low cost and absence of toxicity, especially when compared with standard polymeric resins used for protein adsorption [27].

The ability of kaolinite to attract and hold contaminants can be enhanced by suitable modifications, such as intercalation, grafting of organic groups, acidification, or by thermal treatment [27–31]. Kaolinite surface can be rendered hydrophobic by reacting with organo-functional molecules. Organic derivatives of clays are generally obtained by using silane coupling agents. After surface modification, the organic groups can be attached to clay by chemical bonding, adsorption and coating. Koteja and Matusik used di- and tri-ethanolamine grafted kaolinites to remove the Cd(II), Zn(II), Pb(II) and Cu(II) in equilibrium and kinetic experiments [29]. The ions were attracted by the grafted diethanolamine or tri-ethanolamine, which are N and O-donors and readily form complexes with metals, particularly with Cu(II). Thermal treatment can also change the kaolinite surface through dehydroxylation. Duarte-Silva et al. [27] found that the electrical charge of the kaolinite surface was modified by thermal dehydroxylation, and a secondary pore system made of inter particle voids was created, which led to drastic changes on the adsorbent selectivity, decreasing the retention capacity of bovine serum albumin (10 mg/g sample) and increasing the adsorption of α-lactalbumin (176 mg/g sample). Acid treatment of kaolinite with concentrated inorganic acids is also an important modification method. Such treatments can often replace exchangeable cations with H ions and release Al and other cations. It was reported that acid activation followed by thermal treatment increases the adsorbent capacity to a good extent. Bhattacharyya and Gupta investigated the adsorption of Pb(II) onto kaolinite and its acid activated derivative in aqueous medium [10]. Acid activated kaolinite has higher adsorption capacity compared to the untreated kaolinite due to the increased surface area and pore volume.

2.1.3 Vermiculite

Vermiculite $(Mg, Fe^{2+}, Fe^{3+})_3(Al, Si)_4O_{10}(OH)_2 \cdot 4(H_2O)$ is one of the most known layered silicate mineral with hydrophilic, negatively charged and layered crystalline structure (Fig. 4). Vermiculite has high cation exchange capacity generated as a consequence of the cations with lower valency substituting their main cations, such as Al^{3+} instead of Si^{4+} and Mg^{2+} instead of Al^{3+}. Different with montmorillonite, most of the cation substitutions in vermiculite take place in tetrahedral sheets which limits the potential of this mineral to expand its interlayer space. Vermiculite is one low-cost, high-effective adsorbent for various

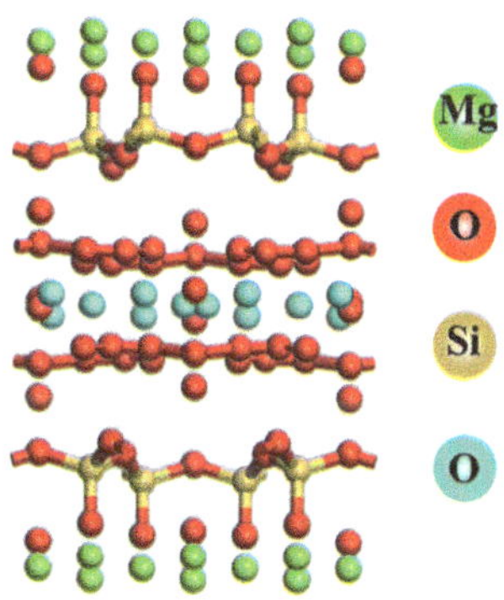

Fig. 4 Schematic vermiculite platelet structure

contaminants, due to their high cation exchange capacity and high specific surface area associated with their small particle sizes.

Cation exchange in the interlayer is mostly considered as the main mechanism for adsorption of metal cations by vermiculite, where ions interact with negative permanent charges. In addition, metal cations also interreact with Si–O and Al–O groups at the mineral particle edges by forming inner- and outer-sphere complexes [32–34]. The adsorption performance of various heavy metal onto vermiculite followed the order of $Pb^{2+} < Cu^{2+} < Cd^{2+} < Zn^{2+} < Ni^{2+} < Mn^{2+}$ [34]. Meanwhile, vermiculite also has been extensively studied to remove dyes from effluents. The adsorption capacity of cationic dyes onto natural vermiculite was dependent on the dyes' molecular structures. Yu et al. [35] compared the adsorptive interactions of methylene blue and crystal violet onto natural vermiculite in aqueous medium. The adsorption isotherm of methylene blue decline apparently and did not meet the Langmuir model. However, the adsorption of crystal violet exhibited a typical Langmuir isotherm under the same initial concentration.

The ability of vermiculite to attract and hold contaminants can be enhanced by suitable modifications. Yu et al. produced chemically activated vermiculite by leaching with HCl or H_2O_2 solution for Pb^{2+} and Cd^{2+} removal [35]. As indicated from Differential Scanning Calorimetry tests, the chemically activated vermiculite show higher thermal stability than that of raw vermiculite treated with the same ions. The Pb^{2+} and Cd^{2+} ions removal efficiency by peroxide activated vermiculite was higher than that of the acid activated vermiculite.

In general, the untreated clay minerals are not suitable to remove hydrophobic pollutants due to their hydrophilicity. The surface properties of clay minerals can be changed from hydrophilic to hydrophobic by exchanging with organic cationic surfactants. Liu et al. [36] synthesized three novel organic vermiculites using different amphoteric surfactants to remove stubborn organic pollutant, e.g., bisphenol A and tetrabromobisphenol A. In addition, compared to dodecyl dimethyl (N-carboxylate) ammonium-Vermiculite and dodecyl dimethyl (3-sulphonate) ammonium-Vermiculite, dodecyl dimethyl (N-phosphate) ammonium-Vermiculite can be preferred used as an inexpensive, renewable and efficient adsorbent for removal of organic contaminants in the environmental treatment.

2.1.4 Illite

The illite mineral $(K, H_3O)(Al, Mg, Fe)_2(Si, Al)_4O_{10}[(OH)_2, (H_2O)]$ is a 2:1 layered silicate mineral with potassium as the interlayer cation (Fig. 5), with considerable ion (isomorphic) substitution being occurred. Different to montmorillonite, the largest charge deficiency of illite is in the tetrahedral sheet rather than in the octahedral sheet. The interlayer potassium ion is not readily exchangeable due to the well fit of potassium in the interlayer position. This structure has a strong interlocking ionic bond to hold the individual layers together and prevents water molecules occupying the interlayer space as it does in the montmorillonite.

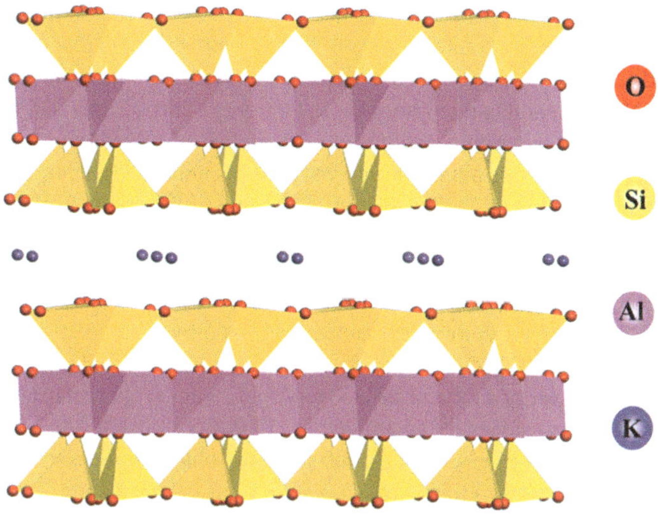

Fig. 5 Schematic illite platelet structure

Illite is often used as a good adsorbent for a variety of pollutants such as heavy metal cations, dyes, and radionuclides [37–42]. Thus most studies focus on using nature illite to remove heavy metal ions from aqueous solutions such as Cu, Sr, Ni, Co, Eu, and Sn, and Cd. Echeverría et al. [40] studied the retention mechanisms of Cd on illite. Below pH 6, ion exchange is the main mechanism for Cd adsorption on illite. Above pH 6, the proton stoichiometry was greater than that at pH values < 6. In addition, an increase in the temperature favored the retention of Cd. Liu et al. investigated the sorption of radionuclide Co(II) on illite as a function of pH, ionic strength, concentration and foreign ions and temperature in the presence and absence of humic acid under ambient conditions by using batch technique [41]. The sorption test indicated that the cost-effective illite is suitable for pre-concentration of Co(II) from large volumes of aqueous solutions. Fil et al. [43] used illite clay as an adsorbent to remove methyl violet dye from solutions. The results from enthalpy change indicated that dye uptake occurred by physical binding. The maximum adsorption capacity of 159.95 mg g^{-1} was found at 60 °C, indicating that illite would be effective good for cationic dye removal. However, Illite drew much less attention compared to other clay minerals primarily due to its low adsorption capacity compared to swelling clay minerals.

2.1.5 Rectorite

Rectorite ($(Na, Ca)Al_4((Si, Al)_8O_{20})(OH)_4 \cdot 2H_2O$) is a kind of uncommon layered silicate mineral that has regularly interstratified structure with alternate pairs of dioctahedral mica-like layer (nonexpansible) and dioctahedral montmorillonite-like layer (expansible) in 1:1 ratio. Due to the presence of smectite layer in rectorite, its adsorption performance would be similar to illite and montmorillonite. The dominant interlayer cation of the mica component is used to subdivide the rectorite group into Na-, K- and Ca-rectorite. Due to their large surface area and high cation exchange capacity, rectorite can be used as potential adsorbents to remove heavy metals and organic contaminants from water, especially after being modified by organic compounds [44–47]. Chang et al. [46] applied the Na-rectorite to removal Ni(II) from aqueous solutions under ambient conditions. The pH value and ionic strength is the main factor for removing Ni(II) and the adsorption capacity increased with increasing pH and decreasing ionic strength. Under acidic conditions, the sorption of Ni(II) on Na-rectorite is mainly dominated by cation exchange, whereas at neutral to alkaline pH values surface complexation is the main sorption mechanism. Rectorite also can be used to remove common pharmaceuticals from groundwater and wastewater. Chang et al. [46] investigated the sorption and intercalation of tetracycline from aqueous solutions onto nature rectorite. Rectorite has high affinity for tetracycline, with a sorption capacity of 140 mg g^{-1} at pH 4–5.

Thus, rectorite modified by cationic surfactant could be potentially used as adsorbents to remove organic contaminants and anionic contaminants from water [48–51]. Hong et al. [48] used stearyl trimethylammonium chloride modified rectorite as adsorbents to remove Cr(VI), showing a sorption capacity of 21 g/kg at pH 4. The uptake of negatively-charged chromate ions in the interlayer regions of stearyl trimethylammonium chloride modified rectorite is probably due to electrostatic force derived from the polarized water molecules and positively-charged groups of the stearyl trimethylammonium cations.

2.1.6 Mica

Micas are widespread in igneous, metamorphic and sedimentary rocks. Their crystal structure accommodates a plethora of elements, leading to a large and diverse mineral group. Micas are subdivided into true micas or brittle micas based on the I cations. Moreover, they may be subdivided into dioctahedral and trioctahedral micas. Micas are hydrated layered silicate mineral, a group of minerals with the extremely variable chemical composition of general crystallochemical formula: (K, Na) $(Al, Fe, Mg)_{2-3}$ $[(Si, Al)_4 O_{10}]$ $(OH, F)_2$. Their crystal structure (Fig. 6) is composed of negatively charged 2:1 layers consisting of two tetrahedral layers of silicon oxide between which one octahedral layer formed by aluminum, magnesium, or iron oxide is sandwiched. In the case of true micas, these interlayer cations are usually univalent, whereas in brittle micas they are divalent.

In addition, numerous studies have been performed to remove metals using mica [52–56]. Charnock and England establish the removal extent of solution containing copper(II), cadmium(II), and lead(II) metal ions using muscovite and biotite [52]. The results indicated that surfaces of muscovite and biotite can strongly interact with particularly Cu(II) and Pb(II) in mildly acidic aqueous solutions. Biotite also strongly interacts with Cd(II) in solution. Paulino et al. [53] tested the effectiveness

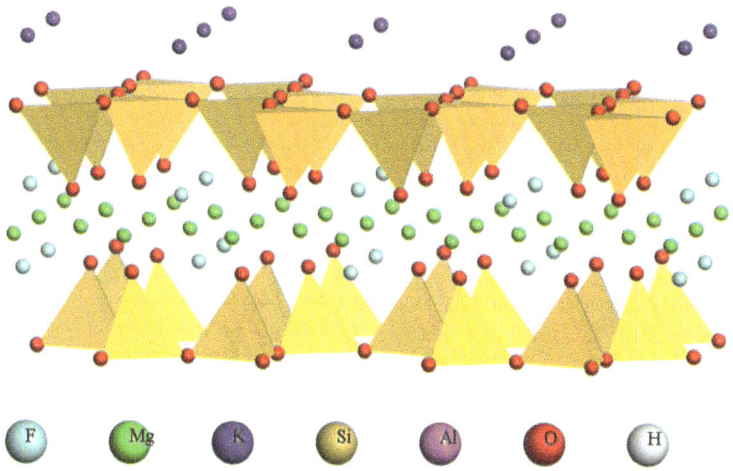

Fig. 6 Schematic mica platelet structure

of biotite (black mica) to remove pesticides from water. The results indicated that more than 80% of all the pesticides were adsorbed, reaching maximum adsorption in acidic conditions (pH 3) at 6 h.

Due to the low cation exchange capacity (CEC), nature micas must be modified as commercial and cost-effective adsorbents to separate radionuclides from groundwater or aqueous nuclear wastes. The low CEC of micas is due to interlayer potassium ions. Komarneni studied the cation exchange selectivity for Cs^+ and Sr^{2+} with K-depleted biotite (Na-biotite) and K-depleted muscovite (Na-muscovite) at room temperature [57]. The results indicated that both K-depleted micas show high selectivity (up to approximately 50% of their theoretical cation exchange capacities) for Cs^+. However, K-depleted biotite shows better selectivity for Cs^+ with high Cs^+ concentrations. These results suggest that K-depleted biotite could be used as an adsorbent to remove radioactive ^{137}Cs as well as ^{90}Sr from groundwater. Yamamoto et al. [58] examined the effect of interlayer cation exchange on the adsorption behavior of Cs^+ using the modified mica in aqueous suspension. The results indicated that phlogopite treated with $BaCl_2$ for a short period (2 h) exhibits equivalent ability to adsorb Cs^+ compared with phlogopite treated with a conventional reagent such as sodium tetraphenylborate (NaTPB). Since $BaCl_2$ is a much cheaper reagent than NaTPB and the treatment period is very short, this finding would be valuable to provide a practical method to remove radioactive Cs^+ from the environment.

2.1.7 Molybdenite

Molybdenite is a typical layered transition metal dichalcogenide (TMD). Each layer consists of two sulfur sheets and one internal molybdenum sheet, forming a S–Mo–S structure (Fig. 7). The thickness of each molybdenite unit layer is around

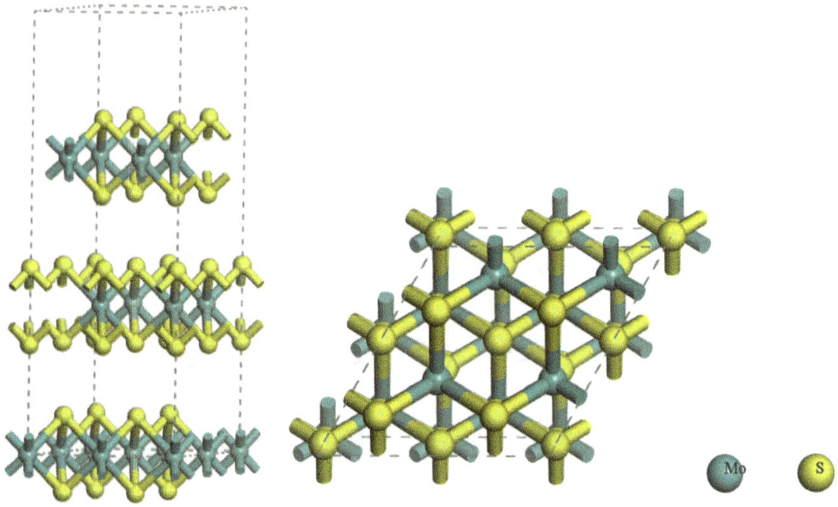

Fig. 7 Schematic molybdenite platelet structure

0.62 nm, leaving a 0.30 nm free spacing between layers [59, 60]. The Mo and S atoms in layers link each other through strong chemical bonds, while each monolayer forming the bulk molybdenite is through weak van der Waals force. According to the computation by Stillinger-Weber (SW) potential, the lattice parameters of molybdenite for the in-plane unit cell is a = 3.09 Å, and the Mo–S bond length is b = 2.39 Å [61]. Molybdenite exhibits three phases (octahedral 1T phase and trigonal prismatic 2H and 3R phases) depends on the atomic arrangement and interlayer stacking. The unit cell exhibits octahedral coordination in 1T phase with one sandwich layer. 2H phase and 3R phase both display trigonal prismatic coordination, i.e., 2H phase has two sandwich layers and 3R phase has three sandwich layers. The 2H phase molybdenite is semiconductor, which is also thermodynamically stable and commonly exists in natural environment, while those with 1T and 3R phases are metallic due to the itinerant electrons, which are rarely found in nature [62, 63]. However, the three phases could be transformed from one to others under special conditions. For example, 1T-MoS_2 and 3R-MoS_2 can be turned into 2H-MoS_2 under annealing [64], while 2H-MoS_2 could also transform into 1T-MoS_2 when introducing Li ions or electron irradiation during the exfoliation of bulk 2H-MoS_2 [65]. With the decreasing of layer numbers, the band gap of molybdenite transforms from indirect band gap to direct band gap, and the energy of the bandgap increases from 1.1 to 1.9 eV [66].

By virtue of the 1.9 eV direct band gap, molybdenite possesses tremendous potentials as field effect transistors, chemical sensors, phototransistors, solar cells, etc. [67, 68]. Furthermore, recent researches have found that molybdenite exhibits excellent performance in the environmental field, which is even much better than graphene-based nanomaterials [69–71]. The superb remedial property of molybdenite as environmental material benefits from its unique two-dimension structure and particular elementary composition [72]. Except for the ultra-high specific surface area, molybdenite has abundant S atoms exposed on the surface and transition metal Mo in the structure. The intrinsic S atoms make molybdenite a promising adsorbent in removal of heavy metals due to the good affinity of S to heavy metals [73–76], while the existence of Mo displays marvelous catalytic performance in degradation of organic contaminants [77].

2.2 *Porous Mineral*

2.2.1 Diatomite

Diatom, distributed widely on Earth, is a single-celled alga with a very small size of only ten to dozens of microns. Deformity of diatomite, such like disciform, acicular, linear, penniform can be found under high-power microscope. Commonly, the main mineral component in diatomite was opal ($SiO_2 \cdot H_2O$) and the content of SiO_2 is supposed to over 60%. It is white or gray for diatomite while it turns to taupe or sepia once the impurity (Al_2O_3, Fe_2O_3, CaO, MgO) increased [78].

Diatomite is widely used as adsorbent material mainly for two reasons, i.e., high surface area with porous structure and abundant surface functional groups [79, 80]. There are interior and exterior two layers to make up the shell of diatom. Studies show three types of pitted perforation on diatomite, i.e., puncta, stria and costa, respectively [81]. The results show that all three pitted perforations are made up of micropores with different size. The puncta refer to the pores with biggish size and there are micropores with certain distance between the biggish pores. Pure and dried diatomite possesses a relative density of 0.4–0.9 g/cm^3 and it has large pore size distribution and high porosity, which can adsorb the liquid that 1.5–4 times weight then itself. Moreover, diatomite has a high chemical stabilization, which make it insoluble in strong acid and strong base except hydrofluoric acid [82]. Meanwhile, diatomite can be regarded as a solid acid due to the abundant silicon hydroxyls and hydrogen bonds existed both in the micropores and surface, which confer upon the particle an overall negative charge. The porous structure and negative charge enable surface adsorption for metal cations.

Due to the high surface area with porous structure and abundant surface functional groups of diatomite, diatomite has been mostly used to deal with dye wastewater [83–86]. The purified diatomite only has a adsorption capacity of approximately 27.86 mg/g for methylene blue [87], while diatomite composited with other material can improve the adsorption capacity markedly [87–89], e.g., diatomite synthesized with activated carbon has a adsorption capacity of methylene blue, reaching 505.1 mg/g [90]. The heavy metal ions were also the important targets for diatomite to deal with [91, 92]. The adsorption capacity of Cu^{2+} and Cd^{2+} on diatomite was 22.7 and 10.05 mg/g, while diatomite grafted with materials possess abundant functional groups have a huge promote on adsorption capacity, such as the adsorption capacity of La^{3+} and Th^{4+} can reach 139.5 and 74.07 mg/g [93, 94]. Except for dye and heavy metal ions, contaminants like arsenic, [95] benzene, [96] phosphate, [97] tetracycline [98–100] and ammonia [101] can also be removed by diatomite.

The adsorption capacity of diatomite is closely related to the high surface area and the silicon hydroxyls in micropores, especially the quantity of hydroxyl [102]. Hence, once the environmental conditions changed, the adsorption capacity of diatomite can be changed. In addition, the hydroxyl in diatomite possesses a certain activity, which can graft some other functional groups and change the adsorptive property of diatomite. This change makes the diatomite to enhance the adsorption capacity by proper chemical modification. Part of the hydroxyls will hydrolyze to lose hydrino, resulting in a negative charge surface of diatomite in the liquid-solid system. For some electropositive colloidal pollutants, negative charge diatomite can make it destabilization by charge neutralization [103]. Excitation and modification can help diatomite obtain higher adsorbing capacity for both organic, inorganic colloidal particles, which can reach 3–4 times weight then itself. Hence, modification for diatomite is necessary, which can realize the destabilization for both electropositive and electronegative colloidal solids and greatly improve the water treatment effect [104].

2.2.2 Zeolite

Zeolite, which gains the name due to the boiling inflation caused by the sharply gasification of water by heating, was found by a Swedish mineralogist from vesicular basalt in 1756. SiO_2 is the main component to constitute the structure of zeolite with the existence of Na^+, Ca^{2+}, Al^{3+} and a small quantity of K^+, Mg^{2+} etc. Zeolite is constituted by silica tetrahedron, in which the Si atom occupies the center with oxygen atom distributed in four vertexes, showed as Fig. 8. The Si^{4+} in tetrahedron can be superseded by Al^{3+} and destroy the electric neutrality, resulting in a negative charge. To neutralize this negative charge, K^+, Na^+, Ca^{2+}, Mg^{2+} etc. are usually adsorbed on the surface of zeolite, which endow it with the ion-exchange property [105]. Moreover, silica (alumina) tetrahedrons can interconnection via oxygen bridge to form various annular shapes in the two-dimension plane. Then these annular shapes will be connected again by oxygen bridge and form plenty of three-dimensional structures with well-regulated shapes (Fig. 8). This structure can extend infinitely in three-dimensional space, resulting in a biggish and porous channel structure [106]. Over 80 types of zeolite have been found, e.g., clinoptilolite, mordenite, analcite, heulandite, erionite, scolecite and natrolite are distribute widely. With the deep study of zeolite, 3000 natural zeolite deposits have been ascertained from approximately 40 countries [107].

The distinctive structure makes the zeolite be a promising adsorbent in sewage treatment. First of all, the cations (Ca^{2+}, Mg^{2+} etc.) in the pore canals are connected with silica skeleton by an unstable Van der Waals force, which can be exchanged by other cations (Na^+, K^+, Ba^{2+}, Cu^{2+} etc.), suggesting that zeolite can be used as an

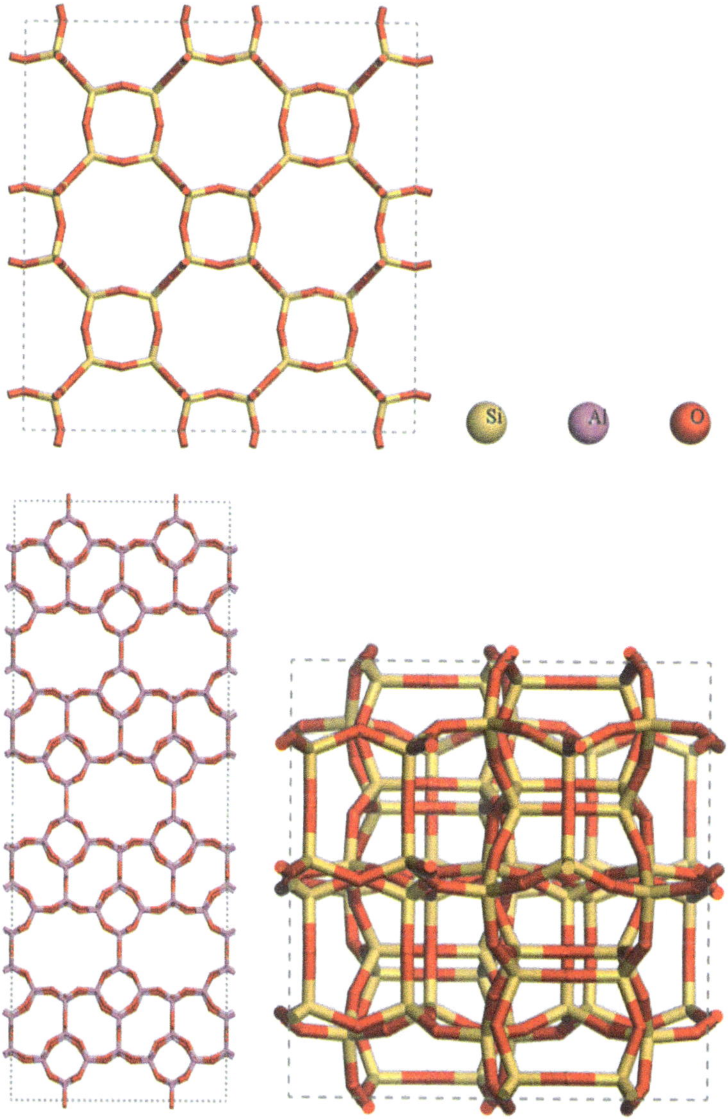

Fig. 8 Schematic structure of zeolite

ideal adsorbent to deal with cationic wastewater. The ion-exchange would not destroy the structure of silica tetrahedron, so it is possible to recycle zeolite by desorbing. Secondly, the diameter of the pore in zeolite is normally 0.3–1.0 nm, which can only allow the adsorbate with diameter below this size to get in and hold back the big ones, indicating that the characteristic adsorption for some adsorbates can be realized [108]. In addition, the exchangeable cations in pore canal will generate a local electric field, which can preferentially adsorb the polar molecules or polarizable molecules. Lastly, zeolite possesses a prodigious internal surface area, which can generate a biggish diffusive force. And there are quantities interconnected pores and aisles with uniform size in the interior of zeolite. This characteristic makes the zeolite have a specific surface area up to 500–1000 m^2/g, providing sufficient surface area for adsorbates to attach and makes the adsorption easier.

Zeolite or its compounds are widely used in wastewater treatment, especially the removal of heavy metal ions, [80, 109–111] organic agents and dyes, [112–114] phosphate and ammonia [115]. Yu et al. systematically studied the adsorption performance of copper cations (Cu(II)) from aqueous solution by natural clinoptilolite-rich zeolite powders modified with a bio-inspired adhesive, polydopamine (PDA). The maximum adsorption capacities of Cu(II) were shown to be 14.93 mg g^{-1} for pristine natural zeolite and 28.58 mg g^{-1} for PDA treated zeolite powders [116]. Kim et al. [117] synthesized a zeolite–nanoscale zero-valent iron composite as an adsorbent to remove Pb(II) from aqueous solution and the capacity of Pb(II) was about 806 mg g^{-1}. Cross-linked beads of activated oil palm ash zeolite/chitosan (Z-AC/C) composite were prepared through the hydrothermal treatment of NaOH activated oil palm ash followed by beading with chitosan, which was used to remove methylene blue and acid blue 29 dyes, showing corresponding maximum adsorption capacities of 199.20 and 270.27 mg g^{-1}, respectively [118]. Liu et al. [119] modified natural zeolites with N,N-dimethyl dehydroabietylamine oxide and used to adsorb anionic dye congo red from aqueous solution, presenting a maximum monolayer adsorption capacity of 69.94 mg/g. Natural zeolite modified by incorporation of hydrated aluminum oxide (HAlO) can be used to remove the phosphate and ammonium simultaneously [115].

2.2.3 Sepiolite

Sepiolite with a transition structure of chainlike and stratiform fiber is one new-fashioned natural mineral absorbent (Fig. 9). The theoretical structural formula of sepiolite can be expressed as $Mg_8Si_{12}O_{30}(OH)_4(OH_2)_4 \cdot 8H_2O$ [120]. The over-all structure of sepiolite is formed by the intertexture of pieces and pore canals extended along the crystal axis of fibers. In addition, every piece constitutes two tetrahedral SiO_2 with one octahedral MgO in the middle [121]. This crystal structure endows sepiolite a high specific surface area (800–900 m^2 g^{-1} in theoretical) and satisfactory chemical and mechanical stability, which shows better properties of adsorption, rheology and catalysis comparing with zeolite, bentonite, vermiculite etc. Hence, sepiolite has been widely applied in the fields of petroleum, chemical industry, medicine, architecture, light textile, environmental protection and traditional agriculture etc. [122].

The connecting channels and voids of sepiolite provides a biggish specific surface area and pore volume, thereby showing a good physical adsorption capacity. Except the physical adsorption, the chemical adsorption also exists and

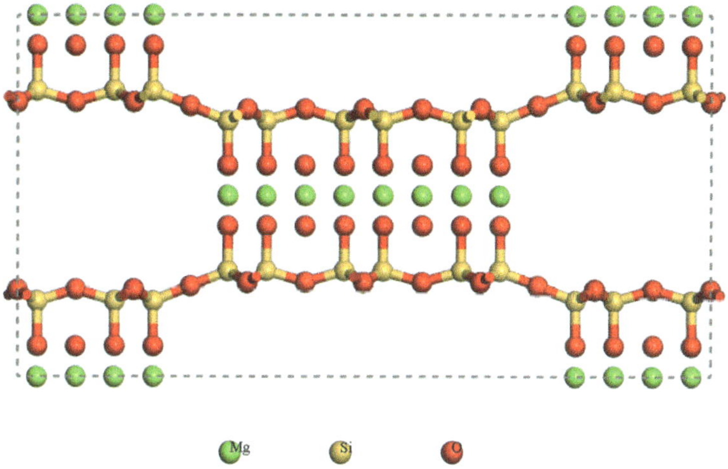

Fig. 9 Schematic sepiolite structure

plays a dominant role. There are three main reasons to adsorb by chemical bond [123]. Firstly, the oxygen atom in the silica tetrahedron layer provides abundant weak charge, which can adsorb impurities by electrostatic attraction. Secondly, the coordinated water molecules of Mg^{2+} on the Lattice edge can form hydrogen bonds with adsorbates. Lastly, the Si–OH generated by the breakage of Si–O–Si bonds on the surface of silica tetrahedron will trigger the complexing reaction with the adsorbates [124]. In addition, Si–OH bond can form covalent bonds with some organic reagents. These three adsorption forms can enhance both the physical and chemical adsorption capacity. Moreover, MgO_6 and SiO_4 on the surface of sepiolite form many alkaline and acid centers, possessing quite strong polarity and adsorbing polar materials easily.

Many researches test the adsorption property of sepiolite for wastewater, indicating that sepiolite could treat the wastewater containing heavy metal ions [125–127], organic pollutants [128, 129] and dyes [130–132]. Yu et al. [133] synthesized a novel magnetic Fe_3O_4/sepiolite composite (MFSC), which had an adsorption capacity of 30.85 mg g^{-1} for Eu(III), via chemical co-precipitation method. Habish et al. [134] used natural and partially acid-activated sepiolites and nanoscale zerovalent iron to prepare composites with an adsorption capacity of approximately 133 mg g^{-1} for Cd(II). The adsorption capacities of sepiolite were 132 mg g^{-1} for marbofloxacin, and 112 mg g^{-1} for Enrofloxacin [135]. The maximum pyrene adsorption capacity of sepiolite was 8.98 mg g^{-1} at 313 K [136]. The monolayer adsorption capacity was 143 mg g^{-1} for malachite green on plasma treated sepiolite by Kaya et al. [137]. Marrakchi et al. [138] prepared cross-linked chitosan/sepiolite composite, which was used as adsorbent to remove methylene blue and reactive orange 16, having the maximum adsorption capacities of 40.986 and 190.965 mg g^{-1}, respectively.

However, the specific surface area and pore size of natural sepiolite are not as large as theoretical calculation and the acidity as well as thermostability are weak, limiting its practical application in a certain extent. Hence, activated modification for sepiolite is necessary to enhance its performance to meet the requirements of various industries. The mainly modification methods for sepiolite include acid activation, ion-exchange activation, hydrothermal activation, high temperature activation and organic modification etc. [139].

2.2.4 Halloysite

Halloysite, named by Berthier in 1826 to commemorate the Belgian scientist Omalius d'Halloy for the discovery of it from carboniferous limestone, is one of the natural clay silicate minerals. Halloysite has a similar structure and chemical composition to kaolinite, the difference is that there are more water molecules in the interlamination of halloysite [140]. Moreover, the kaolinite is commonly lamellate while it is mainly tubular for halloysite, although flake and globularity are also reported. The structural formula of halloysite can be expressed as $Al_2O_3 \cdot 2SiO_2 \cdot 4H_2O$ or $Al_2Si_2O_5(OH)_4 \cdot 2H_2O$, which is obtained from the comparison of halloysite and kaolinite by the results of XRD (X-Ray Diffraction) and TGA (Thermogravimetric Analysis). Pure halloysite is white, while the mineral mixed with impurities (Fe^{3+}, Cr^{3+}, Ti^{4+} ions that may substitute Al^{3+} or Si^{4+} in the crystal texture) will be colored from yellowish to brown [140]. Halloysite has a tubular morphology without end-capping, curly burst or casing, presenting as a natural porous nanocrystalline material. The halloysite nanotubes (HNTs), generally transformed from crystal minerals like feldspar, mica or kaolinite etc., have a length of 0.02–30 μm and a breadth of 0.05–0.2 μm [140]. However, halloysite has a lower density of surface hydroxyl compared with kaolinite.

In halloysite structure, there are Si–O tetrahedron arranged outside and Al–O–OH octahedral inside to form the nanotube with an inner diameter of 10–30 nm [141, 142]. There are Al–OH inside the nanotube while Al–OH and Si–OH distributed on the edge. The functional group outside the nanotube is O–Si–O (ca. 1030 cm^{-1}) certified by FTIR (Fourier Transform Infrared Spectroscopy). The thickness of the bilayers in hydrated mineral is about 1.03 nm while it was 0.72 nm in dehydrated rolls. The dimensions and lengths of the halloysite tubes are unique, depending on the deposit. The external diameter of the tubes is 50–200 nm and the internal diameter varies from 10 to 50 nm [140, 143]. Halloysite is a porous material with abundant hollow tubes, resulting in a huge specific surface area. In addition, the Al–OH on the surface has higher activity. These characteristics endow the halloysite strong adsorption capacity.

Nowadays, halloysite is mainly used to deal with wastewater containing pollutants like ammonia-nitrogen, heavy metal, dye or toxic agents [144–146]. Shu et al. prepared the halloysite-derived mesoporous silica nanotubes with a specific surface area up to 608 $m^2\ g^{-1}$ from the natural halloysite via alkali-treating and acid-etching. The optimized mesoporous silica nanotubes behave favorable monolayer adsorption with a capacity of 618 mg g^{-1} for methylene blue, showing as a potential adsorbent in the future [147]. Massaro et al. [148] used halloysite clay and organic cyclodextrin derivatives prepared the inorganic–organic nanosponge hybrids, presenting an obvious promotion on adsobing dyes. Fard et al. [146] make the amine-terminated dendritic functional groups successfully grow on the surface of Halloysite Nanotubes to improve their adsorption tendency towards the anionic dye molecules. Halloysite nanotubes modified by 3-aminopropyltriethoxysilane can quickly remove Co(II) from solution [149]. Alkaline-assisted pre-activated halloysite has an adsorption capacity of 123.05 and 227.70 mg g^{-1} for Ag^{+} and Pb^{2+}, respectively. New adsorbent, which is prepared by in situ growing Fe_3O_4 nanoparticles on halloysite nanotubes followed by subsequent modification with silane coupling agents, can remove Cr(VI) and Sb(V) Simultaneously [150].

2.2.5 Attapulgite

Attapulgite is one kind water-bearing magnesium-alumina silicate mineral with a catenulate and layered structure. Attapulgite is also called Palygorskite, which was first found from Palygorsk in Russia [151]. Then more of this mineral was found in Attapulays (U.S.A.) and Mormoriron (France) and called attapulgite. Attapulgite possesses a 2:1 (two layers of silica tetrahedron with one alumina/aluminum hydroxide octahedral in the middle) stratiform structure, which combines with sepiolite to form a catenulate and layered structure (Fig. 10) [152]. There is an amphibole-type double chain extending along the c-axis direction, then connected by tetrahedron, forming a circle. The lamination occurs in b-axis direction depending on the periodical inversion of reactive oxygen atom in tetrahedron, resulting in a zeolite-type channel with the size of 0.37 × 0.6 nm paralleled to the chain direction. These channels can be filled with organic matter or water molecules in an ordered

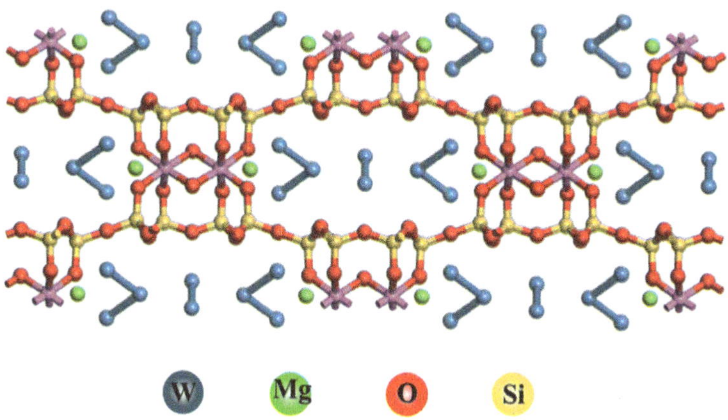

Fig. 10 Crystal structure of palygorskite

arrangement and forming a coordination with Mg^{2+} on the edge of octahedron. The empirical formula of attapulgite is $Mg_5(Si_4O_{10})_2(OH)_2(OH_2)_4 \cdot 4H_2O$ in general [153].

There are abundant microtubules in attapulgite, resulting in a huge internal surface area. Meanwhile, the crystal particles of attapulgite are fine (only dozens of nanometer), so that the external surface area of attapulgite is biggish as well. According to the attapulgite fiber model proposed by Serna in 1978, the internal surface area was about 600 m^2 g^{-1} and external surface area was 300 m^2 g^{-1} theoretically [154]. Due to the huge surface area, attapulgite shows an excellent adsorptive property. In 1978, three kinds of active center of adsorption were classified based on the structure and crystallochemistry characteristic of attapulgite by Serartosa: (i) negative charge caused by the substitution of Si^{4+} by $A1^{3+}$, showing an adsorption driving force to adsorbates, (ii) water molecules coordinating with Mg^{2+} on the edge of octahedron can form hydrogen bond with adsorbates and (iii) the Si–O–Si bond breaking and making the crystal electronegative, then the hydroxyls (–OH) adsorbed on the defects to equilibrate the charge. These Si–OH groups can form covalent bonds with adsorbates on the surface of attapulgite.

The common usage of attapulgite in adsorption field is to remove heavy metal ions from water [155, 156]. Tang et al. fabricated the manganese dioxide/carbon/attapulgite ternary composites via a facile hydrothermal method based on the spent bleaching earth and the prepared composites has a maximum adsorption capacity of 166.64 mg g^{-1} for Pb(II) [157]. A novel rod-like nanocomposite, poly(acrylic acid) brushes-decorated attapulgite, was synthesized by Li et al. [158], and its Ce(III) adsorption capacity reached 295.4 mg g^{-1}. Raw attapulgite (APT) and a novel adsorbent, struvite/attapulgite (MAP/APT) obtained from nutrient-rich wastewater treated by MgO modified APT, were applied as the adsorbent for Cd(II) ion removal from aqueous solution were studied by Wang et al. [159] and the adsorbing capacity of APT and MAP/APT were 10.38 and 121.14 mg g^{-1}, respectively. Some radioactive heavy metal ions like Eu(II), [160] Sr(II) [161] and U(VI) [162, 163] etc. can also be well treated by attapulgite or its compounds.

The adsorption capacity of attapulgite is proportional to its specific surface area. Hence, increasing the specific surface area is a good way to enhance the adsorption capacity of attapulgite. Heat and acid treatments can enlarge the specific surface area of attapulgite significantly. Acid can gradually dissolve the octahedron, which mainly located on the surface of mineral fibers. The acid dissolution forms the uncrystallized SiO_2 as well as the Si–OH group, dissociating the fiber bundles of attapulgite. These increase the microporosity and specific surface area of micropores, thereby enhancing the adsorption capacity of attapulgite. In addition, the method of squeezing and pressing can also enlarge the specific surface area of attapulgite because the shear force is too strong to break the electrostatic attraction force among the fiber bundles and make the fiber bundles smaller.

Sun etc. prepared a composite structured CPA (chitosan-modified purified attapulgite) composed of uniform PA (purified attapulgite) nanorods modified by chitosan under acetic acid conditions, and the maximum adsorption capacity of humic acid could be 112.07 mg g^{-1} [164]. The superparamagnetic attapulgite/Fe_3O_4/polydopamine (APT/Fe_3O_4/PANI) nanocomposites, [165] which was facilely prepared by a one-pot process without the nitrogen protection, can be used as an adsorbent for the removal of methylene blue, and the adsorption ratio toward 100 ppm of methylene blue could reach 95.8%.

2.3 *Other Minerals*

2.3.1 Calcite

Calcite (Fig. 11) is one of the most abundant salts in the environment and is of fundamental importance in both inorganic and biological system. The most stable polymorph of calcium carbonate ($CaCO_3$) has Mohs hardness of 3 and a specific gravity of 2.71. Like most carbonates, calcite can be dissolved in most of acids. It can be either dissolved or precipitated by groundwater, depending on the water temperature, pH, and dissolved ion concentrations. Calcite exhibits a unique characteristic-retrograde solubility which is less soluble in water as the temperature increases.

Fluoride water contamination caused by mineral processing industries or chemical industry has become a worldwide concern due to its severe threat to living organisms, in particular humans [166, 167]. Calcium-based sorbents are one of an effective materials for fluoride removal via adsorption process [168]. Yang et al. [169] used a fixed bed filled with calcite particles to treat the synthetic wastewater containing fluoride and found that fluoride was precipitated as CaF_2. Turner et al.

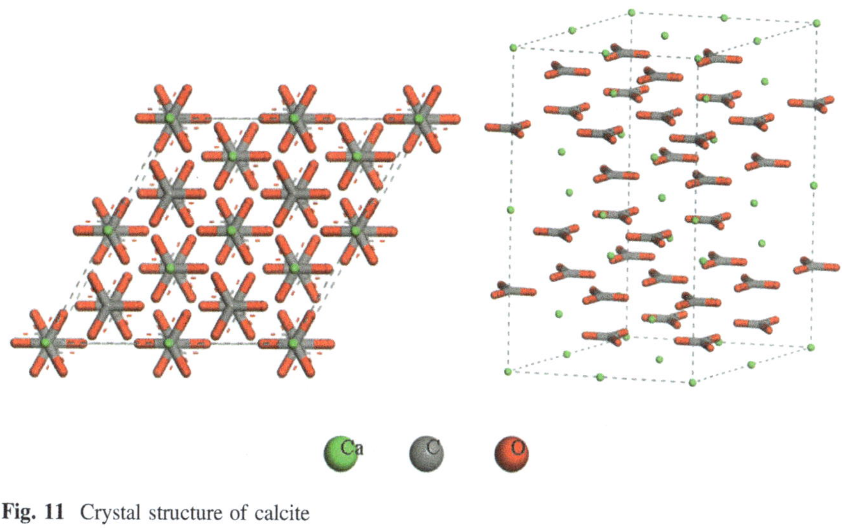

Fig. 11 Crystal structure of calcite

[170] applied calcite ($CaCO_3$) as well as CO_2 addition to remove fluoride from contaminated groundwater. It was found that fluoride removal can be enhanced by injecting CO_2 into the calcite. The mechanisms of calcite dissolution, CaF_2 precipitation and fluoride adsorption have been widely investigated in recent years. Reardon et al. [171] concluded that fluoride removal was firstly achieved by calcite dissolution in water at a certain pH range, which released calcium ions into the water. After that calcium ions interacted with fluoride in water and formed CaF_2 precipitates. Turner et al. [170] reported that the removal of fluoride by calcite was not only due to precipitation but also surface adsorption. Results indicated that CaF_2 precipited at step edges and kinks, whereas fluoride adsorption occurred directly over the entire calcite surface.

In addition to fluoride, calcium carbonate precipitation is sensitive to impurity ions, such as iron and zinc even at trace concentration levels. Zachara et al. [172] reported the adsorption of Zn^{2+} on calcite through exchanging with Ca^{2+} in a surface-adsorbed layer on calcite. Analogously, Zachara et al. [173] also investigated the sorption of seven divalent metal ions (Ba, Sr, Cd, Mn, Zn, Co, and Ni) on calcite. The results indicated that divalent metal ion sorption was dependent on aqueous Ca concentration, and the following selectivity sequence was observed: Cd > Zn $\geq$ Mn > Co > Ni > > Ba = Sr. The sorption was also due to the surface exchange reaction between divalent metal ion and Ca ions on cation-specific surface sites. Calcite was also used to treat wastewater containing heavy metal ions and phosphate [174]. Kobayashi et al. [175] investigated the adsorption of divalent heavy metal ions on calcite surface at 25 °C. Jones et al. investigated the adsorption of As(V), As(III) using calcite as well as the impact of calcium and phosphate. Van Der Weijden et al. [176] studied the influence of phosphate and sulfate on the sorption of cadmium on calcite.

2.3.2 Tourmaline

Tourmaline is one of the research hotspots in the field of nonmetallic mineralogy, particularly its environmental chemistry behavior. Tourmaline belongs to the group of silicate minerals called cyclosilicates. Its Mohs hardness is 7–7.5 and density is 3.02–3.25 g/cm^3. The environmental chemistry behavior of tourmaline are related to its structure, which is characterized by a set of boron triangles, a silicate ring of six tetrahedral (Fig. 12). The general formula of tourmaline is expressed as $XY_3Z_6Si_6O_{18}(BO_3)W_4$, where the X site is usually occupied by a variety of cations, such as Ca, Na and K, Y can be occupied by Fe, Mg, Mn, Al, Li, and Ti, and the Z site is usually occupied by Al (but is also frequently replaced by Fe^{2+}, Fe^{3+}, Ti, Mg, Cr, and V^{3+}. The W site is occupied by OH, but it can also be replaced by F and O [177]. The species of tourmaline can be divided into three categories according to the chemical composition: (1) Schorl, the most common species of tourmaline is schorl, may account for 95% or more of all tourmaline in nature [178]. The formula of schorl is $NaFe_3Al_6[Si_6O_{18}](BO_3)_3(OH_4)$, its color is brownish black to black. (2) Dravite, the sodium magnesium rich tourmaline endmember, with a formula of

$NaMg_3Al_6[Si_6O_{18}](BO_3)_3(OH_4)$ Its color is dark yellow to brownish black. (3) Elbaite (formula $Na(Li, Al)_3Al_6[Si_6O_{18}](BO_3)_3(OH_4)$). Its color is various: red or pinkish-red (rubellite variety), light blue to bluish green (Brazilian indicolite variety) and green (verdelite or Brazilian emerald variety). Tourmaline has spontaneous and permanent poles, which can produce an electrostatic field and a permanent electric dipole as well as adjust solution pH automatically [179]. Therefore, tourmaline is a type of functional material. Under the action of electric field, water molecules are electrolyzed and produced active molecule H_3O^+ and OH^-, which could adsorb impurities, dirt as well as purifying water quality. That is the foundation and driving force for exploitation and application of tourmaline in the field of environmental protection, air purification, printing, as well as wastewater treatment.

Jiang et al. [180] studied the adsorption capacity of heavy metals such as Cu(II), Pb(II), Zn(II) and Cd(II) ions on tourmaline. The efficiency of heavy metal ions removal by tourmaline was in order of Pb(II) > Cu(II) > Cd(II) > Zn(II). Liu et al. [181] used tourmaline to adsorb Cd(II), Zn(II) and Ni(II) from aqueous solutions in binary and ternary component systems. The results indicated that tourmaline is high selectively toward one metal in a two-component or a three-component system with an order of Cd(II) > Zn(II) > Ni(II). Liu et al. [182] investigated the behavior and efficiency of the removal of the diazo Direct Red 23 (DR23) dye from aqueous solution using powdered tourmaline (PT). The results indicated that the OH functional group of PT might be the major surface functional group for dye adsorption. The effective removal for DR23 by PT was achieved through physisorption followed by a stage of slower molecular diffusion. Wang et al. [179] investigated the potential of tourmaline as an adsorbent for removing Pb(II) from acidic water. The adsorption of Pb(II) was strongly dependent on temperature, particle size, and dose of absorbents, but not on pH. Results showed that tourmaline may be explored as a new material for removing pollutants from the environment. After that, Wang et al. [177] investigated the effects of time, temperature, and initial concentration of metal ions on removing heavy metal ions by tourmaline from acidic aqueous solutions.

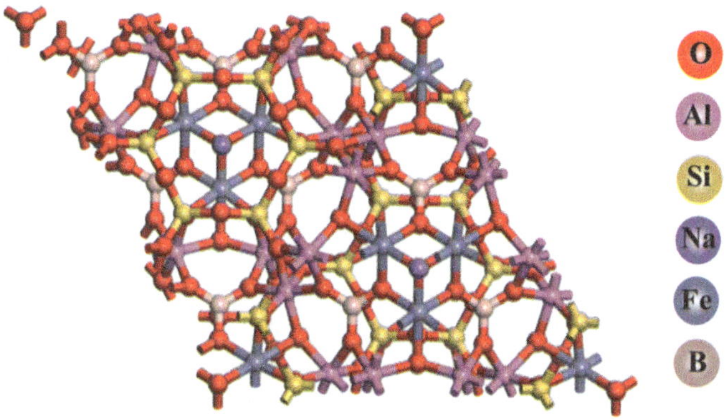

Fig. 12 Crystal structure of tourmaline

The results indicated that the adsorption of Cd(II), Zn(II), and Ni(II) depended significantly on all the above-mentioned parameters.

Furthermore, Li et al. [183] reported that La-modified tourmaline achieved high efficiency for phosphate adsorption. Results showed that over 90.0% of phosphate adsorption occurred within the initial 15 min. Wang et al. [184] synthesized a polyvinyl alcohol semi-IPN poly (acrylic acid)/tourmaline (PVA semi-IPN PAA/Tm) composite as the adsorbent for Pb^{2+} and Cu^{2+} removal. Wang et al. [185] investigated the catalyzed degradation ability of tourmaline on the dyes methylene blue (MB), rhodamine B (RhB), and congo red (CR) as function of pH values. Tokumura et al. [186] studied the photoassisted Fenton discoloration of azo-dye Orange II with tourmaline powder. The efficiency of discoloration increased as dye concentration decreased and the tourmaline concentration increased. Li et al. [187] investigated the effects of different conditions on desorption of Cd(II) from tourmaline at acidic conditions. Results showed that the desorption capacity of tourmaline was too low to cause secondary pollution, indicating tourmaline is quite an effective adsorbent for the removal of heavy metals in the acidic condition.

2.3.3 Rutile/Anatase

Rutile, tetragonal oxidation mineral, is a mineral composed primarily of titanium dioxide (TiO_2). Rutile has a tetragonal unit cel and the titanium cations surrounded by an octahedron of 6 oxygen atoms (Fig. 13). The oxygen anions have a co-ordination number of three resulting in a trigonal planar co-ordination [188]. Brookite and anatase are the two polymorphs of rutile [189, 190].

The photoactivity of TiO_2 is one of the most attractive properties, and TiO_2 is an excellent photocatalyst material for environmental purification [191, 192], due to its unique advantages e.g. biological and chemical resistance, safety, low price and affordability, environmental friendliness and so on [193, 194]. Fujishima and Honda in 1972 discovered that water on TiO_2 electrodes can be photocatalytically split into hydrogen and oxygen [188]. The fundamental processes in enhancing the photocatalytic efficiency of TiO_2 have attracted enormous research interests.

Generation of electron-hole pairs is the main initial process for photocatalysis of organic by photocatalyst. When TiO_2 is under UV irradiation with energy equal to or higher than the band gap energy, usually 3.2 eV (anatase) or 3.0 eV (rutile), electrons were excited from the valence band to the conduction band and the holes were generated at the valance band. The electrons in conduction band are good reducer and the holes in valence band are good oxidizer. The holes react to H_2O or OH^- absorbed at the surface and the hydroxyls with high oxidability are formed. Obviously, the recombination of electrons and holes are detrimental to the photocatalytic efficiency of photocatalyst. The fundamental mechanisms of degradating different surrogate organic compounds (e.g. phenol, chlorophenol, oxalic acid) have been extensively investigated in the photodegradation over TiO_2 surface. Organic contaminants were degraded over TiO_2 surface, resulting in the products of

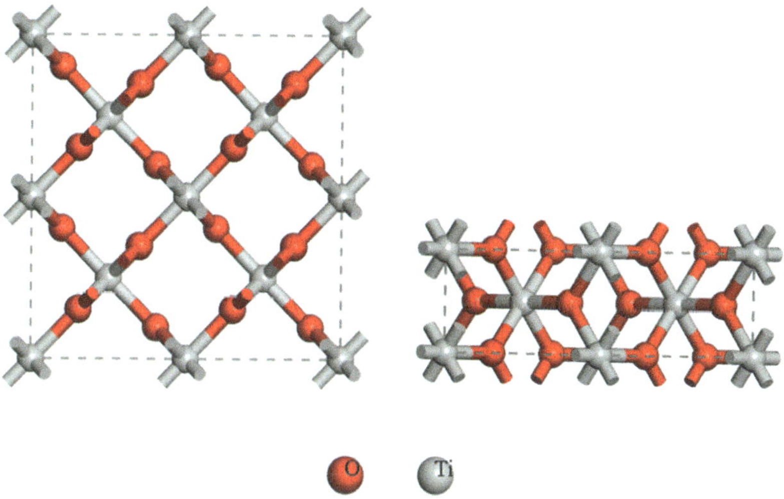

Fig. 13 Crystal structure of rutile

intermediates, and mostly aldehydes and carboxylic acids will be further carboxylated to produce innocuous carbon dioxide and water [192], as shown in Eq. 1:

$$\text{Organic Contaminants} \longrightarrow TiO_2/h\upsilon\text{Intermediate(s)} \rightarrow CO_2 + H_2O \quad (1)$$

The above mentioned photocatalysis reaction (Eq. 1) can be divided into five independent steps:

1. Mass transfer of the organic contaminants in the liquid phase to the TiO_2 surface.
2. Adsorption of the organic contaminants onto the photon activated TiO_2 surface.
3. Photocatalysis reaction for the adsorbed phase on the TiO_2 surface.
4. Desorption of the intermediates from the TiO_2 surface.
5. Mass transfer of the intermediates from the interface region to the bulk fluid.

TiO_2 photocatalysis has been exhibited as a promising technology for complete mineralization of various organic matter in water treatment. Heller [195] reported that TiO_2 is excellent for photocatalytically breaking down organic compounds. Lee et al. [196] investigated the adsorption and photodegradation of humic acids (HA) by homemade nano-structured TiO_2. Results showed that high-efficiency HA adsorption on the TiO_2 surface was achieved at low pH condition and increased cation strength.

Asuha et al. [197] synthesized a mesoporous TiO_2 for the removal of methyl orange (MO) and Cr(VI) from waste waters. It was found that the adsorption for MO is slightly influenced by solution pH, while that for Cr(VI) is strongly dependent on solution pH. López-Munoz et al. [198] evaluated titania and titanate nanotubes as adsorbents for the removal of Hg(II) from aqueous solution.

Light source is one of important factors for influencing the photocatalytic efficiency of TiO_2. Khataee et al. [199] studied the efficiency of TiO_2 in degrading C.I. Basic Blue 3 (BB3) under ultraviolet (UV) light irradiation. In another work, Damodar et al. [200] reported the efficiency PVDF/TiO_2 membrane for the degradation of Reactive Black 5 (RB5) under UV light. Yang et al. [201] successfully synthesized an anatase TiO_2 nanoparticles for the degradation of methyl orange using a sol–gel method. The authors investigated the effects of low pH value and H_2O_2 addition on the photocatalytic degradation of methyl orange in TiO_2 suspensions under UV light. It was found that the photodegradation rate decreased with an increase in pH value of the solution and varied with different amounts of H_2O_2. The obtained pure anatase TiO_2 showed better photocatalytic activity towards methyl orange decolorization than biphase TiO_2.

Girginov et al. [202] studied the photocatalytic efficiency of Ag-doped TiO_2 for the degradation of methyl orange. It was found that silver modified TiO_2 nanoparticles increased the photocatalytic activity in contrast to the standard titania. Besides, Behnajady et al. [203] found that Ag-doped TiO_2 nanoparticles was faster at degrading AR88 photocatalytically than undoped TiO_2. Moreover, Wang et al. [204] reported Au-loaded TiO_2 (Au/TiO_2) as a sonocatalyst for the first time. The results indicated that the Au/TiO_2 has both accelerated discoloration and total organic carbon (TOC) removal of azo dyes such as Orange II, Ethyl Orange and Acid Red G as compared to bare TiO_2 and nano-Au catalyst. Xiu et al. [205] reported that the loading of Cu_2O on the TiO_2 surface could also greatly enhance the photocatalytic effect of the catalyst for the degradation of methyl orange. Jia et al. [206] synthesized a necklace-like MgO/TiO_2 heterojunction structures, which exhibit an excellent removal performance towards toxic As(V) ions combined with good photocatalytic propertiesto organic dyes. Liu et al. [207] proposed one-step method to remove Cr(VI) and Cr(III) with mixture of TiO_2 and titanate nanotubes. Gurboga et al. [208] synthesized a TiO_2–SiO_2 mixed gel spheres for efficient removal of Sr(II) from aqueous and radioactive waste solutions. Nilchi et al. [209] synthesized TiO_2/SiO_2 binary oxide, which is efficient for Pb^{2+} removal from aqueous solutions. Zhou et al. [210] synthesized TiO_2@MoS_2 heterostructure, which showed a high photocatalytic hydrogen production even without the Pt co-catalyst. Moreover, such a heterostructure showed efficient ability for organic dyes removal.

2.3.4 Magnetite

Magnetite (Fig. 14) is one of the main iron ores, with a chemical formula of Fe_3O_4. Magnetite owning black or brownish-black with a metallic luster, has a Mohs hardness of 5–6 and leaves a black streak. It has no cleavage and is of poor reducibility. Magnetite is widely distributed, and there are many causes of formation. Sweden Kiruna is the typical case for the deposit which is formated due to magmatic origin. China Daye iron mine is the typical case for the deposit which is

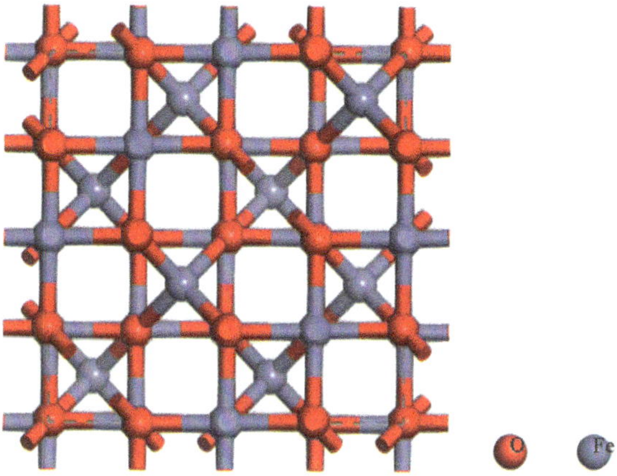

Fig. 14 **a** Face-centred cubic spinel structure of magnetite. **b** Magnification of one tetrahedron and one adjacent octahedron sharing an oxygen atom

formated due to the contact metamorphism. Magnetite is the main raw material for iron smelting.

Magnetite has been widely used in water treatment, such as the removal of heavy metals and dyes in water. Wang et al. [211] investigated the interaction of aqueous As(III) with magnetite during its precipitation from aqueous solution at neutral pH as a function of initial As/Fe ratio. Gao et al. [212] synthesized hierarchical magnetite nanochain assemblies with complex building units, which possess superparamagnetic property and good water-dispersibility. Therefore, the Fe_3O_4 nanochains exhibit excellent ability to remove an organic pollutant from wastewater. Wang et al. [213] successfully synthesized a self-assembled magnetite (Fe_3O_4) peony structures with petal-like nanoslices, showing an high adsorption capacity for Cr(VI) removal from aqueous solution. Magnetite nanoparticles were widely used in the arsenic removal with excellent ability [214]. An et al. [215] prepared and tested a new class of starch-bridged magnetite nanoparticles for removal of arsenate. Results showed that the ability of arsenate adsorption was enhanced compared to bare magnetite particles.

Larraza et al. [216] prepared a hybrid material constituted by MMT, PEI, large Fe_3O_4 NPs for the removal of hexavalent chromium from water. It was found that the hybrid materials might be a potential magnetic adsorbent for Cr(VI) removal from water in a wide range of pH. Shaker et al. [217] synthesized the magnetite nanoparticles for Congo Red removal, which showed a high efficiency of removing Congo Red from wastewater. Ayad et al. [218] synthesized a magnetic chitosan–polypyrrole–magnetite (Cs–PPy–Fe_3O_4), which exhibited a high adsorption capacity for anionic dye removal. Moreover, it is easy to separate Cs–PPy–Fe_3O_4 nanocomposite after adsorption by using an external magnet. Shim et al. [219] synthesized 3D hierarchically structured magnetite/carbon microspheres, which exhibited a superior adsorption capacity of organic pollutants and toxic heavy metals. Huang et al. [220] investigated the molybdate removal using a hybridized zero-valent iron/magnetite/Fe(II) (hZVI) system. The hZVI system could overcome conventional ZVI surface passivation problem, providing an effective chemical platform to harness the reactivity of ZVI for molybdate removal. Sharma and Tiwari [221] prepared a nanomagnetite-loaded poly hydrogel which exhibited effective adsorption capacity for Mn^{2+} removal from contaminated water.

3 Characterization Technology of Mineral Absorbents

3.1 X-Ray Diffraction (XRD)

X-ray diffraction (XRD) is a non-destructive analytic technique primarily used for phase identification of a crystalline, providing information on unit cell dimensions [222]. This technique is based on collecting the diffracted rays of an X-ray beam hitting a sample. A key component of all diffraction is the angle between the incident and diffracted rays.

Following Bragg's description of X-ray diffraction, crystal can be assumed as made of infinite planes containing atoms and whose orientations are ependent from the crystal unit cell parameters. Each family of planes will then diffract according to Bragg's law [223]:

$$\mathrm{n}\lambda = 2d\sin\theta \tag{2}$$

where n is the "order" of the plane, λ is the wavelength of the X-ray, d is the distance between the planes of the family, and θ is the angle between the wave and the plane. Due to their relationship with the spatial arrangements of the atoms inside a periodic structure, the acquisition of the diffracted rays from a material having an internal atomic periodicity can give important insights into the atomic structure of the material itself.

X-ray powder diffraction is most widely used for the identification of unknown crystalline materials such as minerals and inorganic compounds. Determination of unknown solids is critical to studies in geology, environmental science, material science, engineering and biology. Other application includes characterization of crystalline materials, identification of minerals such as clays, determination of unit cell dimensions, measurement of sample purity. Besides, it can be used as quantitative analysis of minerals.

3.2 X-Ray Fluorescence (XRF)

X-ray fluorescence (XRF) is a non-destructive analytical technique used to determine the elemental composition of materials [224]. XRF analyzers determine the chemistry of a sample by measuring the fluorescent (or secondary) X-ray emitted from a sample when it is excited by a primary X-ray source. Each of the elements present in sample produces a set of characteristic fluorescent X-rays that is unique for that specific element, which is why XRF spectroscopy is an excellent technology for qualitative and quantitative analysis of material composition.

A solid or a liquid sample is irradiated with high energy X-rays from a controlled X-ray tube. When an atom in the sample is struck with an X-ray of sufficient energy (greater than the atom's K or L shell binding energy), an electron from one of the atom's inner orbital shells is dislodged. The atom regains stability, filling the vacancy left in the inner orbital shell with an electron from one of the atom's high energy orbital shells. The electron drops to the lower energy state by releasing a fluorescent X-ray. The energy of this X-ray is equal to the specific difference in energy between two quantum states of the electron. The measurement of the energy is the basis of XRF analysis.

XRF is the emission of characteristic "secondary" (or fluorescent) X-rays from a material that has been excited by bombarding with high energy X-rays or gamma rays. The phenomenon is widely used for element analysis and chemical analysis, particularly in the investigation of metals, glass, ceramics and building materials, and for research in geochemistry, forensic science, archaeology and art objects such as paintings and murals [225].

3.3 *ICP-AES*

Inductively coupled plasma atomic emission spectroscopy (ICP-AES), also referred to as inductively coupled plasma optical emission spectrometry (ICP-OES), is an analytical technique used for the detection of trace metals [226]. It is a type of emission spectroscopy that uses the inductively coupled plasma to excite atoms and ions that emit electromagnetic radiation at wavelengths characteristic of a particular element. It is a flame technique with a flame temperature in a range from 6000 to 10,000 K. The intensity of this emission is indicative of the concentration of the element within the sample.

The ICP-AES is composed of two parts: the ICP and the optical spectrometer. And the basic principle of ICP-AES test is as following: An aqueous or organic sample was placed into an analytical nebulizer where it is changed into mist and introduced directly inside the plasma flame. The sample immediately collides with the electrons and charged ions in the plasma and is itself broken down into charged ions. The various molecules break up into their respective atoms which then lose electrons and recombine repeatedly in the plasma, giving off radiation at the characteristic wavelengths of the elements involved. Within the optical chamber(s), the light intensity is measured after the light is separated into its different wavelengths (colours). The intensities of all wavelengths (within the system's range) can be measured simultaneously, allowing the instrument to analyze for every element to which the unit is sensitive all at once. Thus, samples can be analyzed very quickly. The intensity of each line is then compared to previously measured intensities of known concentrations of the elements, and their concentrations are then computed by interpolation along the calibration lines.

The application of ICP-AES includes the determination of metals in wine [227], arsenic in food [228], and trace elements bound to proteins [229]. ICP-AES is often used for trace elements analysis (metal elements and the quantitative part of non-metallic elements) in soil, with low detection limit, good stability, and multi element analysis, high resolution, vertical and horizontal observation etc.

3.4 AFM

The atomic force microscope (AFM), which belongs to a series of scanning probe microscopes with high advanced technique, was invented in 1986 [230]. AFM possesses the properties of high spatial and force resolutions, which can down to sub-nanometer and piconewtons resolution, respectively. Moreover, quantitative information about surface physicochemical properties of minerals such as elasticity, friction or adhesion forces [231–233] can be delivered by AFM. Figure 15 showed the schematic of AFM, a tip mounted to a cantilever spring is used to scan the sample. During scanning, the deflection of the cantilever, which follows the Hooke’s law stating (deflection is proportional to the cantilever spring constant), can deliver the force between the tip and the sample, then the topographic image of the sample can be obtained by plotting the deflection [234]. The deflection of

Fig. 15 Schematic of an atomic force microscope

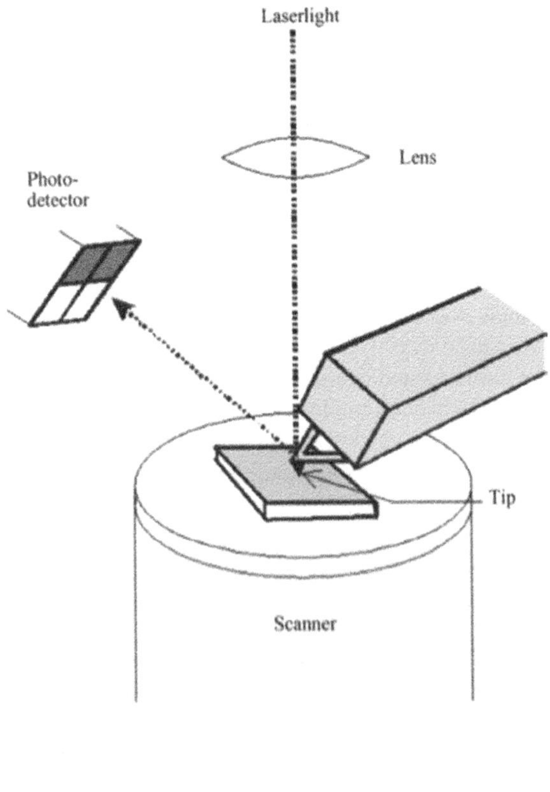

cantilever acts as a function of sample position, which was measured as the difference between voltages, reflecting positions of the sensitive detector. Then the deflection multiplied the factors of photo detector sensitivity and cantilever spring constant transforming into force [235].

The AFM is widely used to observe the surface topography and height of minerals, especially the two-dimensional mineral material [20, 236, 237]. Zhang et al. used AFM to observe (in situ) the molybdenum disulfide (MoS_2) sheet exposed to water to obtain a better understanding for the preparation and application of two-dimensional molybdenum disulfide [237]. Jia et al. [59] investigated the adsorption performance of Hg^{2+} on 2D molybdenum disulfide sheets in water through AFM measurement. A novel method for the determination of the thickness of hydration shell on nanosheets in water has been presented by Zhao etc. The work was realized through the measurements of the viscosity of aqueous nanosheet dispersion as the functions of solid volume concentration and the thickness and size of the nanosheets with AFM [20].

3.5 SEM

Scanning Electron Microscope (SEM) is one large instrument used to measure the surface morphology of samples. Once the energetic incident electron beam bombards the surface of samples, there are single or repeatedly elastic or inelastic collision occurred between electrons and nucleus as well as outer electrons of element [238]. In the process of collision, some electrons were reflected out from the sample surface, others infiltrated inside the sample and gradually lost its energy until stop moving [239]. 99% of the electron energy is transformed into thermal energy and only the remainder 1% successfully generates various signals (Fig. 16), mainly including secondary electron, backscattering electron, absorption electron,

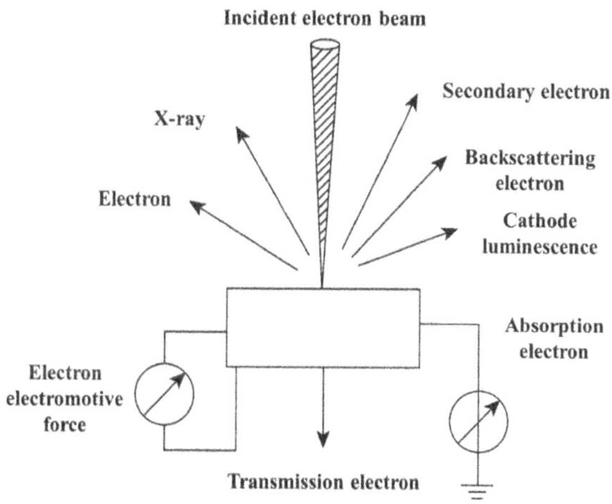

Fig. 16 Schematic of incident electron beam bombards the sample surface

transmission electron, auger electron, electron electromotive force, cathode luminescence, X-ray, etc. [240]. The SEM can receive and use these signals to analyze the samples.

Figure 17 shows the imaging schematic of SEM. Thermions, which came from filament, were accelerated by a voltage (2–30 kV) to form the energetic incident electron beam. Then the incident electron beam does raster scan in certain time and space [238]. During the scanning, secondary electrons deriving from the sample due to the bombardment of incident electron beam, were emitted in all directions and collected by the collector. These secondary electrons were accelerated and shooted on the scintillator. The scintillator transformed the secondary electrons into photo signals, which entered the photomultiplier and transformed into electrical signals followed by amplification in an amplifier. Then the amplified electrical signals were transferred into kinescope grid and the image corresponded to the sample would appeared on the fluorescent screen by modulating the brightness [240].

SEM was widely used in mineral surface morphology analysis. Chang et al. [241] reported a method to fabricate solid-state nanopores using the electron source of a conventional field-emission scanning electron microscope (FESEM). Using a low voltage, high resolution scanning electron microscope (SEM), Harrison etc. examined the topography and underlying morphology of poly(styrene)–poly

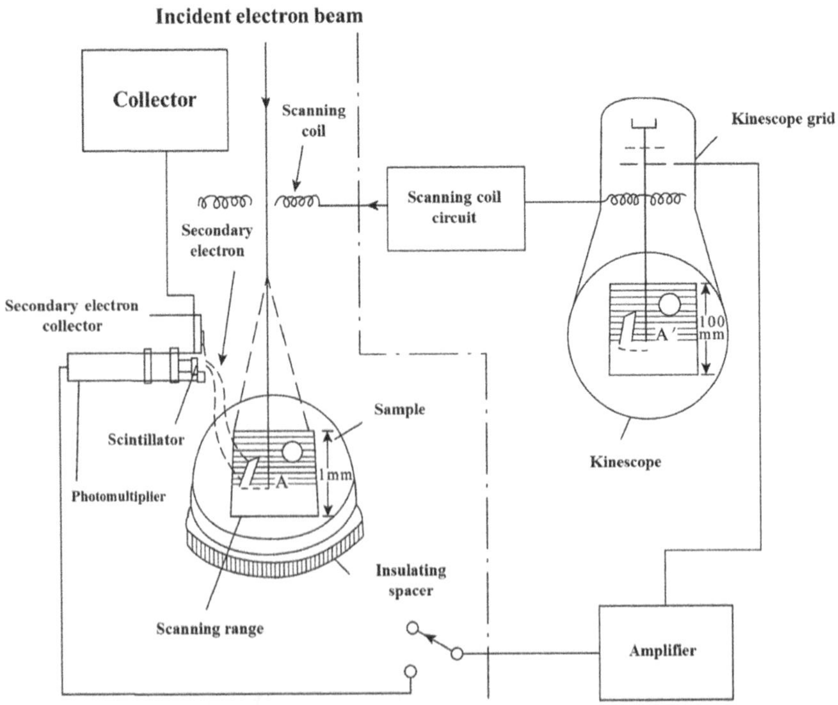

Fig. 17 Imaging schematic of scanning electron microscope

(butadiene) diblock copolymer films [242]. Surface textures of quartz sand grains from several glacial environments at the Feegletscher, Switzerland examined by means of scanning electron microscopy are described by Whalley and Krinsley [243].

3.6 TEM

Transmission electron microscope (TEM) can shoot incident electron beam, which has been accelerated and aggregated, on the very thin sample slice (Fig. 18). The incident electron beam will change the motion direction by the crash with atoms in the sample, resulting in a solid angle scattering. The scattering angle is affected by the density and thickness of sample. Hence, it can form image with different shades, which can appear on the image device (Fluorescent screen, film or light-sensitive coupling components, etc.) via magnification and focusing [244].

The de Broglie wavelength of electron is very small, resulting in tens of thousands to millions of times amplification of sample compared with optical microscope, e.g., a high resolution (0.1–0.2 nm). Hence, TEM can be used to observe the fine structure of sample, even the structure of a column of atoms. When the magnification rate is not high, the contrast of the image is mainly decided by the thickness and component of the sample. The normal image-forming principles of TEM can be divided into three types: (i) absorption image, which was formed mainly on the scattering process, (ii) diffraction image formed according to the diffracted intensity [245], (iii) phase image came from the phase variety [246].

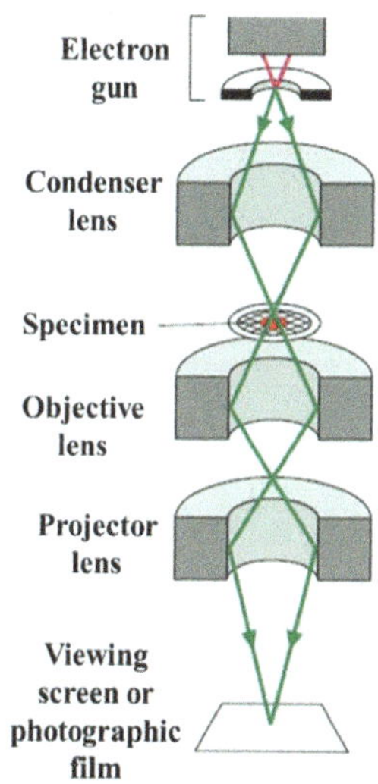

Fig. 18 Imaging schematic of transmission electron microscope

3.7 XPS

The earliest X-ray photoelectron spectroscopy (XPS) was called Electron Spectroscopy for Chemical Analysis (ESCA), which was invented by a Swedish scientist K. Siegbahn. The principle of XPS was based on the ionization of light [247]. When the photon beams were radiated to the sample surface, the electrons on the atomic orbital of a particular element absorb the energy of the photons, freeing the electrons from the bonds of atoms with certain kinetic energy (Fig. 19) [248]. During the process of photo ionization, the binding energy of solid matters can be expressed as follows:

$$E_b = h_\gamma - E_k - \varphi_s \tag{3}$$

where E_k is the kinetic energy of the photon; h_γ is the energy of X-ray; E_b is the ionization energy or binding energy of electron in a particular atomic orbital (the binding energy of an electron is the energy required to transfer electrons from one orbital to the surface Fermi energy level); φ_s is the work function of the spectrograph.

Because the φ_s is decided by the material and state of the spectrograph, it is a constant for the same one spectrograph and not affected by the sample. The average value of φ_s is 3–4 eV, hence, the formula (3) can be simplified as follow:

$$E_b = h_\gamma - E'_k \tag{4}$$

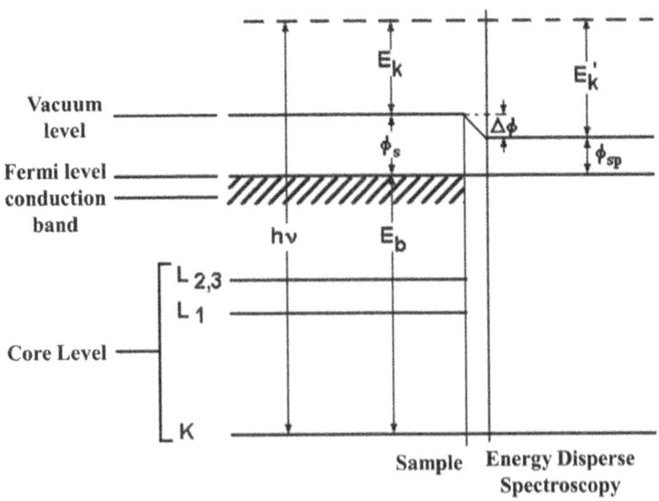

Fig. 19 Schematic of X-ray photoelectron spectroscopy

where E'_k can be detected by the spectrograph, then the E_b can be obtained according to Formula (4). In XPS analysis, it can not only excite valence electrons in atomic orbitals, but also excite the inner orbital electrons due to the high-energy X-ray. The energy of the excited photons is only related to the energy of the incident photons and atomic orbital [249]. Hence, the energy of the photoelectron is characteristic with the specific monochromatic excitation light source and the specific atomic orbital. When fixing the excitation light source energy, the photon energy is only related to the types of element and the ionization excited atomic orbital. For the same element atoms, the electrons in different orbitals is of different binding energy [250]. Hence, the binding energy of photon can be used to determine the type of element.

XPS was widely used in mineral elements and bonds analysis. Katari et al. [251] reported the use of XPS to determine the surface composition of semiconductor nanocrystals. The presence of two sulfur species was detected in XPS studies of thiol and disulfide molecules adsorbed onto gold surfaces reported by Castner et al. [252].

3.8 FTIR

Fourier Transform Infrared Spectroscopy (FTIR) is a kind of selective absorption spectrum, which records the absorption band after selectively absorption of certain frequencies of light by the organic molecules under the irradiation of a certain wavelength infrared. FTIR analysis is an important method to study the relationship between material molecular structure and infrared absorption, which can be effectively applied to the analyse the molecular structure [253]. FTIR analysis is widely used in the determination of high polymer structure and it is one of the basic methods to study the characterization and structural performance of high polymer structures.

FTIR is mainly used to study compounds with dipole moment variation in vibration. In addition to single atoms and homonuclear molecules, almost all organic compounds can absorb the light in the infrared region. The wavelength position and the intensity of infrared absorption band reflect the characteristics of molecular structure, which can be used to identify the structure of unknown objects or determine their chemical groups. Meanwhile, the absorption intensity of absorption band is related to the content of molecular composition or chemical group, which can be used for quantitatively analyse the purity. Due to the strong characteristic infrared spectra analysis, gas, liquid and solid samples can be determined with characteristics of little sample quantity, fast analysing speed and sample without destruction [254]. Therefore, FTIR is often used to identify compounds, determine the molecular structure and provide qualitative and quantitative analysis.

The simplest molecular vibration in FTIR analysis is called harmonic oscillation. The vibration frequency is positively correlated with the interatomic bonding while negatively correlated with the mass, which is so-called basic-frequency absorption. The influences of rotation and interatomic interaction in the actual molecule change the width and position of absorption band. The vibration frequency of the same chemical bonds or groups in different configurations normally do not change significantly, which is called as the characteristic absorption band [255]. FTIR absorption peaks corresponding to different vibration forms of each group in the molecules can provide qualitative and quantitative data of such molecules via the position and intensity of absorption peak.

FTIR has been widely used to identify compounds, determine the molecular structure and provide the qualitative and quantitative analysis. A highly sensitive diethyl ether SnO_2 gas-sensing material has been prepared by a sol–gel method and the surface adsorption and reaction processes between the SnO_2 gas-sensing film and the diethyl ether have been studied by in situ diffuse-reflectance Fourier-transform infrared spectroscopy (DRFT-IR) at different temperatures reported by Wang et al. [256]. Chen et al. [257] employed the subtractively normalized interfacial Fourier transform infrared technique to study the adsorption of benzonitrile (BN) at the Au(111) electrode surface. To provide fundamental knowledge for studying the relative content of vacant sites and exploring the mechanism of interaction between Pb^{2+} and birnessite, FTIR of birnessites with different Mn average oxidation states (AOS) before and after Pb^{2+} adsorption were investigated by Zhao et al. [258]. Tripp and Hair [259] has developed a method to measure the adsorption of polymers on silica particles using in situ FTIR.

3.9 Raman

Raman spectrum was named after an Indian scientist Raman in 1928 due to his findings of Raman scattering. Raman spectrum, as a fingerprint information of matters, is derived from the molecular vibrations and rotation spectra [260]. Hence, Raman spectrum can be used as a powerful tool for authenticating material and analyzing material components. Moreover, the frequency of the peaks in Raman is very sensitive to small changes of the physical structure of matters. Hence, under certain conditions like changing the temperature, pressure or doping characteristics, etc., the structure of the matter will change and the information of environmental factors in different parts of the material can be introduced indirectly by observing the small variation of Raman peaks [261].

Collisions between incident photons and molecules can be elastic or inelastic, in which Raman scattering is an inelastic collision process between photons and molecules. In the process of elastic collision, the frequency of scattered light remains constant, without energy exchange between the molecule and light quantum, which is Rayleigh scattering (Fig. 20a). However, once the inelastic collision happened between molecules and photons, they can have energy exchange

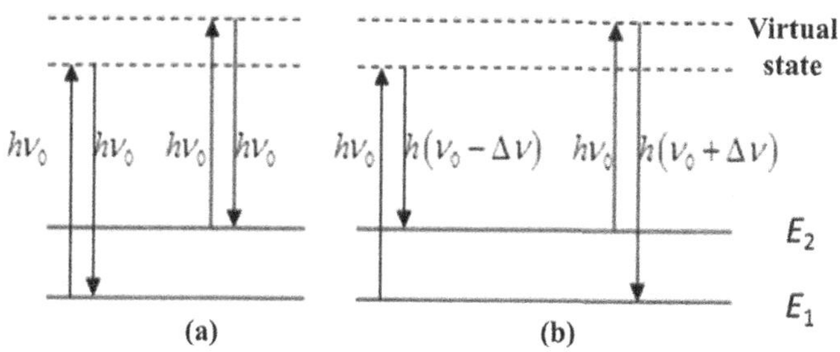

Fig. 20 Energy exchange schematic of elastic collision (**a**) and inelastic collision (**b**)

(Fig. 20b). This energy exchange can be an energy transfer from scattering molecules to photon. E_1 and E_2 in Fig. 20 are the energies of the initial state and the final state of molecules. The energy absorbed or evolved of photon ($\Delta E = E_2 - E_1$) can be the difference between the initial state and the final state of the scattering molecules [262]. If part of the energy from photon was delivered to scattering molecules, the energy of photon will decrease and scatter out with a low frequency ($v' = v_0 - \Delta v$), which is called Stokes line. For scattering molecules, the energy input makes it transite to the excited state E_2. If the scattering molecules are already in the excited state E_2 of vibration or rotation, the incident photon will be able to obtain energy (ΔE, vibration or rotation) from the scattering molecule and scatter out with a higher frequency ($v' = v_0 + \Delta v$), which is called anti-Stokes lines [263].

The frequency distribution of scattered light is showed as Fig. 21 The highest intensity is Rayleigh scattering line, which is located in the central with the frequency of v_0. The high-frequency side is anti-Stokes lines, which has a frequency difference of Δv to Rayleigh scattering line. Meanwhile, the low-frequency side is Stokes lines, which also has a frequency difference of Δv to Rayleigh scattering line. Stokes and anti-Stokes lines that symmetrically distributed on both sides of the Rayleigh line but with much weaker intensity, are normally called Raman lines. Compared with Stokes lines, anti-Stokes lines is weaker due to the ground state of most scattering molecules, so it is hard to find the anti-Stokes lines in Raman

Fig. 21 The frequency distribution of scattered light

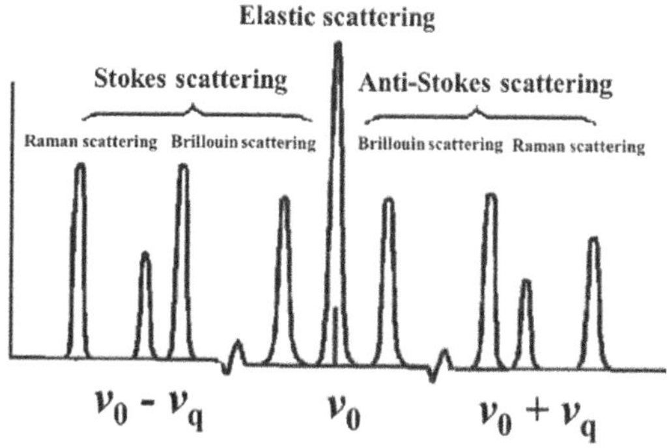

spectra. The frequency of Raman scattering is expressed as $\nu_0 \pm \Delta\nu$, in which the $\Delta\nu$ is called Raman shift and the value depends on the energy levels of vibrational and rotational of the scattered molecules, so the frequency of Raman spectra is not limited by the frequency of excited light. It can identify and analyze the scattered material well by the Raman frequency shift.

Raman spectra can identify and analyze the scattered material well. Raman spectra are reported from single crystals of graphite and other graphite materials and the single crystals of graphite show one single line at 1575 cm^{-1} [264]. The complete Raman spectrum of SnO_2 nanoparticles is presented and analyzed by Diéguez et al. [265]. Prawer et al. [266] used Raman spectrum to characterize the purified nanocrystalline diamond. The energy and polarization characteristics of the one- and two-phonon Raman spectrum have been measured using a 180° backscattering technique reported by Temple et al. [267].

3.10 Nuclear Magnetic Resonance

Nuclear magnetic resonance (NMR) is derived from the precession of the atomic nuclei with the spin angular momentum under an external magnetic field. According to the principle of quantum mechanics, the atomic nuclei also has the spin angular momentum, which is determined by the spin quantum number of the nucleus, as the electron [268]. Researches show that different types of nuclear possess different spin quantum numbers: (i) if the mass number and the proton number in nucleus are both even, the spin quantum number is 0 and (ii) if the mass number is even while the proton number is odd, the spin quantum number is an integer. However, only the nuclear magnetic resonance signal from nucleus with spin quantum number of 1/2 can be used by people. The spin of the nucleus will produce a magnetic moment and the magnetic moment will spin around the external magnetic field. This phenomenon is called procession, which frequency is decided by the strength of the external magnetic field and the property of the nucleus itself. According to the principle of quantum mechanics, the angle between the magnetic moment and the external magnetic field is not continuous, which is decided by the magnetic quantum number of the nucleus. The direction of the magnetic moment of the nucleus can only jump between these magnetic quantum numbers, which can form a series of energy levels. When the nucleus receives energy input from other sources in an external magnetic field, the energy level will jump. To make the precession of the nucleus jump, the energy is required, which is usually provided by an additional radio frequency field [269]. Hence, in a given external magnetic field, a particular nucleus can only absorb the energy provided by a particular frequency field, forming a NMR signal.

NMR is a powerful tool for analyzing the exact structure and how the functional groups are connected. NMR spectrometer can receive the resonant frequency and the corresponding intensity signal of the tested nucleus just like the advanced heterodyne radio. Meanwhile, drawing the NMR graph, which takes the position of resonant peak frequency as the horizontal coordinate and the relative strength of the peak as the vertical coordinate. The state of the electron in a compound molecule is different from that of the electron cloud, resulting in a different shielding effect of the external magnetic field. This means that the actual magnetic field intensity of these nuclei is different. For this reason, they will have resonance absorption at slightly different frequencies. This resonance absorption has a difference frequency of $\Delta\nu$ compared with benchmark nuclear resonant frequency (ν) and the $\Delta\nu/\nu$ is called chemical shift (δ). The types of functional groups in the compounds can be determined according to the characteristics of chemical shift in different specific regions [270]. Interactions between adjacent groups leading to more finely divided peaks can be used to further determine the connection of the adjacent groups within the molecule and infer the chemical structure of the molecule finally.

NMR is widely used in mineral structure and bonds analysis [271]. Silicon-29 magic-angle-spinning NMR spectroscopy has been used to investigate the silicon-aluminum distribution in natural samples of analcite and leucite (before and after heat treatment) as well as a leucite synthesized from a gel [272]. Saidov et al. [273] investigated the nucleation process of sodium sulfate on calcite and quartz particles, respectively, to assess the influence of the substrate chemistry on the crystalline phase being formed and on the nucleation process. NMR proxy for reactive surface area has laid the groundwork for future studies using chemical methods to quantify reactive surface area, which will enable a more accurate prediction of laboratory and field dissolution [274].

3.11 Specific Surface Area and Porosity

Natural minerals play an important role in natural and industrial processes as they can take up organic and inorganic molecules at solid-liquid or solid-gas interfaces. Thus the surface area of minerals is one of their most important properties controlling surface phenomena because almost all the adsorption behaviors take place at solid-liquid or solid-gas interfaces.

Specific surface area (SSA) is a property of solids defined as the total surface area of a material per unit of mass [275]. Values obtained for specific surface area depend on the method of measurement. The SSA was widely measured by adsorption using the Brunauer-Emmett-Teller (BET) isotherm, which is based on gas adsorption, notably nitrogen gas [276]. This has the advantage of measuring the surface of fine structures and deep texture on the particles. However, the results can differ markedly depending on the substance adsorbed.

The porosity of a porous medium (such as rock or sediment) describes the fraction of void space in the material, where the void may contain, for example, air or water. It is defined by the ratio:

$$\phi = \frac{V_V}{V_T} \tag{5}$$

where V_V is the volume of void-space (such as fluids) and V_T is the total or bulk volume of material, including the solid and void components. The mathematical symbols ϕ is used to denote porosity. Porosity is a fraction of the volume of voids over the total volume, between 0 and 1, or as a percentage between 0 and 100%. Strictly speaking, some tests measure the "accessible void", the total amount of void space accessible from the surface. There are many ways to test porosity in a substance or part, such as industrial CT scanning. The term porosity is used in multiple fields including pharmaceutics, ceramics, metallurgy, materials, manufacturing, earth sciences, soil mechanics and engineering.

3.12 Zeta Potential (ξ)

Zeta potential is a scientific term for electrokinetic potential in colloidal dispersions [277]. From a theoretical viewpoint, the zeta potential is the electric potential in the interfacial double layer at the location of the slipping plane relative to a point in the bulk fluid away from the interface. In other words, zeta potential is the potential difference between the dispersion medium and stationary layer of fluid attached to the dispersed particle [278].

The zeta potential is caused by the net electrical charge contained within the region bounded by the slipping plane, and also depends on the location of that plane. Thus it is widely used for quantification of the magnitude of the charge. However, zeta potential is not equal to the Stern potential or electric surface potential in the double layer, because these are defined at different locations. Such assumptions of equality should be applied with caution. Nevertheless, zeta potential is often the only available path to characterize the double-layer properties.

The zeta potential is a key indicator of the stability of colloidal dispersions. The magnitude of the zeta potential indicates the degree of electrostatic repulsion between adjacent, similarly charged particles in dispersion. For molecules and particles that are small enough, a high zeta potential will confer stability, i.e., the solution or dispersion will resist aggregation. When the potential is small, attractive forces may exceed this repulsion and the dispersion may break and flocculate. So, colloids with high zeta potential (negative or positive) are electrically stabilized while colloids with low zeta potentials tend to coagulate or flocculate [279, 280].

References

1. Chen, T., Zhao, Y., Song, S.: Correlation of electrophoretic mobility with exfoliation of montmorillonite platelets in aqueous solutions. Colloids Surf. A **525**, 1–6 (2017)
2. Dent, A.J., Ramsay, J.D.F., Swanton, S.W.: An EXAFS study of uranyl ion in solution and sorbed onto silica and montmorillonite clay colloids. J. Colloid Interface Sci. **150**, 45–60 (1992)
3. Missana, T., Alonso, Ú., García-Gutiérrez, M., Mingarro, M.: Role of bentonite colloids on europium and plutonium migration in a granite fracture. Appl. Geochem. **23**, 1484–1497 (2008)
4. Almeida, C.A.P., Debacher, N.A., Downs, A.J., Cottet, L., Mello, C.A.D.: Removal of methylene blue from colored effluents by adsorption on montmorillonite clay. J. Colloid Interface Sci. **332**, 46–53 (2009)
5. Gürses, A., Doğar, Ç., Yalçin, M., MAçikyildiz, Bayrak, R., Karaca, S.: The adsorption kinetics of the cationic dye, methylene blue, onto clay. J. Hazard. Mater. **131**, 217–228 (2006)
6. Abollino, O., Aceto, M., Malandrino, M., Sarzanini, C., Mentasti, E.: Adsorption of heavy metals on Na-montmorillonite. Effect of pH and organic substances. Water Res. **37**, 1619–1627 (2003)
7. Chen, C., Liu, H., Chen, T., Chen, D., Frost, R.L.: An insight into the removal of Pb(II), Cu (II), Co(II), Cd(II), Zn(II), Ag(I), Hg(I), Cr(VI) by Na(I)-montmorillonite and Ca(II) montmorillonite. Appl. Clay Sci. **118**, 239–247 (2015)
8. Bhattacharyya, K.G., SenGupta, S., Sarma, G.K.: Interactions of the dye, Rhodamine B with kaolinite and montmorillonite in water. Appl. Clay Sci. **99**, 7–17 (2014)
9. Boggs, M.A., Dai, Z., Kersting, A.B., Zavarin, M.: Plutonium(IV) sorption to montmorillonite in the presence of organic matter. J. Environ. Radioact. **141**, 90–96 (2015)
10. Bhattacharyya, K.G., Gupta, S.S.: Pb (II) uptake by kaolinite and montmorillonite in aqueous medium: influence of acid activation of the clays. Colloids Surf. A **277**, 191–200 (2006)
11. Teng, M., Lin, S.: Removal of basic dye from water onto pristine and HCl-activated montmorillonite in fixed beds. Desalination **194**, 156–165 (2006)
12. Aytas, S., Yurtlu, M., Donat, R.: Adsorption characteristic of U (VI) ion onto thermally activated bentonite. J. Hazard. Mater. **172**, 667–674 (2009)
13. Al-asheh, S., Banat, F., Abu-aitah, L.: Adsorption of phenol using different types of activated bentonites. Sep. Purif. Technol. **33**, 1–10 (2003)
14. Tian, S., Jiang, P., Ning, P., Su, Y.: Enhanced adsorption removal of phosphate from water by mixed lanthanum/aluminum pillared montmorillonite. Chem. Eng. J. **151**, 141–148 (2009)
15. Lee, S.Y., Kim, S.J., Chung, S.Y., Jeong, C.H.: Sorption of hydrophobic organic compounds onto organoclays. Chemosphere **55**, 781–785 (2004)
16. Zhu, R., Zhu, L., Zhu, J., Ge, F., Wang, T.: Sorption of naphthalene and phosphate to the CTMAB–Al 13 intercalated bentonites. J. Hazard. Mater. **168**, 1590–1594 (2009)
17. Banat, F., Al-Asheh, S., Al-Anbar, S., Al-Refaie, S.: (2007) Microwave- and acid-treated bentonite as adsorbents of methylene blue from a simulated dye wastewater. Bull Eng. Geol. Environ. **66**.
18. El Mouzdahir, Y., Elmchaouri, A., Mahboub, R., Gil, A., Korili, S.A.: Equilibrium modeling for the adsorption of methylene blue from aqueous solutions on activated clay minerals. Desalination **250**, 335–338 (2010)
19. Bai, H., Zhao, Y., Zhang, X., Wang, W., Zhang, T., Liu, C., Yi, H., Song, S.: Correlation of exfoliation performance with interlayer cations of montmorillonite in the preparation of two-dimensional nanosheets. J. Am. Ceram. Soc. **102**, 3908–3922 (2019)

20. Zhao, Y., Yi, H., Jia, F., Li, H., Peng, C., Song, S.: A novel method for determining the thickness of hydration shells on nanosheets: A case of montmorillonite in water. Powder Technol. **306**, 74–79 (2017)
21. Wang, W., Zhao, Y., Yi, H., Chen, T., Kang, S., Li, H., Song, S.: Preparation and characterization of self-assembly hydrogels with exfoliated montmorillonite nanosheets. Nanotechnology **29**, 025605 (2018)
22. Bhattacharyya, K.G., Sen Gupta, S.: Kaolinite and montmorillonite as adsorbents for Fe(III), Co(II) and Ni(II) in aqueous medium. Appl. Clay Sci. **41**, 1–9 (2008)
23. Zhao, J., He, M.C.: Theoretical study of heavy metal Cd, Cu, Hg, and Ni(II) adsorption on the kaolinite (001) surface. Appl. Surf. Sci. **317**, 718–723 (2014)
24. Srivastava, P., Singh, B., Angove, M.: Competitive adsorption behavior of heavy metals on kaolinite. J. Colloid Interface Sci. **290**, 28–38 (2005)
25. Hamdi Karaoğlu, M., Doğan, M., Alkan, M.: Removal of cationic dyes by kaolinite. Microporous Mesoporous Mater. **122**, 20–27 (2009)
26. Khan, T.A., Dahiya, S., Ali, I.: Use of kaolinite as adsorbent: equilibrium, dynamics and thermodynamic studies on the adsorption of Rhodamine B from aqueous solution. Appl. Clay Sci. **69**, 58–66 (2012)
27. Duarte-Silva, R., Villa-García, M.A., Rendueles, M., Díaz, M.: Structural, textural and protein adsorption properties of kaolinite and surface modified kaolinite adsorbents. Appl. Clay Sci. **90**, 73–80 (2014)
28. Guerra, D.L., Airoldi, C.: The performance of urea-intercalated and delaminated kaolinites-adsorption kinetics involving copper and lead. J. Braz. Chem. Soc. **20**, 19–30 (2009)
29. Koteja, A., Matusik, J.: Di- and triethanolamine grafted kaolinites of different structural order as adsorbents of heavy metals. J. Colloid Interface Sci. **455**, 83–92 (2015)
30. Bhattacharyya, K.G., Sen Gupta, S.: Removal of Cu(II) by natural and acid-activated clays: An insight of adsorption isotherm, kinetic and thermodynamics. Desalination **272**, 66–75 (2011)
31. Bhattacharyya, K.G., Sen, S.: Gupta calcined tetrabutylammonium kaolinite and montmorillonite and adsorption of Fe(II), Co(II) and Ni(II) from solution. Appl. Clay Sci. **46**, 216–221 (2009)
32. Kongchum, M., Hudnall, W.H., DeLaune, R.D.: Relationship between sediment clay minerals and total mercury. J. Environ. Sci. Health. A. Tox. Hazard. Subst. Environ. Eng. **46**, 534–539 (2011)
33. Brigatti, M.F., Colonna, S., Malferrari, D., Medici, L., Poppi, L.: Mercury adsorption by montmorillonite and vermiculite: a combined XRD. TG-MS, and EXAFS study, Appl Clay Sci. **28**, 1–8 (2005)
34. Abollino, O., Giacomino, A., Malandrino, M., Mentasti, E.: Interaction of metal ions with montmorillonite and vermiculite. Appl. Clay Sci. **38**, 227–236 (2008)
35. Yu, X., Wei, C., Wu, H.: Effect of molecular structure on the adsorption behavior of cationic dyes onto natural vermiculite. Sep. Purif. Technol. **156**, 489–495 (2015)
36. Liu, S., Wu, P., Chen, M., Yu, L., Kang, C., Zhu, N., Dang, Z.: Amphoteric modified vermiculites as adsorbents for enhancing removal of organic pollutants: bisphenol A and tetrabromobisphenol A. Environ. Pollut. **228**, 277–286 (2017)
37. Turan, N.G., Elevli, S., Mesci, B.: Adsorption of copper and zinc ions on illite: determination of the optimal conditions by the statistical design of experiments. Appl. Clay Sci. **52**, 392–399 (2011)
38. Zhao, D., Chen, C.: Effect of fulvic acid on the sorption and desorption of $^{90}Sr^{2+}$ on red earth. J. Radioanal. Nucl. Chem. **270**, 445–452 (2006)
39. Bradbury, M.H., Baeyens, B.: Sorption modelling on illite. Part II: Actinide sorption and linear free energy relationships, Geochim Cosmochim Acta **73**, 1004–1013 (2009)
40. Echeverría, J.C., Churio, E., Garrido, J.J.: Retention mechanisms of Cd on illite. Clays Clay Miner. **50**, 614–623 (2002)

41. Liu, X., Guo, Z., Chen, Y., Cai, L., Chi, Y.: Effect of pH, ionic strength, foreign ions, humic acid and temperature on sorption of radionuclide ^{60}Co(II) on illite. J. Radioanal. Nucl. Chem. **291**, 1357–1366 (2012)
42. Zhang, L., Yu, X., Zhang, H., Chen, E., Dong, T.: Impact of environmental conditions on the sorption behavior of ^{60}Co(II) on illite. J. Radioanal. Nucl. Chem. **295**, 1473–1485 (2013)
43. Fil, B.A., Korkmaz, M., Özmetin, C.: Application of nonlinear regression analysis for methyl violet (MV) dye adsorption from solutions onto illite clay. J. Dispers. Sci. Technol. **37**, 991–1001 (2016)
44. Zhang, G., Liu, Y., Xie, Y., Yang, X., Hu, B., Ouyang, S.: Zinc adsorption on Na-rectorite and effect of static magnetic field on the adsorption. Appl. Clay Sci. **29**, 15–21 (2005)
45. Zhang, G., Yang, X., Liu, Y., Jia, Y., Yu, G., Ouyang, S.: Copper (II) adsorption on Ca-rectorite, and effect of static magnetic field on the adsorption. J. Colloid Interface Sci. **278**, 265–269 (2004)
46. Chang, P., Wang, X., Yu, S., Wu, W.: Sorption of Ni (II) on Na-rectorite from aqueous solution: Effect of pH, ionic strength and temperature. Colloids Surf. A Physicochem. Eng. Asp. **302**, 75–81 (2007)
47. Chang, P., Jean, J., Jiang, W., Li, Z.: Mechanism of tetracycline sorption on rectorite. Colloids Surf. A **339**, 94–99 (2009)
48. Hong, H., Jiang, W.T., Zhang, X., Tie, L., Li, Z.: Adsorption of Cr(VI) on STAC-modified rectorite. Appl. Clay Sci. **42**, 292–299 (2008)
49. Huang, Y., Ma, X., Liang, G., Yan, H.: Adsorption of phenol with modified rectorite from aqueous solution. Chem. Eng. J. **141**, 1–8 (2008)
50. Zhao, X., Zhu, L., Zhang, Y., Yan, J., Lu, X., Huang, Y., Tang, H.: Removing organic contaminants with bifunctional iron modified rectorite as efficient adsorbent and visible light photo-Fenton catalyst. J. Hazard. Mater. **215–216**, 57–64 (2012)
51. Huang, Y., Ma, X., Liang, G., Yan, Y., Wang, S.: Adsorption behavior of Cr(VI) on organic-modified rectorite. Chem. Eng. J. **138**, 187–193 (2008)
52. Farquhar, M.L., Vaughan, D.J., HughesCR, Charnock J.M., England, K.E.R.: Experimental studies of the interaction of aqueous metal cations with mineral substrates: lead, cadmium, and copper with perthitic feldspar, muscovite, and biotite. Geochim. Cosmochim. Acta **61**, 3051–3064 (1997)
53. Paulino, L., Ceolin, W., Rebello, T., Junior, A., Morais, M.M., Rosado, J., Veloso, A.D., Paulino, B.F., Martins, L.L.: Biotite (black mica) as an adsorbent of pesticides in aqueous solution, water, air. Soil Pollut. **8**, 231 (2015)
54. Yang, J., Lee, J.Y., Park, Y., Baek, K., Choi, J.: Adsorption of As (III), As (V), Cd (II), Cu (II), and Pb (II) from aqueous solutions by natural muscovite. Sep. Sci. Technol. **45**, 814–823 (2010)
55. Komarneni, S., Rustum, R.: A cesium-selective ion sieve made by topotactic leaching of phlogopite mica. Science **239**, 1286–1289 (1988)
56. Connecticut, T., Experiment, A., Haven, N.: Potassium and cesium ion selectivity in relation to clay mineral structure. Clays Clay Miner. **18**, 47–52 (1970)
57. Cho, Y., Komarneni, S.: Cation exchange equilibria of cesium and strontium with K-depleted biotite and muscovite. Appl. Clay Sci. **44**, 15–20 (2009)
58. Yamamoto, T., Fujimura, T., Shimada, T., Takagi, S.: Preparation of modified mica as an effective adsorbent to remove Cs^+ from water. Chem. Lett. **43**, 860–861 (2014)
59. Jia, F., Wang, Q., Wu, J., Li, Y., Song, S.: Two-dimensional molybdenum disulfide as a superb adsorbent for removing Hg^{2+} from water. ACS Sustain. Chem. Eng. **5**, 7410–7419 (2017)
60. Ganatra, R., Zhang, Q.: Few-layer MoS_2: a promising layered semiconductor. ACS Nano (2014)
61. Jiang, J.W.: Graphene versus MoS_2: a short review. Front Phys. (2015)
62. Wang, Z., Mi, B.: Environmental applications of 2D molybdenum disulfide (MoS_2) nanosheets. Environ. Sci. Technol. (2017)

63. Zhao, Z.Y., Liu, Q.L.: Study of the layer-dependent properties of MoS_2 nanosheets with different crystal structures by DFT calculation. Catal. Sci. Technol. **8**, 1867–1879 (2018)
64. Eda, G., Yamaguchi, H., Voiry, D., Fujita, T., Chen, M., Chhowalla, M.: Photoluminescence from chemically exfoliated MoS_2. Nano Lett. **11**, 5111–5116 (2011)
65. Zhao, X., Ma, X., Sun, J., Li, D., Yang, X.: Enhanced catalytic activities of surfactant-assisted exfoliated WS_2 nanodots for hydrogen evolution. ACS Nano (2016)
66. Kumar, A., umar, P.K., Late, D.J., Kumar, A., Patel, S., Singh, J.: 2D layered transition metal dichalcogenides (MoS_2): synthesis, applications and theoretical aspects. Appl. Mater. Today. **13**, 242–270 (2018)
67. Li, T., Galli, G.: Electronic properties of MoS_2 nanoparticles. J. Phys. Chem. C. (2007)
68. Kim, S., Konar, A., Hwang, W.S., Lee, J.H., Lee, J., Yang, J., Jung, C., Kim, H., Yoo, J.B., Choi, J.Y., Jin, Y.W., Lee, S.Y., Jena, D., Choi, W., Kim, K.: High-mobility and low-power thin-film transistors based on multilayer MoS_2 crystals. Nat. Commun. (2012)
69. Heiranian, M., Farimani, A.B., Aluru, N.R.: Water desalination with a single-layer MoS_2 nanopore. Nat. Commun. (2015)
70. Gash, A.E., Spain, A.L., Dysleski, L.M., Flaschenriem, C.J., Kalaveshi, A., Dorhout, P.K., Strauss, S.H.: Efficient recovery of elemental mercury from Hg(II)-contaminated aqueous media using a redox-recyclable ion-exchange material. Environ. Sci. Technol. **32**, 1007–1012 (1998)
71. Jia, F., Sun, K., Yang, B., Zhang, X., Wang, Q., Song, S.: Defect-rich molybdenum disulfide as electrode for enhanced capacitive deionization from water. Desalination **446**, 21–30 (2018)
72. Jia, F., Zhang, X., Song, S.: AFM study on the adsorption of Hg^{2+} on natural molybdenum disulfide in aqueous solutions. Phys. Chem. Chem. Phys. (2017)
73. Chiriță, P.: Aqueous oxidation of iron monosulfide (FeS) by molecular oxygen. Miner. Process Extr. Metall. Rev. (2016)
74. Pearson, R.G.: Hard and soft acids and bases, HSAB, part 1: fundamental principles. J. Chem. Educ. (1968)
75. Aditya, A., Suresh, A., Sriramoju, S.K., Dash, P.S., Pati, S., Padmanabhan, NPH.: Optimization study of sodium hydroxide consumption in the coal demineralization process. Miner. Process Extr. Metall. Rev. (2018)
76. Uçar, D.: Sequential precipitation of heavy metals using sulfide-laden bioreactor effluent in a pH controlled system. Miner. Process Extr. Metall. Rev. (2017)
77. Lin, L., Miao, N., Huang, J., Zhang, S., Zhu, Y., Horsell, D.D., Ghosez, P., Sun, Z., Allwood, D.A.: A photocatalyst of sulphur depleted monolayered molybdenum sulfide nanocrystals for dye degradation and hydrogen evolution reaction. Nano Energy **38**, 544–552 (2017)
78. Flower, R.J.: Diatomites: Their Formation, Distribution, and Uses, 2nd edn. Elsevier (2013)
79. Nenadovic, S., Nenadovic, M., Kovacevic, R., Matovic, L., Matovic, B., Jovanovic, Z., Novakovic, G.J.: Influence of diatomite microstructure on its adsorption capacity for Pb(II). Sci. Sinter. **41**, 309–317 (2009)
80. Yurekli, Y.: Removal of heavy metals in wastewater by using zeolite nano-particles impregnated polysulfone membranes. J. Hazard. Mater. **309**, 53–64 (2016)
81. Selim, Q., Ibrahim, S.S.: Microscopic evaluation of diatomite for advanced applications: case study. Microsc. Sci. Technol. Appl. Educ. 2174–2181 (2010)
82. Round, B.F.E.: The composition of some diatomites from the southery. J. Roy. Microscopical Soc. **80**, 59–69 (1961)
83. Kim, C., Zhang, Z., Wang, L., Sun, T., Hu, X.: Core-shell magnetic manganese dioxide nanocomposites modified with citric acid for enhanced adsorption of basic dyes. J. Taiwan Inst. Chem. Eng. **67**, 418–425 (2016)
84. Hethnawi, A., Nassar, N.N., Manasrah, A.D., Vitale, G.: Polyethylenimine-functionalized pyroxene nanoparticles embedded on diatomite for adsorptive removal of dye from textile wastewater in a fixed-bed column. Chem. Eng. J. **320**, 389–404 (2017)

85. Darvishi Cheshmeh Soltani, R., Safari, M., Maleki, A., Godini, H., Mahmoudian, M.H., Pordel, M.A.: Application of nanocrystalline iranian diatomite in immobilized form for removal of a textile dye. J. Dispers. Sci. Technol. **37** (2016)
86. Dang, T.D., Banerjee, A.N., Tran, Q.T., Roy, S.: Fast degradation of dyes in water using manganese-oxide-coated diatomite for environmental remediation. J. Phys. Chem. Solids **98**, 50–58 (2016)
87. Zhang, J., Ping, Q., Niu, M., Shi, H., Li, N.: Kinetics and equilibrium studies from the methylene blue adsorption on diatomite treated with sodium hydroxide. Appl. Clay Sci. **83–84**, 12–16 (2013)
88. Yuan, W., Yuan, P., Liu, D., Yu, W., Laipan, M., Deng, L., Chen, F.: In situ hydrothermal synthesis of a novel hierarchically porous TS-1/modified-diatomite composite for methylene blue (MB) removal by the synergistic effect of adsorption and photocatalysis. J. Colloid Interface Sci. **462**, 191–199 (2016)
89. Zhang, Y., Jing, Z., Li, Y., Fan, J., Kan, W.: Hydrothermal synthesis of tobermorite from diatomite and its adsorption performance for methylene blue. Mater. Res. Innov. **19** (2015)
90. Liu, D., Yuan, W., Yuan, P., Yu, W., Tan, D., Li, Hu, He, H.: Physical activation of diatomite-templated carbons and its effect on the adsorption of methylene blue (MB). Appl. Surf. Sci. **282**, 838–843 (2013)
91. Zhu, J., Wang, P., Lei, M.J., Zhang, W.L.: Polyhydroxyl-aluminum pillaring improved adsorption capacities of Pb^{2+} and Cd^{2+} onto diatomite. J. Cent. South Univ. **21**, 2359–2365 (2014)
92. Yang, Z., Lu, Y., Li, Y.: The adsorption of zinc(ii) from aqueous solution on modified diatomite. Appl. Mech. Mater. **438–439**, 211–214 (2013)
93. Zhou, Q., Yang, H., Yan, C., Luo, W., Li, X., Zhao, J.: Synthesis of carboxylic acid functionalized diatomite with a micro-villous surface via UV-induced graft polymerization and its adsorption properties for Lanthanum(III) ions. Colloids Surf. A Physicochem. Eng. Asp. **501**, 9–16 (2016)
94. Yusan, S., Bampaiti, A., Erenturk, S., Noli, F., Aslani, M.A.A., Aytas, S.: Sorption of Th (IV) onto ZnO nanoparticles and diatomite-supported ZnO nanocomposite: kinetics, mechanism and activation parameters. Radiochim. Acta **104**, 635–647 (2016)
95. Du, Y., Fan, H., Wang, L., Wang, J., Wu, J., Dai, H.: Alpha-Fe_2O_3 nanowires deposited diatomite: highly efficient absorbents for the removal of arsenic, J. Mater. Chem. A. **1**, 7729–7737 (2013)
96. Yu, W., Deng, L., Yuan, P., Liu, D., Yuan, W., Chen, F.: Facile preparation of hierarchically porous diatomite/MFI-type zeolite composites and their performance of benzene adsorption: the effects of NaOH etching pretreatment Wenbin. Chem. Eng. J. **270**, 450–458 (2015)
97. Chen, J., Yan, L.G., Yu, H.Q., Li, S., Qin, L.L., Liu, G.Q., Li, Y.F., Du, B.: Efficient removal of phosphate by facile prepared magnetic diatomite and illite clay from aqueous solution. Chem. Eng. J. **287**, 162–172 (2016)
98. Guo, Y., Mi, X., Li, G., Chen, X.: Sonocatalytic degradation of antibiotics tetracycline by Mn-modified diatomite. J. Chem. (2017)
99. Meng, X., Liu, Z., Deng, C., Zhu, M., Wang, D., Li, K., Deng, Y., Jiang, M.: Microporous nano-MgO/diatomite ceramic membrane with high positive surface charge for tetracycline removal. J. Hazard. Mater. **320**, 495–503 (2016)
100. Chen, Y., Liu, K.: Preparation and characterization of nitrogen-doped TiO_2/diatomite integrated photocatalytic pellet for the adsorption-degradation of tetracycline hydrochloride using visible light. Chem. Eng. J. **302**, 682–696 (2016)
101. Xia, P., Wang, X., Wang, X., Song, J., Wang, H., Zhang, J.: Struvite crystallization combined adsorption of phosphate and ammonium from aqueous solutions by mesoporous MgO–loaded diatomite. Colloids Surf. A **506**, 220–227 (2016)
102. Cheng, L., Kang, Y., Li, G.: Effect factors of benzene adsorption and degradation by nano-TiO_2 immobilized on diatomite. J. Nanomater. (2012)
103. Aksoy, S., Tekba, M., Engin, G., Bekta, N.: Removal of direct orange-46 from aqueous solutions using MN-diatomite (2011)

104. Erdem, E., Çölgeçen, G., Donat, R.: The removal of textile dyes by diatomite earth. J. Colloid Interface Sci. **282**, 314–319 (2005)
105. van Bekkum, H., Flanigen, E., Jansen, J.: In: Introduction to Zeolite Science and Practice (1991)
106. Guth, J.L., Kessler, H., Wey, R.: In: New Developments in Zeolite Science and Technology (1986)
107. Du Scott, M., Auerbach, P.K., Carrado, K.A.: In: Handbook of Zeolite Science and Technology (2003)
108. Jacobsen, C.J.H., Madsen, C., Houzvicka, J., Schmidt, I., Carlsson, A.: Mesoporous zeolite single crystals. J. Am. Chem. Soc. **122**, 7116–7117 (2000)
109. Nguyen, T.C., Loganathan, P., Nguyen, T.V., Vigneswaran, S., Kandasamy, J., Naidu, R.: Simultaneous adsorption of Cd, Cr, Cu, Pb, and Zn by an iron-coated Australian zeolite in batch and fixed-bed column studies. Chem. Eng. J. **270**, 393–404 (2015)
110. Malamis, S., Katsou, E.: A review on zinc and nickel adsorption on natural and modified zeolite, bentonite and vermiculite: Examination of process parameters, kinetics and isotherms. J. Hazard. Mater. **252–253**, 428–461 (2013)
111. Arancibia-Miranda, N., Baltazar, S.E., García, A., Muñoz-Lira, D., Sepúlveda, P., Rubio, M. A., Altbir, D.: Nanoscale zero valent supported by zeolite and montmorillonite: template effect of the removal of lead ion from an aqueous solution. J. Hazard. Mater. **301**, 371–380 (2016)
112. Bhatnagar, A., Sillanpää, M.: Removal of natural organic matter (NOM) and its constituents from water by adsorption—a review. Chemosphere **166**, 497–510 (2017)
113. Khanday, W.A., Marrakchi, F., Asif, M., Hameed, B.H.: Mesoporous zeolite–activated carbon composite from oil palm ash as an effective adsorbent for methylene blue. J. Taiwan Inst. Chem. Eng. **70**, 32–41 (2017)
114. Alver, E., ü Metin, A.: Anionic dye removal from aqueous solutions using modified zeolite: adsorption kinetics and isotherm studies. Chem. Eng. J. **200–202**, 59–67 (2012)
115. Guaya, D., Valderrama, C., Farran, A., Armijos, C., Cortina, J.L.: Simultaneous phosphate and ammonium removal from aqueous solution by a hydrated aluminum oxide modified natural zeolite. Chem. Eng. J. **271**, 204–213 (2015)
116. Yu, Y., Shapter, J.G., Popelka-Filcoff, R., Bennett, J.W., Ellis, A.V.: Copper removal using bio-inspired polydopamine coated natural zeolites. J. Hazard. Mater. **273**, 174–182 (2014)
117. Kim, S.A., Kamala-Kannan, S., Lee, K.J., Park, Y.J., Shea, P.J., Lee, W.H., Kim, H.M., Oh, B.T.: Removal of Pb(II) from aqueous solution by a zeolite-nanoscale zero-valent iron composite. Chem. Eng. J. **217**, 54–60 (2013)
118. Khanday, W.A., Asif, M., Hameed, B.H.: Cross-linked beads of activated oil palm ash zeolite/chitosan composite as a bio-adsorbent for the removal of methylene blue and acid blue 29 dyes. Int. J. Biol. Macromol. **95**, 895–902 (2017)
119. Liu, S., Ding, Y., Li, P., Diao, K., Tan, X., Lei, F., Zhan, Y., Li, Q., Huang, B., Huang, Z.: Adsorption of the anionic dye Congo red from aqueous solution onto natural zeolites modified with N, N-dimethyl dehydroabietylamine oxide. Chem. Eng. J. **248**, 135–144 (2014)
120. Özcan, A.S., Erdem, B., Özcan, A.: Adsorption of Acid Blue 193 from aqueous solutions onto Na-bentonite and DTMA-bentonite. J. Colloid Interface Sci. **280**, 44–54 (2004)
121. Rytwo, G., Tropp, D., Serban, C.: Adsorption of diquat, paraquat and methyl green on sepiolite: experimental results and model calculations. Appl. Clay Sci. **20**, 273–282 (2002)
122. Galan, E.: In: Palygorskite-Sepiolite. Occurrences, Genesis and Uses (1984)
123. Çoruh, S., Geyikçi, F., Elevli, S.: Adsorption of neutral red dye from an aqueous solution onto natural sepiolite using full factorial design. Clays Clay Miner. **59**, 617–625 (2011)
124. Lemić, J., Tomašević-Čanović, M., Djuričić, M., Stanić, T.: Surface modification of sepiolite with quaternary amines. J. Colloid Interface Sci. **292**, 11–19 (2005)
125. Mahmoud, M.R., Rashad, G.M., Metwally, E., Saad, E.A., Elewa, A.M.: Adsorptive removal of $^{134}Cs^{+}$, $^{60}Co^{2+}$ and $^{152+154}Eu^{3+}$ radionuclides from aqueous solutions using sepiolite: Single and multi-component systems. Appl. Clay Sci. **141**, 72–80 (2017)

126. Bahabadi, F.N., Farpoor, M.H., Mehrizi, M.H.: Removal of Cd, Cu and Zn ions from aqueous solutions using natural and Fe modified sepiolite, zeolite and palygorskite clay minerals. Water Sci. Technol. **75**, 340–349 (2017)
127. Doğan, M., Turhan, Y., Alkan, M., Namli, H., Turan, P., Demirbaş, Ö.: Functionalized sepiolite for heavy metal ions adsorption. Desalination **230**, 248–268 (2008)
128. Zhang, K., Pan, J., Tang, C., Shen, J., Peng, G., Qiu, D.: Study on adsorption of octenylsuccinate by sepiolite. Desalin. Water Treat. **57**, 8765–8775 (2016)
129. Sabah, E., Ouki, S.: Mechanistic insight into pyrene removal by natural sepiolites. Environ. Sci. Pollut. Res. **24**, 21680–21692 (2017)
130. Moreira, M.A., Ciuffi, K.J., Rives, V., Vicente, M.A., Trujillano, R., Gil, A., Korili, S.A., de Faria, E.H.: Effect of chemical modification of palygorskite and sepiolite by 3-aminopropyltriethoxisilane on adsorption of cationic and anionic dyes. Appl. Clay Sci. **135**, 394–404 (2017)
131. Chaudhary, R.P., Pawar, P.B., Vaibhav, K., Saxena, S., Shukl, S.: Quantification of adsorption of azo dye molecules on graphene oxide using optical spectroscopy. Miner. Met. Mater. Soc. **69**, 236–240 (2016)
132. Wang, Q., Ren, G., Jia, F., Song, S.: Preparation and characterization of nanoscale zero-valent iron-loaded porous sepiolite for decolorizing methylene blue in aqueous solutions. JOM **69**, 699–703 (2017)
133. Yu, S., Liu, X., Xu, G., Qiu, Y., Cheng, L.: Magnetic Fe_3O_4/sepiolite composite synthesized by chemical co-precipitation method for efficient removal of Eu(III). Desalin. Water Treat. **3994**, 1–12 (2015)
134. Habish, A.J., Lazarević, S., Janković-Častvan, I., Jokić, B., Kovač, J., Rogan, J., Janaćković, Đ., Petrović, R.: Nanoscale zerovalent iron (nZVI) supported by natural and acid-activated sepiolites: the effect of the nZVI/support ratio on the composite properties and Cd^{2+} adsorption. Environ. Sci. Pollut. Res. **24**, 628–643 (2017)
135. Sturini, M., Speltini, A., Maraschi, F., Profumo, A., Tarantino, S., Gualtieri, A.F., Zema, M.: Removal of fluoroquinolone contaminants from environmental waters on sepiolite and its photo-induced regeneration. Chemosphere **150**, 686–693 (2016)
136. Kara, M., Yuzer, H., Sabah, E., Celik, M.S.: Adsorption of cobalt from aqueous solutions onto sepiolite. Water Res. **37**, 224–232 (2003)
137. Kaya, M., Dilekoğlu, M.F., Şahin, Ö., Saka, C.: Plasma treated sepiolite: a new adsorbent for removal of malachite green from contaminated water. Plasma Chem. Plasma Process. **36**, 1417–1430 (2016)
138. Marrakchi, F., Khanday, W.A., Asif, M., Hameed, B.H.: Cross-linked chitosan/sepiolite composite for the adsorption of methylene blue and reactive orange 16. Int. J. Biol. Macromol. **93**, 1231–1239 (2016)
139. d'Espinose de la Caillerie, J., Gruver, V., Fripiat, J.: Modification of the surface properties of natural phyllosilicate sepiolite by secondary isomorphic substitution. J. Catal. **151**, 420–430 (1995)
140. Joussein, E., Petit, S., Churchman, J., Theng, B., Righi, D., Delvaux, B.: Halloysite clay minerals—a review. Clay Miner. **40**, 383–426 (2005)
141. Abdullayev, E., Lvov, Y.: Halloysite clay nanotubes for controlled release of protective agents. ACS Nano. **11**, 10007–10026 (2011)
142. Bates, A.S.T., Hildebrand, F.: Morphology and structure of endellite and halloysite. Am. Mineral. **35**, 463–483 (1950)
143. Price, R.R., Gaber, B.P., Lvov, Y.: In-vitro release characteristics of tetracycline HCl, khellin and nicotinamide adenine dineculeotide from halloysite; a cylindrical mineral. J. Microencapsul. **18**, 713–722 (2001)
144. Riahi-Madvaar, R., Taher, M.A., Fazelirad, H.: Synthesis and characterization of magnetic halloysite-iron oxide nanocomposite and its application for naphthol green B removal. Appl. Clay Sci. **137**, 101–106 (2017)

145. Kim, T., Nguyen, L., Cao, X.T., Park, C., Lim, K.T.: Preparation, characterization and application of magnetic halloysite nanotubes for dye removal. Mol. Cryst. Liq. Cryst. **644**, 153–159 (2017)
146. Fard, F.S., Akbari, S., Pajootan, E., Arami, M.: Enhanced acidic dye adsorption onto the dendrimer-based modified halloysite nanotubes. Desalin. Water Treat. **57**, 26222–26239 (2016)
147. Shu, Z., Chen, Y., Zhou, J., Li, T., Sheng, Z., Tao, C., Wang, Y.: Preparation of halloysite-derived mesoporous silica nanotube with enlarged specific surface area for enhanced dye adsorption. Appl. Clay Sci. **132–133**, 114–121 (2016)
148. Massaro, M., Colletti, C.G., Lazzara, G., Guernelli, S., Noto, R., Riela, S.: Synthesis and characterization of halloysite-cyclodextrin nanosponges for enhanced dyes adsorption. ACS Sustain. Chem. Eng. **5**, 3346–3352 (2017)
149. Wang, X., Chen, Y., Zhang, W., He, W., Wang, J., Chen, B.: Rapid adsorption of cobalt (II) by 3-aminopropyltriethoxysilane modified halloysite nanotubes. Korean J. Chem. Eng. **33**, 3504–3510 (2016)
150. Zhu, K., Duan, Y., Wang, F., Gao, P., Jia, H., Ma, C., Wang, C.: Silane-modified halloysite/Fe_3O_4 nanocomposites: simultaneous removal of Cr(VI) and Sb(V) and positive effects of Cr (VI) on Sb(V) adsorption. Chem. Eng. J. **311**, 236–246 (2017)
151. Frost, R.L., Locos, O.B., Ruan, H., Kloprogge, J.T.: Near-infrared and mid-infrared spectroscopic study of sepiolites and palygorskites. Vib. Spectrosc. **27**, 1–13 (2001)
152. Leboda, R., Chodorowski, S., Skubiszewska-Ziba, J., Tarasevich, Y.I.: Effect of the carbonaceous matter deposition on the textural and surface properties of complex carbon-mineral adsorbents prepared on the basis of palygorskite. Colloids Surf. A **178**, 113–128 (2001)
153. Frost, R.L., Cash, G.A., Kloprogge, J.T.: 'Rocky Mountain leather', sepiolite and attapulgite—an infrared emission spectroscopic study. Vib. Spectrosc. **16**, 173–184 (1998)
154. Serna, C., Vanscoyoc, G.: Infrared study of sepiolite and palygorskite surfaces. Dev. Sedimentol. **27**, 197–206 (1979)
155. Wang, X., Wang, C.: Chitosan-poly(vinyl alcohol)/attapulgite nanocomposites for copper(II) ions removal: pH dependence and adsorption mechanisms. Colloids Surf. A **500**, 186–194 (2016)
156. Feng, Y., Wang, Y., Liu, S., Jiang, J., Cao, C., Yao, J.: Simple fabrication of easy handling millimeter-sized porous attapulgite/polymer beads for heavy metal removal. J. Colloid Interface Sci. **502**, 52–58 (2017)
157. Wang, H., Liu, Y., Li, M., Huang, H., Xu, H.M., Hong, R.J., Shen, H.: Fabrication of manganese dioxide/carbon/attapulgite composites derived from spent bleaching earth for adsorption of Pb(II) and brilliant green. Optoelectron. Adv. Mater. Rapid Commun. **4**, 1166–1169 (2010)
158. Li, X.J., Yan, C.J., Luo, W.J., Gao, Q., Zhou, Q., Liu, C., Zhou, S.: Exceptional cerium(III) adsorption performance of poly (acrylic acid) brushes-decorated attapulgite with abundant and highly accessible binding sites. Chem. Eng. J. **284**, 333–342 (2017)
159. Wang, H., Wang, X., Ma, J., Xia, P., Zhao, J.: Removal of cadmium (II) from aqueous solution: a comparative study of raw attapulgite clay and a reusable waste–struvite/attapulgite obtained from nutrient-rich wastewater. J. Hazard. Mater. **329**, 66–76 (2017)
160. Chen, Z., He, J., Chen, L., Lu, S.: Sorption and desorption properties of Eu(III) on attapulgite. J. Radioanal. Nucl. Chem. **307**, 1093–1104 (2016)
161. Liu, D., Zheng, H.: Enhanced adsorption of radioactive strontium ions from aqueous solution by H_2O_2-modified attapulgite. J. Radioanal. Nucl. Chem. **311**, 1883–1890
162. Pan, D., Fan, Q., Fan, F., Tang, Y., Zhang, Y., Wu, W.: Removal of uranium contaminant from aqueous solution by chitosan@attapulgite composite. Sep. Purif. Technol. **177**, 86–93 (2017)
163. Han, H., Cheng, C., Hu, S., Li, X., Wang, W., Xiao, C., Xu, Z., Shao, D.: Facile synthesis of gelatin modified attapulgite for the uptake of uranium from aqueous solution. J. Mol. Liq. **234**, 172–178 (2017)

164. Sun, N., Zhang, Y., Ma, L., Yu, S., Li, J.: Preparation and characterization of chitosan/purified attapulgite composite for sharp adsorption of humic acid from aqueous solution at low temperature. J. Taiwan Inst. Chem. Eng. **78**, 96–103 (2016)
165. Mu, B., Kang, Y., Zheng, M., Wang, A.: One-pot preparation of superparamagnetic attapulgite/Fe_3O_4/polydopamine nanocomposites for adsorption of methylene blue. Times Polym. Compos. **020147**, 020147 (2016)
166. Wu, P., Wu, J., Xia, L., Liu, Y., Xu, L., Song, S.: Adsorption of fluoride at the interface of water with calcined magnesium-ferri-lanthanum hydrotalcite-like compound. RSC Adv. **7**, 26104–26112 (2017)
167. Ben Nasr, A., Walha, K., Puel, F., Mangin, D., Ben Amar, R., Charcosset, C.: Desalination and water treatment precipitation and adsorption during fluoride removal from water by calcite in the presence of acetic acid. Desalin. Water Treat. **52**, 2231–2240 (2014)
168. Meenakshi, Maheshwari R.C.: Fluoride in drinking water and its removal. J. Hazard. Mater. **137**, 456–463 (2006)
169. Yang, M.I.N., Hashimoto, T., Hoshi, N., Myoga, H.: Fluoride removal in a fixed bed packed with granular calcite. Water Res. **33**, 3395–3402 (1999)
170. Turner, B.D., Binning, P.J., Sloan, S.W.: A calcite permeable reactive barrier for the remediation of Fluoride from spent potliner (SPL) contaminated groundwater. J. Contam. Hydrol. **95**, 110–120 (2008)
171. Reardon, E.J.: A limestone reactor for fluoride removal from wastewaters. Environ. Sci. Technol. **34**, 3247–3253 (2000)
172. Zachara, J.M., Kittrick, J.A., Harsh, J.B.: The mechanism of Zn^{2+} adsorption on calcite. Geochim. Cosmochim. Acta **52**, 2281–2291 (1988)
173. Zachara, J.M., Cowan, C.E., Resch, C.T.: Sorption of divalent metals on calcite. Geochim. Cosmochim. Acta **55**, 1549–1562 (1991)
174. Freeman, J.S., Rowell, D.L.: The absorption and precipitation of phosphate onto calcite. Eur. J. Soil Sci. **32**, 75–84 (2010)
175. Kobayashi, H., Satoh, K., Sawada, K.: Adsorption of divalent heavy metal ions on calcium carbonate (calcite). Bunseki Kagaku **53**, 101–107 (2004)
176. Van Der Weijden, R.D., Meima, J., Comans, R.N.J.: Sorption and sorption reversibility of cadmium on calcite in the presence of phosphate and sulfate. Mar. Chem. **57**, 119–132 (1997)
177. Wang, C., Liu, J., Zhang, Z., Wang, B., Sun, H.: Adsorption of Cd (II), Ni (II), and Zn (II) by tourmaline at acidic conditions : kinetics, thermodynamics, and mechanisms. Ind. Eng. Chem. Res. **51** (2012)
178. Lameiras, F.S., Nunes, E.H.M., Leal, J.M.: Backgrounds for the industrial use of black tourmaline based on its crystal structure characteristics. Ferroelectrics **377**, 107–119 (2011)
179. Wang, C.P., Wu, J.Z., Sun, H.W., Wang, T., Liu, H.B., Chang, Y.: Adsorption of Pb (II) ion from aqueous solutions by tourmaline as a novel adsorbent. Ind. Eng. Chem. Res. **50**, 8515–8523 (2011)
180. Kan, J., Tie-heng, S.U.N., Li-na, S.U.N., Hai-bo, L.I.: Adsorption characteristics of copper, lead, zinc and cadmium ions by tourmaline. J. Environ. Sci. **18**, 1221–1225 (2016)
181. Liu, H., Wang, C., Liu, J., Wang, B., Sun, H.: Competitive adsorption of Cd (II), Zn (II) and Ni (II) from their binary and ternary acidic systems using tourmaline. J. Environ. Manag. **128**, 727–734 (2016)
182. Liu, N., Wang, H., Weng, C., Hwang, C.: Adsorption characteristics of Direct Red 23 azo dye onto powdered tourmaline. Arab. J. Chem. (2016, In press)
183. Li, G., Chen, D., Zhao, W., Zhang, X.: Efficient adsorption behavior of phosphate on La-modified tourmaline. J. Environ. Chem. Eng. **3**, 515–522 (2015)
184. Zheng, Y., Wang, A.: Removal of heavy metals using polyvinyl alcohol semi-IPN poly (acrylic acid)/tourmaline composite optimized with response surface methodology. Chem. Eng. J. **162**, 186–193 (2010)
185. Wang, C., Zhang, Y., Yu, L., Zhang, Z., Sun, H.: Oxidative degradation of azo dyes using tourmaline. J. Hazard. Mater. **260**, 851–859 (2013)

186. Tokumura, M., Znad, H.T., Kawase, Y.: Modeling of an external light irradiation slurry photoreactor: UV light or sunlight-photoassisted Fenton discoloration of azo-dye Orange II with natural mineral tourmaline powder. Chem. Eng. Sci. **61**, 6361–6371 (2006)
187. Li, N., Li, X., Wang, C., Shi, X., Liu, J.: Journal of environmental chemical engineering desorption of Cd (II) from tourmaline at acidic conditions: kinetics, equilibrium and thermodynamics. J. Environ. Chem. Eng. **4**, 30–36 (2016)
188. Diebold, U.: The surface science of titanium dioxide. Surf. Sci. Rep. **48**, 53–229 (2003)
189. Veréb, G., Gyulavári, T., Pap, Z., Baia, L., Mogyorósi, K., Dombi, A., Hernádi, K.: Visible light driven photocatalytic elimination of organic- and microbial pollution by rutile-phase titanium dioxides: new insights on the dynamic relationship between morpho-structural parameters and photocatalytic performance. Rsc. Adv. **5**, 66636–66643 (2015)
190. Tijani, J.O., Fatoba, O.O., Madzivire, G., Petrik, L.F.: A review of combined advanced oxidation technologies for the removal of organic pollutants from water. Water Air Soil Pollut. **225**, 2102 (2010)
191. Qi, K., Zasada, F., Piskorz, W., Indyka, P., Gryboś, J., Trochowski, M., Buchalska, M., Kobielusz, M., Macyk, W., Sojka, Z.: Self-sensitized photocatalytic degradation of colorless organic pollutants attached to rutile nanorods—experimental and theoretical DFT+D studies. J. Phys. Chem. C **120**, 5442–5456 (2010)
192. Chong, M.N., Jin, B., Chow, C.W., Saint, C.: Recent developments in photocatalytic water treatment technology: a review. Water Res. **44**, 2997–3027 (2010)
193. Chanathaworn, J., Bunyakan, C., Wiyaratn, W., Chungsiriporn, J.: Photocatalytic decolorization of basic dye by TiO_2 nanoparticle in photoreactor. Songklanakarin J. Sci. Technol. **34**, 203–210 (2012)
194. Dolat, D., Quici, N., Kusiak-Nejman, E., Morawski, A.W., Puma, G.L.: One-step, hydrothermal synthesis of nitrogen, carbon co-doped titanium dioxide (N, C TiO_2) photocatalysts. Effect of alcohol degree and chain length as carbon dopant precursors on photocatalytic activity and catalyst deactivation. Appl. Catal. B Environ. **115–116**, 81–89 (2012)
195. Heller, A.: Conversion of sunlight into electrical power and photoassisted electrolysis of water in photoelectrochemical cells. Cheminform **14**, 154–162 (2006)
196. Lee, P.F., Sun, D.D., Leckie, J.O.: Adsorption and photodegradation of humic acids by nano-structured TiO_2 for water treatment. J. Adv. Oxid. Technol. **10**, 72–78 (2006)
197. Asuha, S., Zhou, X.G., Zhao, S.: Adsorption of methyl orange and Cr (VI) on mesoporous TiO_2 prepared by hydrothermal method. J. Hazard. Mater. **181**, 204–210 (2010)
198. López-mu, M., Arencibia, A., Cerro, L., Pascual, R., Melgar, Á.: Applied surface science adsorption of Hg (II) from aqueous solutions using TiO_2 and titanate nanotube adsorbents. Appl. Surf. Sci. **367**, 91–100 (2010)
199. Khataee, A.R., Fathinia, M., Aber, S., Zarei, M.: Optimization of photocatalytic treatment of dye solution on supported TiO_2 nanoparticles by central composite design: intermediates identification. J. Hazard. Mater. **181**, 886–897 (2010)
200. Damodar, R.A., You, S.J., Chou, H.H.: Study the self cleaning, antibacterial and photocatalytic properties of TiO_2 entrapped PVDF membranes. J. Hazard. Mater. **172**, 1321–1328 (2006)
201. Yang, H., Ke, Z., Shi, R., Li, X., Dong, X., Yu, Y.: Sol-gel synthesis of TiO_2 nanoparticles and photocatalytic degradation of methyl orange in aqueous TiO_2 suspensions. J. Alloy. Compd. **413**, 302–306 (2006)
202. Girginov, C., Stefchev, P., Vitanov, P., Dikov, H.: Silver doped TiO_2 photocatalyst for methyl orange degradation. J. Eng. Sci. Technol. Rev. **5**, 14–17 (2009)
203. Behnajady, M.A., Modirshahla, N., Shokri, M., Rad, B.: Enhancement of photocatalytic activity of TiO_2 nanoparticles by Silver doping: photodeposition versus liquid impregnation methods. Glob. Nest. J. **10**, 1–7 (2009)
204. Wang, Y., Zhao, D., Ma, W., Chen, C., Zhao, J.: Enhanced sonocatalytic degradation of azo dyes by Au/TiO_2. Environ. Sci. Technol. **42**, 6173–6178 (2009)

205. Xiu, F.R., Zhang, F.S.: Preparation of nano-Cu_2O/TiO_2 photocatalyst from waste printed circuit boards by electrokinetic process. J. Hazard. Mater. **172**, 1458–1463 (2009)
206. Jia, Y., Yu, X.Y., Luo, T., Jin, Z., Sun, B., Liu, J.H., Huang, X.J.: Necklace-like mesoporous MgO/TiO_2 heterojunction structures with excellent capability for water treatment. Dalt. Trans. **43**, 2348–2351 (2014)
207. Liu, W., Ni, J., Yin, X.: Synergy of photocatalysis and adsorption for simultaneous removal of Cr(VI) and Cr(III) with TiO_2 and titanate nanotubes. Water Res. **53**, 12–25 (2014)
208. Gürboğa, G., Tel, H.: Preparation of TiO_2-SiO_2 mixed gel spheres for strontium adsorption. J. Hazard. Mater. **B120**, 135–142 (2005)
209. Nilchi, A., Garmarodi, S.R., Darzi, S.J.: Adsorption behavior of nano sized sol-gel derived TiO_2-SiO_2 binary oxide in removing Pb^{2+} metal ions. Sep. Sci. Technol. **45**, 801–808 (2010)
210. Zhou, W., Yin, Z., Du, Y., Huang, X., Zeng, Z., Fan, Z., Liu, H., Wang, J., Zhang, H.: Synthesis of few-layer MoS_2 nanosheet-coated TiO_2 nanobelt heterostructures for enhanced photocatalytic activities. Small **9**, 140–147 (2013)
211. Wang, Y., Morin, G., Ona-Nguema, G., Menguy, N., Juillot, F., Aubry, E., Guyot, F., Calas Jr., G., Brown, G.E.: Arsenite sorption at the magnetite–water interface during aqueous precipitation of magnetite: EXAFS evidence for a new arsenite surface complex. Geochim. Cosmochim. Acta **72**, 2573–2586 (2008)
212. Gao, M.R., Zhang, S.R., Jiang, J., Zheng, Y.R., Tao, D.Q., Yu, S.H.: One-pot synthesis of hierarchical magnetite nanochain assemblies with complex building units and their application for water treatment. J. Mater. Chem. **21**, 16888–16892 (2011)
213. Wang, J., Xia, T., Wu, C., Feng, J., Meng, F., Shi, Z., Meng, J.: Self-assembled magnetite peony structures with petal-like nanoslices: one-step synthesis, excellent magnetic and water treatment properties. Rsc. Adv. **2**, 4220–4227 (2012)
214. Yang, W., Zhao, N., Zhang, N., Chen, W., Kan, A.T., Tomson, M.B.: Time-dependent adsorption and resistant desorption of arsenic on magnetite nanoparticles: kinetics and modeling. Desalin. Water Treat. **44**, 100–109 (2012)
215. An, B., Liang, Q., Zhao, D.: Removal of arsenic(V) from spent ion exchange brine using a new class of starch-bridged magnetite nanoparticles. Water Res. **45**, 1961–1972 (2011)
216. Larraza, I., López-Gónzalez, M., Corrales, T., Marcelo, G.: Hybrid materials: magnetite-polyethylenimine-montmorillonite, as magnetic adsorbents for Cr(VI) water treatment. J. Colloid Interf. Sci. **385**, 24–33 (2012)
217. Shaker, S., Zafarian, S., Chakra, C.S., Rao, K.V.: Preparation and characterization of magnetite nanoparticles by sol-gel method for water treatment. Int. J. Innov. Res. Sci. Eng. Technol. **2**, 2069–2973 (2015)
218. Ayad, M., Salahuddin, N., Fayed, A., Bastakoti, B.P., Suzuki, N., Yamauchi, Y.: Chemical design of a smart chitosan-polypyrrole-magnetite nanocomposite toward efficient water treatment. Phys. Chem. Chem. Phys. Pccp. **16**, 21812–21819 (2015)
219. Shim, H.W., Park, S., Song, H.J., Kim, J.C., Jang, E., Hong, K.S., Kim, T.D., Kim, D.W.: Biomineralized multifunctional magnetite/carbon microspheres for applications in Li-ion batteries and water treatment. Chem. (Easton) **21**, 4655–4663 (2015)
220. Huang, Y.H., Tang, C., Zeng, H.: Removing molybdate from water using a hybridized zero-valent iron/magnetite/Fe(II) treatment system. Chem. Eng. J. **200–202**, 257–263 (2012)
221. Sharma, N., Tiwari, A.: Nanomagnetite-loaded poly (acrylamide-co-itaconic acid) hydrogel as adsorbent for effective removal of Mn from contaminated water. Desalin. Water Treat. **57**, 5654–5672 (2016)
222. Maqsood, A., Iqbal, K.: Materials characterization by non-destructive methods. J. Pakistan Mater. Soc. **4**, 31–38 (2010)
223. Zhang, X., Yi, H., Zhao, Y., Min, F., Song, S.: Study on the differences of Na- and Ca-montmorillonites in crystalline swelling regime through molecular dynamics simulation. Adv. Powder Technol. **27**, 779–785 (2016)
224. Rydberg, J.: Wavelength dispersive X-ray fluorescence spectroscopy as a fast, non-destructive and cost-effective analytical method for determining the geochemical composition of small loose-powder sediment samples. J. Paleolimnol. **52**, 265–276 (2014)

225. De Viguerie, L., Sole, V.A., Walter, P.: Multilayers quantitative X-ray fluorescence analysis applied to easel paintings. Anal. Bioanal. Chem. **395**, 2015–2020 (2009)
226. Ramesh, A., Mohan, K.R., Seshaiah, K., Jeyakumar, N.D.: Determination of trace elements by inductively coupled plasma-atomic emission spectrometry (ICP-AES) after preconcentration on a support impregnated with piperidine dithiocarbamate. Anal. Lett. **34**, 219–229 (2002)
227. Aceto, M., Abollino, O., Bruzzoniti, M.C., Mentasti, E., Sarzanini, C., Malandrino, M.: Determination of metals in wine with atomic spectroscopy (flame-AAS, GF-AAS and ICP-AES); a review. Food Addit. Contam. **19**, 126–133 (2002)
228. Benramdane, L., Bressolle, F., Vallon, J.J.: Arsenic speciation in humans and food products: a review. J. Chromatogr. Sci. **37**, 330–344 (2017)
229. Ma, R., Mcleod, C.W., Tomlinson, K., Poole, R.K.: Speciation of protein-bound trace elements by gel electrophoresis and atomic spectrometry. Electrophoresis **25**, 2469–2477 (2017)
230. Zemła, J., Danilkiewicz, J., Orzechowska, B., Pabijan, J., Seweryn, S., Lekka, M.: Atomic force microscopy as a tool for assessing the cellular elasticity and adhesiveness to identify cancer cells and tissues. Semin. Cell Dev. Bio. **l**, 1–10 (2017)
231. Smith, J.R., Tsibouklis, J., Nevell, T.G., Breakspear, S.: AFM friction and adhesion mapping of the substructures of human hair cuticles. Appl. Surf. Sci. **285**, 638–641 (2009)
232. Guo, Q., Xia, Y., Sandig, M., Yang, J.: Characterization of cell elasticity correlated with cell morphology by atomic force microscope. J. Biomech. **45**, 304–309 (2009)
233. Sirghi, L., Bretagnol, F., Mornet, S., Sasaki, T., Gilliland, D., Colpo, P., Rossi, F.: Atomic force microscopy characterization of the chemical contrast of nanoscale patterns fabricated by electron beam lithography on polyethylene glycol oxide thin films. Ultramicroscopy **109**, 222–229 (2009)
234. Butt, H.J., Cappella, B., Kappl, M.: Force measurements with the atomic force microscope: technique, interpretation and applications. Surf. Sci. Rep. **59**, 1–152 (2005)
235. Cappella, B., Dietler, G.: Force-distance curves by atomic force microscopy. Surf. Sci. Rep. **34**, 1–104 (1999)
236. Peng, W., Li, H., Liu, Y., Song, S.: Effect of oxidation degree of graphene oxide on the electrochemical performance of CoAl-layered double hydroxide/graphene composites. Appl. Mater. Today **7**, 201–211 (2017)
237. Zhang, X., Jia, F., Yang, B., Song, S.: Oxidation of molybdenum disulfide sheet in water under in situ atomic force microscopy observation. J. Phys. Chem. C **121**, 9938–9943 (2017)
238. Joy, D.C., Newbury, D.E., Davidson, D.L.: Electron channeling patterns in the scanning electron microscope. J. Appl. Phys. **53** (1982)
239. Kanemaru, T., Hirata, K., Takasu, S., Isobe, S., Mizuki, K., Mataka, S., Nakamura, K.: A fluorescence scanning electron microscope. Ultramicroscopy **109**, 344–349 (2009)
240. Seiler, H.: Secondary electron emission in the scanning electron microscope. J. Appl. Phys. **54** (1983)
241. Chang, H., Iqbal, S.M., Stach, E.A., King, A.H., Zaluzec, N.J., Bashir, R.: Fabrication and characterization of solid-state nanopores using a field emission scanning electron microscope. Appl. Phys. Lett. **88**, 2004–2007 (2006)
242. Harrison, C., Park, M., Chaikin, P.M., Register, R.A., Adamson, D.H., Yao, N.: Layer by layer imaging of diblock copolymer films with a scanning electron microscope. Polymer (Guildf) **30**, 2733–2744 (1998)
243. Whalley, W., Krinsley, D.: A scanning electron microscope study of surface textures of quartz grains from glacial environments. Sedimentology **21**, 87–105 (1968)
244. Crewe, A.V., Wall, J., Welter, L.M.: A high-resolution scanning transmission electron microscope. J. Appl. Phys. **39**, 5861–5868 (1968)
245. Albina, S.J.P., Borisevich, Y., Andrew, Lupini R.: Depth sectioning with the aberration-corrected scanning transmission electron microscope. Somat. Cell Genet. **103**, 3044–3048 (1984)
246. Williams, D.B., Carter, C.B.: In: Transmission Electron Microscopy (2009)

247. (1984) Peak fitting methods and parameter settings in XPS analysis for organic sulfur in coal. Chem. Phys. Lett. **220**, 7–10
248. Oelhafen, P.: Practical surface analysis by auger and X-ray photoelectron spectroscopy. Int. J. Adhes. Adhes. **34**, 203 (1984)
249. Briggs, D., Grant, J.T.: Surface analysis by Auger and Photoelectron Spectroscopy 2014 (2003)
250. Chen, M., Wang, X., Yu, Y.H., Pei, Z.L., Bai, X.D., Sun, C., Huang, R.F., Wen, L.S.: X-ray photoelectron spectroscopy and auger electron spectroscopy studies of Al-doped ZnO films. Appl. Surf. Sci. **158**, 134–140 (2000)
251. Katari, J.E.B., Colvin, V.L., Alivisatos, A.P.: X-ray photoelectron spectroscopy of CdSe nanocrystals with applications to studies of the nanocrystal surface. J. Phys. Chem. **98**, 4109–4117 (1994)
252. Castner, D.G., Hinds, K., Grainger, D.W.: X-ray photoelectron spectroscopy sulfur 2p study of organic thiol and disulfide binding interactions with gold surfaces. Langmuir **12**, 5083–5086 (1996)
253. Susi, H., Byler, D.M.: In: Resolution-Enhanced Fourier Transform Infrared Spectroscopy of Enzymes (1986)
254. Books OUR: In: Fundamentals of Fourier Transform Infrared Spectroscopy (1996)
255. Bates, J.B.:In: Fourier Transform Infrared Spectroscopy (2016)
256. Wang, N., Huang, K., Song, J.: Gas-sensing properties and in-situ diffuse-reflectance fourier-transform infrared spectroscopy study of diethyl ether adsorption and reactions on SnO_2 film. Mater. Sci. **35**, 265–274 (2017)
257. Chen, A., Richer, J., Roscoe, S.G., Lipkowski, J.: Electrochemical and fourier transform infrared spectroscopy studies of benzonitrile adsorption at the Au (111) electrode. Langmuir **13**, 4737–4747 (1997)
258. Zhao, W., Liu, F., Feng, X., Tan, W., Qiu, G., Chen, X.: Fourier transform infrared spectroscopy study of acid birnessites before and after Pb^{2+} adsorption. Clay Miner. **47**, 191–204 (1993)
259. Tripp, C.P., Hair, M.L.: Measurement of polymer adsorption on colloidal silica by in S u transmission fourier transform infrared spectroscopy. Society, 3523–3529 (1993)
260. Schweiger, G.: Observation of morphology-dependent resonances caused by the input field in the Raman spectrum of microdroplets. J. Raman Spectrosc. **21**, 165–168 (1990)
261. Agullo-Rueda, F., Moreno, J.D., Montoya, E., Guerrero-Lemus, R., Martin-Palma, R.J., Martinez-Duart, J.M.: Selection rules in the Raman spectrum of porous silicon. Mater. Res. Soc. Symp. Proc. **452**, 571–575 (1997)
262. Marconi, G.: On the nature of the lowest excited states of styrene: calculation of the resonance Raman spectrum and pre-resonance profiles. 3–6 (1983)
263. Bennink, R.S., Marino, A.M., Wong, V., Stroud, J.R.C.R., Boyd, R.W., Narducci, F.A.: Electromagnetically induced absorption and transparency in a single Raman spectrum. Phys. Rev. A At Mol. Opt. Phys. **4**, 1–2 (2003)
264. Tuinstra, F., Koenig, J.L.: Raman spectrum of graphite. J. Chem. Phys. **53**, 1126–1130 (1970)
265. Diéguez, A., Romano-Rodríguez, A., Vilà, A., Morante, J.R.: The complete Raman spectrum of nanometric SnO_2 particles. J. Appl. Phys. **90**, 1550–1557 (2000)
266. Prawer, S., Nugent, K., Jamieson, D., Orwa, J., Bursill, L., Peng, J.: The Raman spectrum of nanocrystalline diamond. Chem. Phys. Lett. **332**, 93–97 (2000)
267. Temple, P.A., Hathaway, C.E.: Multiphonon Raman spectrum of silicon. Phys. Rev. B **7**, 3685–3697 (1973)
268. Callaghan, P.: Principles of nuclear magnetic resonance microscopy (1991)
269. Rutledge, D.: Encyclopedia of nuclear magnetic resonance (1996)
270. Dunn, K.J., Bergman, D.J., Latorraca, G.: In: Nuclear Magnetic Resonance: Petrophysical and Logging Applications (2002)
271. Mcwhinnie, W.: In: Electron Spin Resonance and Nuclear Magnetic Resonance Applied to Minerals (1985)

272. Murdoch, J.B., Stebbins, J.F., Carmichael, I.S.E., Pines, A.: A silicon-29 nuclear magnetic resonance study of silicon-aluminum ordering in leucite and analcite. Phys. Chem. Miner. **15**, 370–382 (1988)
273. Saidov, T.A., Espinosa-Marzal, R.M., Pel, L., Scherer, G.W.: Nucleation of sodium sulfate heptahydrate on mineral substrates studied by nuclear magnetic resonance. J. Cryst. Growth **338**, 166–169 (2012)
274. Sanders, R.L.: Clay mineral reactivity across scales utilizing solid-state nuclear magnetic resonance (2011)
275. Okolo, G.N., Everson, R.C., Neomagus, H.W.J.P., Roberts, M.J., Sakurovs, R.: Comparing the porosity and surface areas of coal as measured by gas adsorption, mercury intrusion and SAXS techniques. Fuel **141**, 293–304 (2015)
276. Zhang, L., Li, H., Chen, F., Zhang, D., Wu, M., Pan, B., Xing, B.: New insights provided by solvent relaxation NMR measured surface area in liquids to explain phenolics sorption on silica nanoparticles. Environ. Sci. Nano. **4**, 577–584 (2017)
277. Hunter, R.J., Ottewill, R.H., Rowell, R.L.: In: Zeta Potential in Colloid Science, pp. 219–257. Acad Press (1981)
278. Delf, L.B., Jelvehgari, M., Zakeri-Milani, P., Valizadeh, H.: Statistical optimization of oral Vancomycin-Eudragit RS nanoparticles using response surface methodology. Iran. J. Pharm. Res. IJPR **11**, 1001 (2012)
279. O'Brien, R.W., Midmore, B.R., Lamb, A., Hunter, R.J.: Electroacoustic studies of moderately concentrated colloidal suspensions. Faraday Discuss. Chem. Soc. **90**, 301–312 (1990)
280. Barner, A., Myers, M.: The effects of carboxylic acids on the aqueous dispersion and electrophoretic deposition of ZrO_2. J. Eur. Ceram. Soc. **32**, 235–244 (2013)

Surface Chemistry of Mineral Adsorbents

Shaoxian Song, Weijun Peng, and Hongqiang Li

Abstract Surface chemistry of mineral adsorbents dominates the interaction of adsorbent and adsorbates, and thus the adsorption. The surface chemistry related to adsorption at solid/liquid interfaces including surface areas, surface charges, surface interactions, adsorption thermodynamics and kinetics, etc. are described in the chapter. The roles of surface area, surface charge, and surface interaction in the adsorption of contaminants have been deep discussed. In addition, the chapter also reviewed the characterization technology of adsorption including component, structural and morphology, surface and interface properties, and other properties. The chapter provides a fundamental understanding of the surface chemistry of mineral adsorbents for the selection, development of efficient mineral based adsorbents.

Keywords Surface chemistry · Solid/liquid interfaces · Adsorption thermodynamics and kinetics · Characterization technologies of adsorption

---

S. Song (✉)
Hubei Key Laboratory of Mineral Resources Processing and Environment, Wuhan University of Technology, Luoshi Road 122, Wuhan 430070, Hubei, China
e-mail: ssx851215@whut.edu.cn

S. Song
School of Resources and Environmental Engineering, Wuhan University of Technology, Luoshi Road 122, Wuhan 430070, Hubei, China

W. Peng
School of Chemical Engineering, Zhengzhou University, Zhengzhou 450001, Henan, People's Republic of China

H. Li
School of Xingfa Mining, Wuhan Institute of Technology, Wuhan 430205, Hubei, China

S. Song and B. Li (eds.), *Adsorption at Natural Minerals/Water Interfaces*, Engineering Materials,
https://doi.org/10.1007/978-3-030-54451-5_2

1 Introduction

Surface chemistry of mineral adsorbents plays a predominant role in the interaction of adsorbent and adsorbates, and thus the adsorption. In this chapter, the basic knowledge of surface chemistry related to adsorption at solid/liquid interfaces are described, including surface charges, surface areas, surface structures, porous structure, surface interaction, adsorption thermodynamics and kinetics, etc. The structure, chemical composition, exchangeable ion type and small crystalsize of mineral adsorbents are responsible for several unique properties, including a high cation exchange capacity, a large chemically active surface area, interlamellar surfaces having unusual hydration characteristics leading to a low hydraulic conductivity and adsorption capacities. These properties enable them very useful for adsorption. For instance, surface area provides numerous adsorption sites for various adsorbates, while the surface charge drives the heterogeneous charged adsorbent and adsorbate to integration via electrostatic attraction. In addition, the surface interaction also plays a main role in the adsorption of contaminants, and the common mechanisms of interaction for the mineral adsorbents, including ion exchange, electrostatic interactions, hydrophobic/hydrophilic interactions, ligand exchange, cation bridge, water bridge, etc., have been summarized in the chapter. The effects of the surface characteristics on the adsorption at mineral/water interfaces are discussed. Furthermore, the adsorption thermodynamics, adsorption kinetics and characterization technologies of adsorption have also been detailed described.

2 Internal and External Surfaces

Surface area of mineral adsorbents is very important for the adsorption. It determines the contact areas between adsorbent and adsorbate. The vast internal surface (between the interlayer) provides rich available sites for adsorption and a high cation exchange capacity. While the hydrolysable surface hydroxyl sites on the external surface of mineral adsorbents offer a strong binding to a cation contaminants by complexation. In addition, both the internal and external surfaces play a dominant role in adsorption.

2.1 Characteristics

Most mineral adsorbents hold great internal and external surface areas due to their cage-like structures (as shown in Fig. 1). Montmorillonite (MMT) is a low charge expandable phyllosilicate mineral in which Si-tetrahedral sheets and Al-octahedral sheets are combined in a 2:1 ratio. As reported previously, MMT has a large number of Si sites due to its interlayer expansion, compared with the number of Al

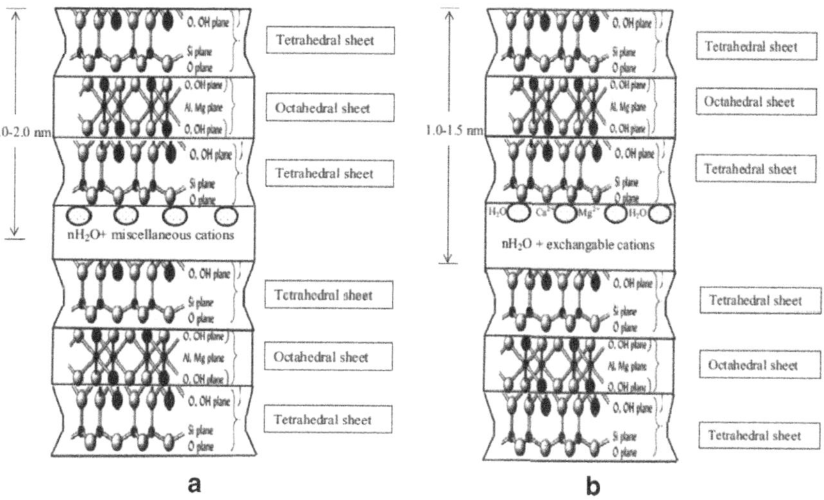

Fig. 1 Diagrammatic representation of crystal structures **a** MMT (smectite) and **b** Vermiculite [2]

sites at its edges. The internal surface area accounts for 97% of the total surface area. As a first approximation, Si sites may be said to account for over 97% of the total number of available adsorption sites of MMT [1].

Moreover, it has been confirmed that Cu(II) species strongly bounded on the external surfaces of MMT at basic pH conditions, and the amount of strongly bond Cu(II) increased with increasing pH. As the adsorption of metal cation at the surface hydroxyl site is coupled with a release of H^+ to the solution, it is strongly pH dependent and favoured by high pH values. Consequently, the pH dependent sorption mechanism can be properly interpreted by considering it as a strong binding to a hydrolysable surface hydroxyl site on the external surface of MMT particles. Hyun et al. [1] found that the ion exchange to the internal surface and complexation to the external surface contributed to the overall sorption to a varying degree at intermediate pH conditions in both electrolyte conditions (as shown in Table 1).

Table 1 Summary of the proposed mechanisms of U(VI) sorption on MMT

Proposed mechanism	Ion exchange	Surface complexation
Binding site	Interlayer site	Edge site
Location of site	Internal surface	External surface
Charge character of site	Fixed	Variable
Supposed nature of binding	Electrostatic attraction (weak)	At least partially chemical interaction (stronger)
Prevalence condition	Low ionic strength	High pH
	Low pH	

2.2 Measurements

Several methods including Brunauer, Emmitt and Teller (BET) gas adsorption technique, methylene blue, and cation exchange capacities by conventional titration procedure have been proposed to measure the internal and external surface area of mineral adsorbents.

(1) BET gas adsorption technique

Kohayashi and Owada [3] have examined the accessibility of the surface of homoionic MMTs to nitrogen and carbon dioxide adsorbate at 77 and 195 K, respectively, using a dynamic measuring system. From variation in the surface area obtained by applying BET theory to these measurements, these authors considered that there was a certain degree of penetration of both nitrogen and carbon dioxide between the unit platelets or lamellae forming the crystals of MMT. Their experimental data indicated that the extent of penetration was time dependent and was a function of the interlayer forces that governed by the size and charge of the replaceable cations.

Researchers have recently pointed out that in the dry state MMT forms complexly interwoven martrix in which one lamella may conceivably pass through several crystalline regions. It is considered that the term "quasi-crystalline" may be the most appropriate description here since in these regions the lamellae are stacked in parallel array but not necessarily in perfect crystalline order. In these circumstances the definition of apparent crystal size must be somewhat arbitrary. It was considered that the area determined by nitrogen adsorption at low temperatures was essentially a measure of the surface area external to these quasi-crystalline regions. The difference in the degree of association of the lamellae in aqueous suspensions and the subsequent statistical arrangement of the units on drying, and partly to variation in accessibility of areas of overlap of quasi-crystalline regions with size of the exchangeable cations all had effect on the measurements. Besides, the surface areas of homoionic illite and MMTs were measured by the application of the BET theory to the nitrogen and carbon dioxide adsorption isotherms, and the results shown in Table 2.

Table 2 BET surface areas from nitrogen and carbon dioxide adsorption on $<2\mu$ fraction of homoionic illite and MMTs

Clay	Surface area (m^2/g)	
	Nitrogen adsorption	Carbon dioxide adsorption
Sodium Fithian illite	105	72.4
Sodium Willalooka illite	195	162
Caesium Willalooka illite	184	159
Sodium Wyoming bentonite	39.8	40.6
Caesium Wyoming bentonite	99.1	189
Sodium Redhill montmorillonite	147	132
Caesium Redhill montmorillonite	158	215

The BET area obtained for a MMT was particularly dependent on the efficiency of the outgassing procedure used. The specific surface area of Ca-Wyoming bentonite was increased to a value of about 120 m^2 g^{-1} with decreasing hydration due to the accessibility to nitrogen of surface propped apart by water molecules. Similar effects have been noted in the detailed studies of the surface areas of both illite and MMTs with small amounts of presorbed water. Invariably the specific surface area increased to a maximum with the gradual removal of adsorbed water from the clay surface, followed by a decrease due to the collapse of regions between surfaces propped apart by the last remaining water molecules. In conclusion, the distinction between internal and external surfaces is essentially one of alignment of the lamellae to the adsorbate.

(2) Methylene blue adsorption

The adsorption of methylene blue dye by mineral adsorbents is currently used for determining either their cation exchange capacities or their surface areas. The amount of methylene blue adsorbed by or exchanged on to a mineral adsorbent from an aqueous solution is usually measured using a colorimeter or spectrophotometer. This measurement is complicated by the dimerization of methylene blue when the concentration exceeds about 7×10^{-6} mol/L, corresponding to an optical density higher than about 0.6. The change of colour with concentration is related to the use of methylene blue as a field test for identifying mineral adsorbents. The adsorption of methylene blue by glass surfaces is another possible source of error.

Hang and Brindley used methylene blue to measure the surface area of florida kaolinite, delaminated kaolinite, Oswego illite and Na montmorillonite [4]. The specific surface areas of the clays are calculated from the amounts of methylene blue adsorbed when flocculation is optimum by taking the area per adsorbed molecule at 130 Å, which corresponds to the molecules lying flat on the clay mineral surfaces. The amounts of methylene blue absorbed and the resulting areas obtained are listed in Table 3.

As shown in Table 3, the calculated surface area of Na-MMT is 750–800 m^2 g^{-1} where the range of values arises from the uncertain contribution from the edges of particles. Moreover, the agreement between the surface areas determined by methylene blue absorption and gas adsorption or calculation is everywhere. It is thought that the slightly higher values obtained by methylene blue absorption for delaminated kaolinite and illite may arise from a small amount of a swelling mineral in which, under gas absorption methods, would collapse and contribute little to the total area.

Table 3 Comparison of results by methylene blue adsorption and by other methods

	Methylene blue absorbed at optimum flocculation (meq/100 g clay)	Surface area, m^2/g of clay		Cation exchange capacity meq/100 g clay	
		BET meas.	MB absorption	Titration	MB absorption
Florida kaolinite	4.50 ± 0.25	33.0 ± 2.0	35.2 ± 2.0	7.3 ± 0.2	6.5 ± 0.4 (low conc.)
					7.2 ± 0.3 (high conc.)
Delaminated kaolinite	1.10 ± 0.05	5.5 ± 0.6	8.6 ± 0.6	1.87 ± 0.03	1.95 ± 0.2 (high conc.)
Oswego illite	10.00 ± 0.25	68 ± 2	78 ± 2	13.9 ± 0.1	11.5 ± 0.4 (low conc.)
					13.2 ± 0.25 (high conc.)
Na montmorillonite	95.0 ± 0.5	*	743 ± 4	124.9 ± 1	126 ± 2 (high conc.)

*Obtained by calculation to be 750–800 m^2/g of clay

3 Surface Charges

3.1 Characteristics

Mineral adsorbents contain layered particles or their crystallites composed of many layers. The individual layers have a negative surface charge, which is the sum of variable and permanent charges. The variable charge occurred at the edges of mineral layers is dominated by the athydroxyl groups exhibiting acid/base properties. The variable surface charge, which is due to the protonation and/or deprotonation of some surface sites, is mainly a function of the pH and ionic strength of the aqueous solution. In acidic solution the edge of mineral adsorbents is expected to carry a positive charge, while in alkaline solution the edge is negatively charged, as shown for kaolinite in Fig. 2.

The surface charge is frequently determined by electrophoretic measurements. Traditional electrophoretic measurements are compromised by the shape and anisotropic character of layered silicate particles. In fact, it is not exactly clear what the measured electrophoretic mobility represents. In any event, electrophoretic mobility measurements have been made and apparent points of zero charge (PZC) have been reported for mineral adsorbents particles.

The permanent charge is due to the presence of non-equivalent, isomorphic substitutions of the central atoms within the octahedral and/or tetrahedral sheets, or vacancies. Commonly, Al^{3+} substitutes for Si^{4+} in tetrahedral coordination, and Mg^{2+} substitutes for Fe^{3+} or Al^{3+} in octahedral coordination. Zhang et al. [5] used a specific organic ion (benzyltrimethylammonium ions, $BTMA^+$) to determine the substitution sites of mica, and the atomic force microscopy imaging of the

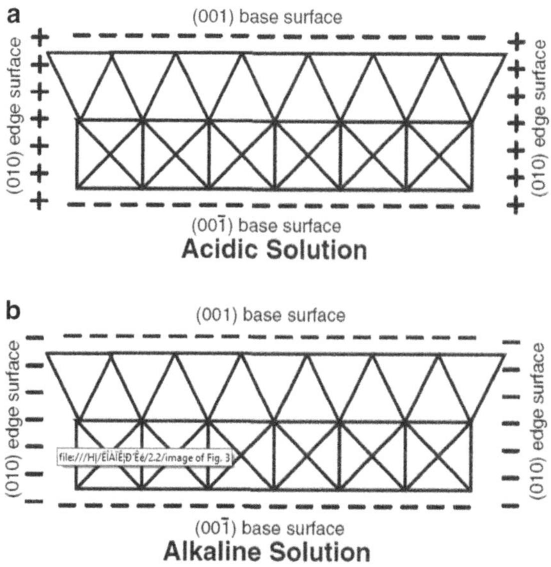

Fig. 2 Surface charge of kaolinite at its face surfaces and edge surfaces in **a** acidic and **b** alkaline solution

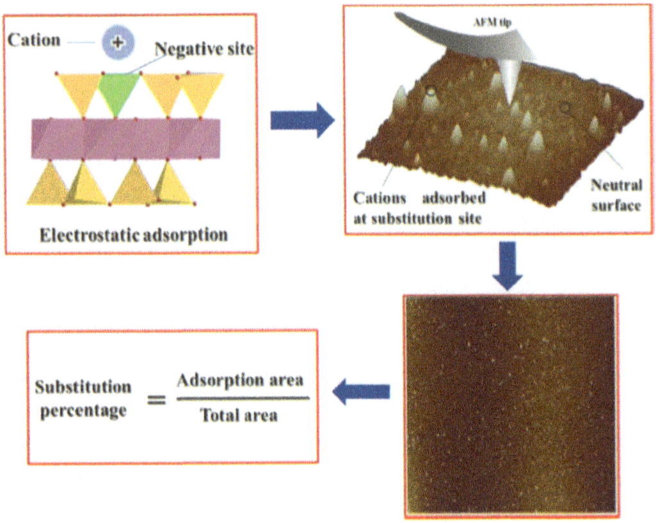

Fig. 3 Schematic diagram of the method for determining the isomorphous substitution degree [5]

adsorption. By means of an image processing software, the AFM images was analyzed and the areas as well as area percentages of the substitution sites on the clay mineral surfaces could be determined (as shown in Fig. 3). A high degree of isomorphous substitution means a stronger electronegative surface. Therefore, the permanent charge is independent of the pH of the surrounding aqueous solution.

Details of typical structural formula and charge characteristics of some common layer mineral adsorbents are given in Table 4. The structural charge can vary from practically zero to >200 cmol kg^{-1}. Therefore, while the structures of all 2:1 layer-silicate surfaces, for example, are superficially similar, the surface chemical properties can be greatly different depending on the position and quantity of structural charge. The structural charge must be balanced by cations at or near the mineral surface. The spatial distribution of these exchangeable "counterions" in the vicinity of the surface greatly affects the colloidal behaviour of mineral adsorbents.

3.2 *Auxiliary Categories of Surface Charge*

Surface charge of mineral adsorbents particles can develop as a result of proton or other ion complexation at the particle surface and from isomorphic ion substitution within the crystal structure. Net total particle surface charge, σ_P, (in units of moles of charge per kilogram of solid [mol_c kg^{-1}]) is the sum of four components:

Table 4 Chemical composition and charge characteristics of representative layered mineral adsorbents

Minerals	Chemical formula	Charge per half unit cell		Structural charge ($cmol_c\ kg^{-1}$)
		Tetrahedral	Octahedral	
2:1 Dioctahedral				
Pyrophyllite	$Si_4Al_2O_{10}(OH)_2$	0	0	0
Montmorillonite	$Ca_{0.165}Si(Al_{1.67}Mg_{0.33})O_{10}(OH)_2$	0	−0.33	92
Beidellite	$Ca_{0.25}(Si_{3.5}Al_{0.5})Al_2O_{10}(OH)_2$	−0.5	0	135
Nontronite	$Ca_{0.25}(Si_{3.5}Al_{0.5})Fe_2^{3+}O_{10}(OH)_2$	−0.5	0	117
Muscovite	$K_{0.94}(Si_{3.11}Al_{0.89})(Al_{1.95}Mg_{0.05})O_{10}(OH)_2$	−0.89	−0.05	237
2:1 Trioctahedral				
Talc	$Si_4Mg_3O_{10}(OH)_2$	0	0	0
Hectorite	$Ca_{0.165}Si_4(Mg_{2.67}Li_{0.33})O_{10}(OH)_2$	0	−0.33	89
Saponite	$Ca_{0.165}(Si_{3.67}Al_{0.33})Mg_3O_{10}(OH)_2$	−0.33	0	87
Vermiculite	$Mg_{0.31}(Si_{3.15}Al_{0.85})(Mg_{2.69}Fe^{3+}_{0.23}Fe^{2+}_{0.08})O_{10}(OH)_2$	−0.85	+0.23	157
Phlogopite	$K(Si_3Al)Mg_3O_{10}(OH)_2$	−1.0	0	240
1:1 Dioctahedral				
Kaolinite	$Si_2Al_2O_5(OH)_4$	0	0	0
1:1 Trioctahedral				
Serpentine	$Si_2Mg_3O_5(OH)_4$	0	0	0

$$\sigma_P = \sigma_O + \sigma_H + \sigma_{IS} + \sigma_{OS} \quad (1)$$

where σ_O is the net permanent structural-charge density resulting from isomorphic substitutions; σ_H is the net proton surface-charge density resulting from the difference in surface excess (q) of H^+ and OH^- ions($q_H - q_{OH}$); σ_{IS} is the net inner-sphere complex charge density resulting from total charge of ions (excluding H^+ and OH^-) bound into inner-sphere surface coordination; and σ_{OS} is the net outer-sphere complex charge density resulting from total charge of ions (excluding H^+ and OH^-) bound into outer-sphere surface coordination. If σ_P is nonzero, then it must be balanced by charge of ions adsorbed in the diffuse layer, σ_D:

$$\sigma_P = -\sigma_D \quad (2)$$

The net charge of ions (excluding H^+ and OH^-) adsorbed into surface complexes and the diffuse ion swarm is given by the surface excess of cation minus anion charge (Δq, $mol_c\ kg^{-1}$):

$$\Delta q = (q_+ - q_-) = \sigma_{IS} + \sigma_{OS} + \sigma_D \tag{3}$$

The point of zero net charge (PZNC) is defined as the pH value where Δq is equal to zero.

The surface charge balance requires that the intrinsic charge of mineral adsorbents particles, σ_{in}, including that resulting from both isomorphic substitutions (permanent charge) and proton adsorption (variable charge) must be balanced by adsorption from solution of background cation and anion:

$$\sigma_{in} = \sigma_0 + \sigma_H = -\Delta q \tag{4}$$

Therefore, by Eq. 4, this is the pH value where $\Delta q = -\sigma_0$. Experimental methods for measuring surface charge components in Eq. 4 include proton titration for σ_H, preferential Cs^+ adsorption for σ_0, and background ion adsorption for Δq. Whereas σ_0 is constant for a given mineral adsorbent, σ_H and σ_q vary with pH and ionic strength.

These surface charge components must be measured independently to assess conformation of the data to Eq. 4. This is particularly important in whole mineral adsorbents can diminish the accuracy of charge measurements. Adsorption and dissolution reactions both consume aqueous H^+ or OH^-; therefore, protontitration data must account for the stoichiometry of dissolution side reactions. Furthermore, re-adsorption of charged dissolution products (e.g., Al^{3+}) can impact the surface excess of index ions used to measure adsorbed ion charge.

3.3 Surface Charge of Trilayer Mineral Adsorbents

For trilayer mineral adsorbents, an octahedral layer was sandwiched by two silica tetrahedral layers. The surface charge may be affected by the substitution in the silica tetrahedral sheet and in the alumina or the magnesia octahedral sheet. Taking MMT as an example, the basal surface of MMT particles have a negative and permanent surface charge whereas the edge surfaces have a pH-dependent surface charge because of the presence of amphoteric surface sites that can exchange protons with the surrounding water [6].

MMT particles are made of a succession of TOT layers constituted of two inward-pointing tetrahedral (T) sheets with a central octahedral (O) sheet (2:1 layer, Fig. 4). Each TOT layer has excess negative permanent charge at its basal surface, e.g., the isomorphic substitution of Si^{4+} by Al^{3+} and Fe^{3+} ions in the tetrahedral sheets, Al^{3+} and Fe^{3+} ions by Mg^{2+} and Fe^{2+} ions in the octahedral sheet. In MMT, most of the isomorphic substitutions occur in the octahedral sheet and the negative

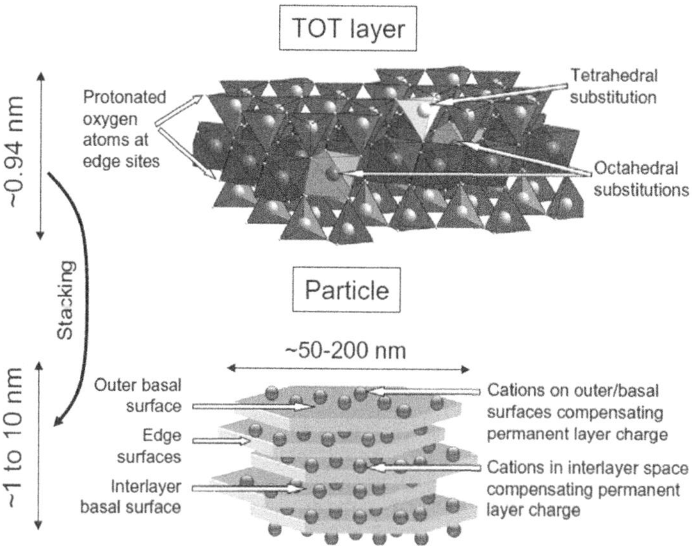

Fig. 4 TOT layer and particle structure of MMT [7]

surface charge is compensated by adsorbed (partially or fully hydrated) cations (e.g. Na^+, K^+, Ca^{2+} or Mg^{2+} ions). These cations can be located in the interlayer space, sandwiched between two TOT layers, or on the outer basal surfaces (Fig. 4). The chemical formula of a Na^+-MMT with both tetrahedral and octahedral isomorphic substitution is Na_{x+y} $[Si^{4+}_{8-y}(Al^{3+}, Fe^{3+})_y]^{IV}$ $[(Al^{3+}, Fe^{3+})_{4-x}(Mg^{2+}, Fe^{2+})_x]^{VI}$ $O_{20}(OH)_4$ (omitting water), with a molecular weight of ~ 750 g mol^{-1} [8].

Zhao et al. reported that the long-range forces at a mica surface were monotonically repulsive for pH 6–10 and the measured forces were pH-independent, thereby confirming that mica basal planes have a permanent surface charge from isomorphous substitution of lattice elements [9]. In contrast, Assemi et al. [10] reported that the basal plane surfaces of mica carry a pH-independent surface charge. They found the surface charge dependency of the mica face surface was a function of pH in their surface force analysis between a silica particle and mica substrate. More detailed research is needed in order to better understand the surface charge properties of the layered silicate minerals. In addition, potentiometric titration is considered to be a better technique to obtain the surface charge information for layered silicates. It should be noted that titration techniques are not affected by particle shape and non-uniform surface charge densities. For example, Christianah et al. employed the potentiometric and mass titration techniques to study the surface properties of MMT, the point of zero charge (PZC) and point of zero net proton charge (PZNPC) of MMT edges at different ionic strengths present pH_{PZC} and pH_{PZNPC} to be 3.4 ± 0.2 (as shown in Fig. 5) [11].

However, since titration can only provide an overall surface charge density of the particles, other techniques are still needed to determine the surface charge

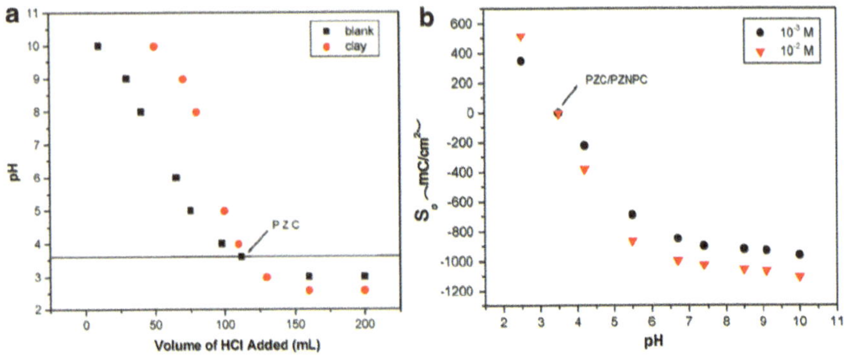

Fig. 5 Potentiometric titration curves for blank distilled water and MMT in distilled water (**a**), and surface charge density of MMT as a function of pH for different KCl concentrations (**b**) [11]

properties at face and edge surfaces. Atomic force microscopy has been used to determine the edge surface charges of mineral adsorbents by surface force measurement. The results showed that the edge surfaces for these two layered silicates are pH-dependent. The isoelectric point of the edge surface for muscovite (~pH 7.5) is slightly lower than that for the talc edge surface (~pH 8.1), apparently due to the difference in the strength of the Al–OH and Mg–OH surface acid groups.

Furthermore, researchers have found that micas and illites hold variable external layer charge, which is varied with the solution. Additionally, some micas and illites have smectite-like external surfaces with lower surface charge, resulting in asymmetrical 2:1 layer charges. This would result from differing degrees of tetrahedral substitution on either side of the outmost 2:1 layer [12]. The magnitude of charge and its origin (tetrahedral oroctahedral) should determine the degree of selectivity of the silicate surface for different metal ions. In the cases of tetrahedrally-derived charge, where ionic interactions act over short distances, even slight differences in ionic potential of adsorbed metal species might be expected to result in adsorption selectivity.

3.4 Surface Charge of Bilayer Clay Minerals

With respect to 1:1 type layered mineral adsorbents, it should be noticed that the electrophoresis gives a reasonable estimation of surface charge densities of spheres having uniform surface charge densities. This technique however does not define sufficiently the charging of platy shaped particles having non-uniform surface charge densities on their face and edge surfaces. The surface chemistry study of bilayer mineral adsorbents can be appropriately considered beyond the planar structures such as kaolinite, to include the tubular halloysite with the same composition as kaolinite, the corresponding magnesium silicates antigorite (planar structure) and chrysotile (tubular structure) that recently discussed in the literature [24]. As might be expected the electrokinetic behavior of halloysite is quitesimilar to the behavior of silica, i.e., the PZC of both halloysite and silica is less than pH 3, since the silica tetrahedral sheet is exposed at the surface of the halloysite tubes. Less clear is why the electrokinetic behaviour of the kaolinite planar structure is similar to silica.

It seems that the aluminum octahedral sheet (gibbsite sheet) of kaolinite has little influence even though equivalent exposure is expected for the planar kaolinite structure. On the other hand, the tubular chrysotile behaviour is similar to the behavior of brucite (PZC at pH 11) since the magnesia octahedral sheet of chrysotile is exposed. The planar antigorite seems to exhibit electrokinetic behavior intermediate between brucite and silica, indicating equal importance of both the magnesia octahedral sheet and the silica tetrahedral sheet.

A few studies have reported surface charge densities of layered silicates such as kaolinite, by titration. The points of zero charge for kaolinite particles as determined by potentiometric titrations were pH 4.3 [36], pH 4.5 [42] and pH 4.6 [43], significantly higher than that measured by electrophoresis. It is concluded that titration may better in describing the surface charge properties of layered mineral adsorbents. However, titration gives the overall surface charge density of the particles at any particular pH, and still does not describe the surface charge properties of the face (basal planes) and edge surfaces.

Gupta and Miller [13] interrogated the interaction forces between a siliconnitride tip and each of these surfaces using AFM surface force measurements. For the first time, they reported that the silica tetrahedral face of kaolinite is negatively charged at pH > 4, whereas the alumina octahedral face of kaolinite is positively charged at pH < 6, and negatively charged at pH > 8. These results contradict the generally accepted view that the face surfaces of kaolinite carry a permanent negative charge due to minor substitution of Al^{3+} for Si^{4+} in the silica tetrahedral sheet.

3.5 Surface Charge for Adsorption

Mineral adsorbents possess two kinds of electrical charges a variable (pH dependent) charge resulting from proton adsorption/desorption reaction of surface sites M_{surf}–OH and a structural negative charge from X-sites resulting from isomorphous substitution in the structure. Adsorption of ions on charged surfaces can be described to involve both a coordination reaction at specific surface sites and an electrostatic interaction between adsorbing ions and the charged surface [14]. Previously studies found that cation exchange at permanent negatively charged sites on the siloxane faces including interlayer regions of MMT, with the interlayer region accounting for the relatively slow sorption, and inner-sphere surface complex formation at variable charge surface hydroxyl groups (M_{surf}–OH) at the crystal edges. Metal ions exchange may have occurred not only on external faces, but also within the interlayer regions. As the surface of MMT is more protonated at low pH, H^+ competes with metal ion resulting in active sites less able to retain heavy metal ions, this may be explained by the surface complexation reactions that influenced by the electrostatic attraction between negatively charged groups at MMT surface and the metal ions. The sorption of Cu^{2+} and Ni^{2+} by MMT at higher pH is similar to metal ion uptake onto oxide surfaces [15] that is, sorption increases to near 100% over a narrow pH range.

For many 2:1 type mineral adsorbents, the permanent negative charge is balanced by cationic counterions occupying interlayer space. The counterions can be exchanged by other organic/inorganic cations, making them efficient for cationic contaminants. On the other hand, when special cations (e.g., organic cations, hydroxymetal cations) were exchanged into the interlayers, the modified clay minerals can be further used as adsorbents for a wide range of contaminants. As for cationic contaminants, the ion exchange to the fixed charge site at the internal surface and complexation to the variable charge site at the external surface dominated the whole process.

4 Surface Interactions

4.1 Water-Mineral Adsorbents Surface Interaction

The interaction of H_2O with mineral adsorbents was the subject of continuous study over many years. Understanding the interaction of mineral adsorbents with water is fundamental to many natural and industrial processes. A quantitative description of microscopic dynamics of water adsorbed in the mineral adsorbents is not trivial because various interactions determines the diffusion mobility of water molecules: water-cation, water-clay surface, and water-water. In order to control their influence, their relative role should be elucidated. The quasielastic neutron scattering (QENS) [16–18] and other methods [19] have been used to study the water microscopic dynamics in the mineral adsorbents interlayer. Recent investigations [20–22] have discovered that the intermediate scattering function I(Q, t) of water molecules in mineral adsorbents is not exponential but can be described by the stretched-exponential function. The stretched behavior of I(Q, t) can be interpreted as an evidence for a broad distribution of relaxation times, presumably because the water molecules have widely different local environments (as shown in Fig. 6).

Moreover, previous experimental and theoretical studies pointed out an inherent hydrophobicity of the siloxane surface in the absence of layer charge. Neutral-layer phyllosilicates such as talc and pyrophyllite are known to be rather hydrophobic [23]. The interaction of water molecules with siloxane surface of charged-layer mineral adsorbents is influenced by the location of isomorphous substitutions in the mineral structure. The A1–O surface oxygens in tetrahedrally-substituted phyllosilicates would be expected to form strong hydrogen bonds [24]. Besides, the expansion of basal spacing occurred due to incorporation of water molecules bound to each other as well as on the surfaces of mineral adsorbents by inter-molecular hydrogen bonding (as shown in Fig. 7) [25].

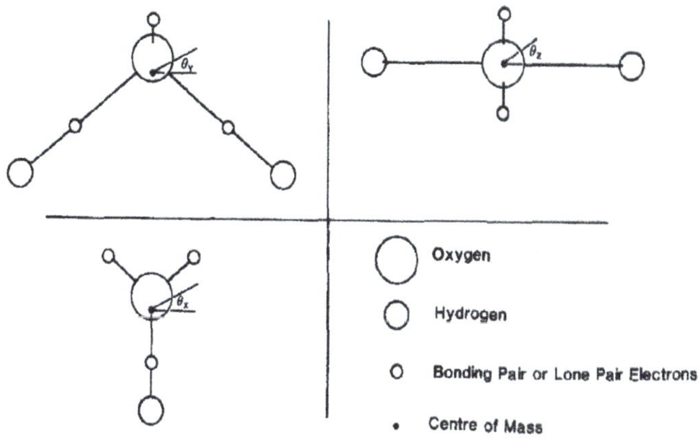

Fig. 6 Interaction sites in the QPEN model of a water molecular

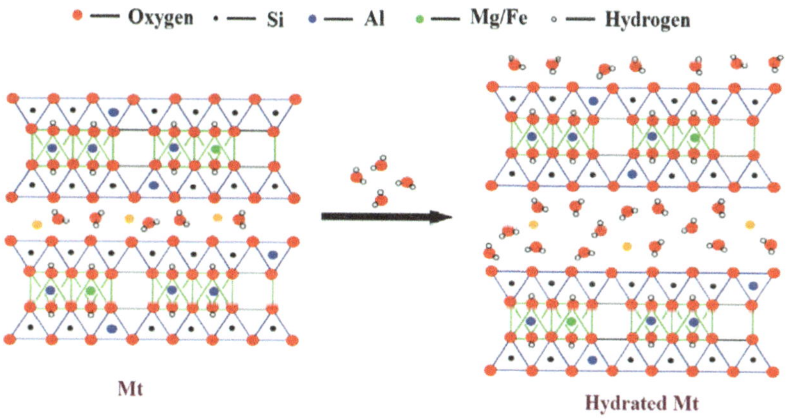

Fig. 7 Schematic of the hydration of Mt [25]

It was assumed that if the charge deficit occurs in the octahedral sheet, the negative charge on the surface oxygen atoms is delocalized and the adsorbed water molecules form only weak hydrogen bonds with these surface atoms. The strong hydration of metal exchange cations obscures the inherent hydrophilicity or hydrophobicity of the siloxane surfaces. The nature of the siloxane surface in smectites was measured by the adsorption of aromatic hydrocarbons from water by organo-clays. The siloxane surfaces of smectites can effectively adsorb aromatic hydrocarbons from water, if the hydrophilic, inorganic exchange cations are replaced with small, hydrophobic organic cations [26]. They also used the Li-fixation method [27] to create reduced-charge MMTs to study the effects of layer charge on organic-mineral interactions. The sorption of aromatic hydrocarbons was directly proportional to the surface area and inversely related to the layer charge.

These results supported the hydrophobicity of the siloxane surfaces in smectites which increases as the layer charge decreases.

On the other hand, numerous studies contradict the hypothesis that mineral adsorbent surfaces are inherently hydrophobic. Earlier infrared spectroscopic studies indicated that interlayer water is strongly coupled to the mineral adsorbents surface [28]. Researchers [29] made use of the oxidation state of the structural Fe as the method for varying the layers charge: reduction of Fe(III) to Fe(II) increases their negative charge. They compared the water holding capacity and swelling of reduced and oxidized smectite clays saturated with either Na or hydrophobic organic cations. It was found that reduction of the structural Fe increases the hydration energy of smectite basal surfaces, and the main conclusion was that smectite interlayer surfaces are hydrophilic. Another evidence for the hydrophilic nature of smectite interlayer surfaces is provided by neutron diffraction experiments performed with vermiculite [30] and Wyoming MMT [31]. Furthermore, the separation between different interactions affecting the molecular dynamics and structure of the adsorbed water can be done by means of molecular modeling. Several computation studies have been performed in order to investigate the microscopic properties of water adsorbed on hydrophobic and hydrophilic surfaces of different mineral adsorbents [32–34]. Sobolev et al. [35] investigated the interaction between water molecules and internal mineral adsorbent surfaces by means of neutron diffraction and quasielastic neutron scattering (as shown in Fig. 8). A hydrophobic cation, TMA^+ (NC_4H_{12}), was used to saturate the interlayer space of nontronite NAu-1 in order to reduce hydration of interlayer cations that could hinder the effects related to the mineral adsorbent-water interactions. The water content was low in order to reduce hydrogen bonding between water molecules. It was found that water molecules form strong hydrogen bonds with surface oxygen atoms of nontronite. The diffusion activation energy value Ea = 29 ± 3 kJ/mol was obtained for water molecules hydrating the mineral adsorbent surface, which indicates that the surfaces of smectite with tetrahedral substitutions are hydrophilic.

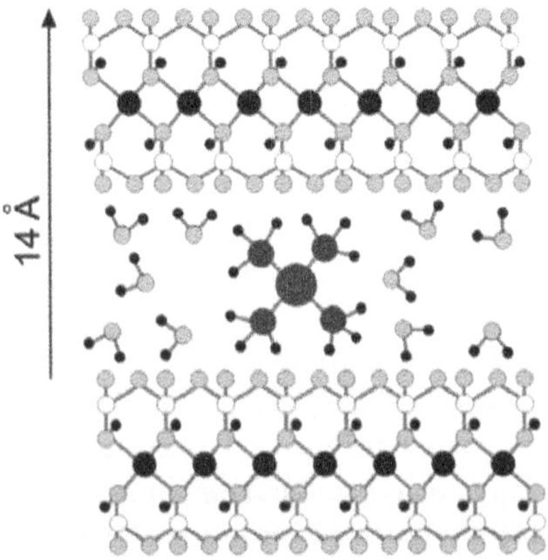

Fig. 8 Schematic view of two nontronite mineral layers [the layers are formed by tetrahedrally coordinated Si atoms (white circles), octahedrally coordinated Fe atoms (large black circles), oxygen atoms (light gray circles) and hydrogen atoms (small black circles)] [35]

4.2 *Surface Interaction for Adsorption*

Mineral adsorbents have been widely used for extensive variety contaminants, including hydrophobic organic contaminants, cationic/anionic dyes, heavy metal cations, oxyanions, radioactive nuclides, etc. According, a wide variety of mechanisms are involved for the uptake of these contaminants, such as ion exchange, electrostatic interactions, hydrophobic/hydrophilic interactions, ligand exchange, cation bridge, water bridge, hydrogen bond and van der Waals forces.

The related mechanisms for uptaking contaminants mainly include surface adsorption, partition, surface precipitation, and structural incorporation (Fig. 9) [36, 37].

Surface adsorption refers to the concentrate of contaminants onto or near the surface or pores of an adsorbent, which includes physical adsorption (driven by London–van der Waals forces) and chemisorption (involving the formation of chemical bonds) [38]. Adsorption of hydrophobic organic contaminants (HOC) onto activated carbon is a typical physical adsorption process [39]; while adsorption of heavy metal cations and oxyanions on metal (hydr) oxides always involves chemisorption [37, 40].

Besides, the 'ion-exchange' process, referring to the exchange of ionic contaminants with the pre-adsorbed ions on the adsorbents, has been regarded as one type of adsorption process as well in some publications [37]. Electrostatic interaction is the dominant interaction force between the adsorbed ions and the adsorbent, and adsorption of heavy metal cations by MMt is primarily an ion-exchange

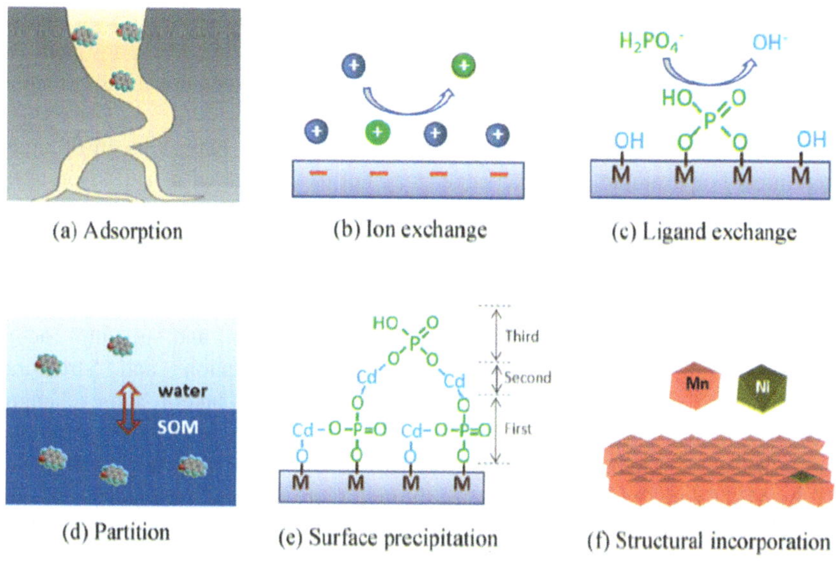

Fig. 9 Schematic illustrations showing the main mechanisms for the uptake of contaminants on various adsorbents

controlled process [41]. An analog adsorption process to 'ion-exchange' is the 'ligand-exchange', in which the contaminants exchange the ligands on the adsorbents during the adsorption process. However, a ligand can be an ion or a molecule binding to a metal ion on the adsorbent, and this adsorption process always involves a highly specific chemisorption [42]. The adsorption of oxyanions on metal (hydr) oxides has been regarded as a typical ligand-exchange process [43].

'Partition' refers to the distribution of contaminants between two phases (e.g., organic phase-bulk water, organic phase-air, bulk water air), which means that the contaminants will penetrate into the entire network of a bulk phase, rather than concentrate onto the surface of the adsorbents [44, 45]. A typical partition-dominated process is the uptake of HOC by soil organic matters (SOM), and one can generally observe relatively linear isotherm, non-competitive uptaking of multi-contaminants, and only small exothermic heats [44, 45].

'Surface-precipitation' involves the formation of precipitates on the surface of the adsorbents, which generally needs relatively high concentrations of cations and anions (but still below the formation of precipitates in aqueous solution) [46–48]. The adsorbents concentrate contaminants on their surface by adsorption/ ion-exchange at first, which is followed by the formation of precipitates because of the over-saturation of cations and anions on the surface [37, 47, 48]. The formation of surface-precipitates, therefore, requires relatively high surface concentration of cations and anions. A typical adsorption process that involves the surface-precipitation is the co-adsorption of cations (e.g., Cd^{2+}, Zn^{2+}) and oxyanions (e.g., phosphate, arsenate) on the surface of metal (hydr)oxides [37, 49].

'Structural incorporation' refers to the incorporating of ions into the solid phase of adsorbents [50, 51], e.g., sequestration of metal cations into the crystal structure of minerals by isomorphous substitution [37]. This adsorption process always follows the surface adsorption process and generally has a low adsorption rate. As the contaminants are incorporated into the bulk phase of the adsorbents, they can be well sequestrated by this type of adsorption [37, 51].

However, the categorization of adsorption mechanisms may differ in different literatures. For example, 'ion-exchange' may not be considered as an 'adsorption' process; 'surface-precipitation' may be regarded as one form of 'structural incorporation' [37]. Moreover, in many cases the uptake of contaminants on adsorbents is indeed operated by more than one type of mechanisms (as shown in Table 5). For example, the uptake of heavy metal cations on metal (hydr) oxides always simultaneously involves surface adsorption, surface precipitation, and structural incorporation [37, 51, 52]. As both the 'surface precipitation' and 'structural incorporation' processes should start from the 'adsorption' and/or 'ion-exchange' process, completely differentiate these processes may be difficult under some circumstances [47]. Medhi and Bhattacharyya found that the adsorption process of Cu (II) by MMT and TMA-MMT occurred through more than one mechanism, among which chemical, mainly electrostatic, interactions play a great role in the adsorption over the physical types of interactions [25]. Wang et al. [53] used organo-MMT to remove cationic organic dyes, and concluded that the removal of MB was proposed

Table 5 Some MMT based mineral adsorbents and their related adsorption mechanisms

Adsorbent	Contaminants	Adsorption mechanisms	References
MMT	Heavy metal cations, radioactive nuclides, cationic dye	Cation exchange, specific adsorption, surface precipitates	[54–56]
AMMT	Heavy metal cations, cationic dye, radioactive nuclides, anionic dyes	Cation exchange, specific adsorption; anion exchange	[57–59]
TMMT	Heavy metal cations, radioactive nuclides	Cation exchange	[54, 57]
Type I OMMT(TMA-MMT)	HOC	Adsorption	[60, 61]
Type II OMMT (HDTMA-MMT)	HOC, anionic dyes, oxyanions	Partition, adsorption; anion exchange	[62–65]
Type III OMMT (polymer-MMT)	Heavy metal cations, anionic dyes, oxyanions, HOC	Specific adsorption; anion exchange; partition, surface adsorption	[63–66]
IMMT (hydroxyiron-MMT)	Heavy metal cations, oxyanions F−, oxyanions + heavy metal cations	Cation exchange, specific adsorption, surface precipitation; LIGAND-exchange, anion exchange, surface precipitation; specific adsorption, surface precipitation	[67–72]
IMMT (NZVI-MMT)	Heavy metal cations, oxyanions	Adsorption, reduction	[73–75]
IOMMT (hydroxyiron-HDTMA-MMT)	HOC, oxyanions, HOC + oxyanions	Partition, adsorption; ligand exchange, anion exchange; combination of above four mechanisms	[76–80]

to be a synergistic effect of ion exchange, partition adsorption, hydrogen bonding and electrostatic interactions, as shown in Fig. 10.

5 Adsorption Thermodynamics

Adsorption depends on the existence of a force field at the surface of a solid, which reduces the potential energy of an adsorbed molecule below that of the ambient fluid phase. For the purpose of estimating the performance and predicting the

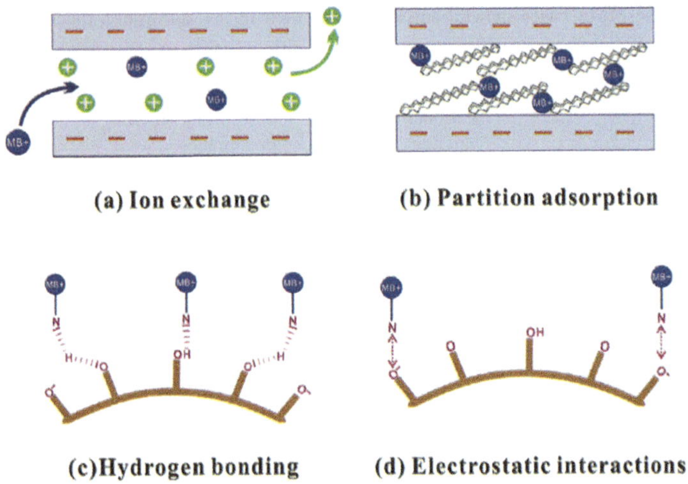

Fig. 10 Schematic illustration of the mechanism for the adsorption of MB onto OMMT [53]

mechanism of an adsorption separation process and the characterization and optimization of an adsorption process, the value of the adsorption thermodynamics parameters such as activation energy, activation parameters, Gibb's free energy change, enthalpy, entropy, and isosteric heat of adsorption including are required [81, 82]. From these thermodynamics parameters, it is possible to verify if the adsorption is favorable, spontaneous, endothermic, or exothermic. It is possible to infer about the adsorption nature, i.e., physisorption or chemisorption, and verify if the operation is controlled by enthalpy or entropy. Moreover, it is possible to obtain information regarding the disorder in the solid–liquid interface during the adsorption [83].

5.1 *Adsorption Reaction Process*

According to Langmuir, the adsorption process can be depicted as follows:

$$A + B \leftrightarrow AB \tag{5}$$

In which A represents free adsorptive solute molecules, B is vacant sites on the adsorbent, and AB is the occupied sites. When this reaction attains the thermodynamic equilibrium, the chemical potentials in the liquid phase (μl) and in the solid–liquid interface (μs-l) are equal, and the Gibbs free energy change (ΔG) tends to zero, leading to Eq.

$$\Delta G = \mu_{s-l} - \mu_l = \Delta G^0 + RT\ \ln(K_e) = 0 \tag{6}$$

then

$$\Delta G^0 = -RT\ \ln(K_e) \tag{7}$$

where R is the universal gas constant, T is the temperature, and K_e is the equilibrium thermodynamic constant.

The relationship of ΔG^0 with ΔH^0 and ΔS^0 can be expressed as follow:

$$\Delta G^0 = \Delta H^0 - T\Delta S^0 \tag{8}$$

Substituting Eq. (4) in Eq. (3), Eq. (3) could be transformed to the following relation:

$$\ln(K_e) = -\frac{\Delta H^0}{RT} + \frac{\Delta S^0}{R} \tag{9}$$

The plot of ln K_e against 1/T theoretically yields a straight line that allows calculation of ΔH^0 and ΔS^0 from the respective slope and interception of Eq. (9) [84]. The graph is known as the Van't Hoff plot.

The use of Van't Hoff plot is relatively simple but is dependent of the correct calculation of the equilibrium thermodynamic constant K_e, a reasonable mean to find K_e is presented. In a solid–liquid adsorption system, the equilibrium thermodynamic constant is given by Eq. (10):

$$K_e = \frac{\text{activity of occupied sites}}{(\text{activity of vacant sites})(\text{activity of adsorbate in solution})} \tag{10}$$

Assuming that the activity of the occupied and unoccupied sites is the same, Eq. (10) could be rewritten as:

$$K_e = \frac{\theta}{(1-\theta)\alpha_e} \tag{11}$$

In which θ_e is the fraction of the surface covered at equilibrium and α_e is the activity of the adsorbate in solution at equilibrium.

For the Langmuir model, θ is given by Eq. (12)

$$\theta = \frac{q_e}{q_m} \tag{12}$$

For other isotherms, q_m can be replaced by the parameter relative to the maximum adsorption capacity in mol/g.

5.2 Activation Energy

Activation energy is an important parameter in a thermodynamic study as it determines the temperature dependence of the reaction rate. The activation energy of a reaction is usually denoted by E_a, and given in units of kJ mol^{-1}. In adsorption separation, it is defined as the energy that must be overcome by the adsorbate ion/molecule to react/interact with the functional groups on the surface of the adsorbent. According to the Arrhenius equation Eq. (13), the activation energy (E_a) for the adsorption of an adsorbate ion/molecule onto an adsorbent surface in an adsorption process can be determined from experimental measurements of the adsorption rate constant at different temperatures:

$$\ln k = \ln A - \frac{E_a}{RT} \tag{13}$$

where k is the adsorption rate constant, A is a constant called the frequency factor, E_a is the activation energy (kJ mol^{-1}), R is the gas constant (8.314 J mol^{-1} K^{-1}) and T is the temperature (K). By plotting ln k versus 1/T and from the slope and the intercept, values of E_a and A can be obtained [85, 86].

The magnitude of activation energy may give an idea about the type of adsorption. Two main types of adsorption may occur, physical and chemical. The forces of physical adsorption consist of the ubiquitous dispersion–repulsion forces (van der Waals forces), which are a fundamental property of all matter, supplemented by various electrostatic contributions (polarization, field–dipole and field gradient–quadrupole interactions), which can be important or even dominant for polar adsorbents. The forces involved in chemisorption are much stronger and involve a substantial degree of electron transfer or electron sharing, as in the formation of a chemical bond [87, 88]. The activation energy for physisorption is usually not greater than 4.2 kJ mol^{-1}. While, two kinds of chemisorptions are encountered, activated and less frequently nonactivated. Activated chemisorption means that the rate varies with temperature according to finite activation energy (between 8.4 and 83.7 kJ mol^{-1}) in the Arrhenius equation (high E_a). However, in some systems, a non-activated chemisorption occurs very rapidly, suggesting the activation energy is near zero.

5.3 Thermodynamic Parameters

Thermodynamic considerations of an adsorption process are necessary to conclude whether the process is spontaneous or not. For example, the negative values of ΔG^0 show a spontaneous and favorable process. The higher the ΔG^0 magnitude, the more favorable and spontaneous the adsorption. For significant adsorption to occur, the Gibb's free energy change of adsorption, ΔG^0, must be negative. A decrease in the negative value of ΔG^0 with an increase in temperature indicates that the adsorption process is more favourable at higher temperatures. On the contrary, an increase in the negative value of ΔG^0 with an increase in temperature implies that lower temperature makes the adsorption easier.

Standard enthalpy change (ΔH^0) is the enthalpy change that occurs in an adsorption process. In an endothermic process ($\Delta H^0 > 0$), the adsorbate species has to displace more than one water molecule for their adsorption and this results in the endothermicity of the adsorption process. In an exothermic process ($\Delta H^0 > 0$), the total energy absorbed in bond breaking is less than the total energy released in bond making between adsorbate and adsorbent, resulting in the release of extra energy in the form of heat. The magnitude of ΔH^0 can give an idea about the interactions that occur between the adsorbent and adsorbate. Physisorption, such as van der Waals interactions, is usually lower than 20 kJ mol^{-1}, and electrostatic interaction ranges from 20 to 80 kJ mol^{-1}. Chemisorption bond strengths can be from 80 to 450 kJ mol^{-1} [85, 86]. Chemisorptions were highly specific and the adsorption energies are generally substantially greater than those for physical adsorption (see Table 6).

In relation to the ΔS^0, negative values show that the randomness decreases at the solid solution interface during the adsorption, and positive values suggest the possibility of some structural changes or readjustments in the adsorbate–adsorbent complex. The positive ΔS^0 value also corresponds to an increase in the degree of freedom of the adsorbed species. A negative value of ΔS^0 suggests that the adsorption process is enthalpy driven. A negative value of ΔS^0 also implies a decreased disorder at the solid/liquid interface during the adsorption process causing the adsorbate ions/molecules to escape from the solid phase to the liquid phase.

Table 6 Physical adsorption and chemisorptions [87, 88]

Physical adsorption	Chemisorption
Low heat of adsorption (1.0–1.5 times latent heat of evaporation)	High heat of adsorption (>1.5 times latent heat of evaporation)
Nonspecific	Highly specific
Monolayer or multilayer	Monolayer only
Adsorption of dodecylamine hydrochloride on graphene oxide in water	May involve dissociation
Only significant at relatively low temperatures	Possible over a wide range of temperatures
Rapid, nonactivated, reversible	Activated, may be slow and irreversible
No electron transfer, although polarization of sorbate may occur	Electron transfer leading to bond formation between sorbate and surface

6 Adsorption Kinetics

Adsorbate residence time and the reactor dimensions are controlled by the adsorption kinetics. Thus, predicting the rate at which adsorption takes place for a given system is probably the most important factor in adsorption system design.

The basis for kinetics study is the kinetic isotherm, which is obtained experimentally by tracking the adsorbed amount against time; kinetic investigations develop a model to describe the adsorption rate. Ideally, the model should, with minimal complexity, enable one to identify operating conditions with minimal mass transfer resistance and predict adsorbent performance [89]. The kinetics is dependent on material factors, such as adsorbent and adsorbate types, and experimental factors, such as temperature and pH.

According to the present state of knowledge [90], the adsorption process can be described by the following consecutive steps:

(i) Transport of the adsorbate (ions in case of solution) from the bulk to the external surface of the adsorbent. However, it seems implausible to the authors that this step is the slowest step;
(ii) Diffusion of adsorbate across the so-called liquid film surrounding adsorbent particles (liquid film diffusion);
(iii) Diffusion of adsorbate in the liquid contained in the pores of sorbate particle and along the pore walls (intraparticle diffusion);
(iv) Adsorption and desorption of adsorbate molecules on/from the sorbent surface(surface reaction).

Any of the above steps may be the slowest step determines the overall rate of the interactions and hence the kinetics of the adsorption process [91].

6.1 *Surface Reaction*

The most widely used are those assuming that step (iv) makes a significant contribution in the kinetics of a process, referring to the "surface reaction". Many expressions related to the concept of a chemical reaction occurring on the surface and its order, such as the Lagergren pseudo-first order (PFO), pseudo-second order (PSO) or Elovich equations have been applied for correlating kinetic data measured in many different systems [92].

(1) Lagergren pseudo first order model

The Lagergren equation is probably the earliest known example describing the rate of adsorption in the liquid-phase systems; which is one of the most widely used kinetic equations and the earliest known one describing the adsorption rate based on the adsorption capacity [93]. Equation (5) is one of the most used equations particularly for pseudo first order kinetics:

$$dq_t/dt = k_1(q_e - q_t) \tag{14}$$

where q_e and q_t are the values of amounts adsorbed per unit mass at equilibrium and at any time t. k_1 (min^{-1}) is the pseudo first order adsorption rate coefficient, the value of k_1 is expected to be influenced by experimental conditions, such as pH and temperature.

The integrated form of the Eq. (14) for the boundary conditions of $t = 0$, $q_t = 0$ and $t = t$, $q_t = q_t$,

$$\ln(q_e - q_t) - \ln q_e - k_1 t \tag{15}$$

According to Eq. (15), the values of k_1 and q_e can be obtained from the slope of the linear plot of ln ($q_e - q_t$) versus t. For many adsorption processes, the Lagergren pseudo first order model is found suitable only for the initial 20–30 min of interaction and not fit for the whole range of contact time [94]. The real test of the validity of Eq. (15) arises from a comparison of the experimentally determined q_e values and those obtained from the plots of ln ($q_e - q_t$) versus t. If this test is not valid, then higher order kinetic models are to be tested with respect to the experimental results [95].

(2) Pseudo-second order model

In recent years, the pseudo-second-order (PSO) rate expression has been widely applied to the adsorption of pollutants from aqueous solutions. Most environmental kinetic adsorption can be modeled well by PSO, thereby indicating its superiority to other models. It has been successfully applied to the adsorption of metal ions, dyes, herbicides, oils, and organic substances from aqueous solutions. The second order kinetics may be tested on the basis of the equation:

$$dq_t/d_t = k_2(q_e - q_t)^2 \tag{16}$$

where k_2 is the second order rate coefficient, the effects of pH and temperature on k_2 are not well studied because of difficulties that arise from the effects on equilibrium isotherm shapes. q_e and q_t are the values of amount adsorbed per unit mass at equilibrium and at any time t.

The integral form of the model is shown in Eq. (17),

$$t/q_t = 1/k_2 q_e^2 + (1/q_e)\Delta t \tag{17}$$

The ratio of the time/adsorbed amount should be a linear function of time. The pseudo-second order rate coefficient, k_2, can be determined experimentally from the slope and intercept of a plot of t/q_t against t.

The initial adsorption rate, h, of a second order process as $t \to 0$ is defined as,

$$h = k_2 q_e^2 \tag{18}$$

The pseudo-second-order equation has the following advantages, .e.g., it does not have the problem of assigning an effective adsorption capacity, i.e., the adsorption capacity, the rate constant of pseudo-second-order. The initial adsorption rate can be determined from the equation without knowing any parameter beforehand.

(3) Elovich equation

The Elovich equation is the rate equation, proposed by Roginsky and Zeldovich in 1934 to describe the kinetics of adsorption of carbon monoxide on manganese dioxide. The Elovich equation assumes that the actual solid surfaces are energetically heterogeneous and that neither desorption nor interactions between the adsorbed species could substantially affect the kinetics of adsorption at low surface coverage, it is known to describe chemisorption well [96]. The Elovich equation has been used in Eq. 19.

$$\mathrm{d}q_t/\mathrm{d}t = \alpha \exp(-\beta q_t) \tag{19}$$

The Elovich coefficients, α and β, represent the initial adsorption rate (g mg^{-1} min^{-2}) and the desorption coefficient (mg g^{-1} min^{-1}) respectively. Assuming $\alpha\beta t \gg 1$, and qt = 0 at t = 0 and qt = qt at t = t, the linear form of the Eq. (19) is as following:

$$q_t = \beta \ln(a\beta) + \beta \ln t \tag{20}$$

Thus, the Elovich coefficients α and β could be computed from the plots of q_t versus ln t. In practice, the applicability of the Elovich equation is restricted to the initial part of the adsorbate–adsorbent interaction process, when the values of the fractional surface coverage lower than about 0.7, the system is relatively far from equilibrium.

6.2 *Intra-particle Diffusion*

For porous adsorbents, in many cases the diffusion of the adsorbate molecules or ions into the pores may control the rate of uptake of an adsorbate.

A typical intraparticle diffusion model is the so-called homogeneous solid diffusion model (HSDM), which can describe mass transfer in an amorphous and homogeneous sphere. The pore diffusion kinetics model is represented by the following familiar expression:

$$q_t/q_e = 1 - (6/\pi^2)\Sigma(1/n^2)\exp(-n^2\pi^2 D_c t/r^2) \tag{21}$$

In Eq. 8, the ratio, q_t/q_e, giving the fractional approach to equilibrium, D_c = intra-crystalline diffusivity, r = particle radius, t = reaction time, and the summation is carried out from n = 1 to n = α.

For short times, or more precisely for $q_t/q_e < 0.3$, the Eq. (21) could be simplified by the following expression:

$$\ln q_t = (1/2)\ln(t) + \ln(k) \tag{22}$$

with

$$k = 6(D_c/\pi r^2)^{1/2} \tag{23}$$

For long duration, or more precisely for qt/qe > 0.85, the Eq. (21) can be rewritten in the following simplified form:

$$\ln(1 - q_t/q_e) = (-\pi^2 D_c/r^2)t + \ln(6/\pi^2) \tag{24}$$

and

$$k' = \pi^2 D_c/r^2 \tag{25}$$

Therefore, the plot of ln ($1 - q_t/q_e$) versus t should be linear with a slope of ($-\pi^2 Dc/r^2$), which is known as the diffusion time constant, and k′ is the overall rate constant. This simplified Crank long times equation (Eq. 24) is also known as the Boyd equation [97].

In recent years, HSDM has been applied to different kinds of adsorption systems, such as the adsorption of salicylic acid and 5-sulfosalicylic acid from aqueous solutions by hypercross linked polymeric adsorbent NDA-99 and NDA-101.

6.3 *Liquid Film Diffusion*

When the flow of the reactant through the liquid film surrounding the adsorbent particles is the slowest process, liquid film diffusion of the adsorbate control the kinetics of the rate processes. In liquid/solid adsorption systems the rate of solute accumulation in the solid phase is equal to that of solute transfer across the liquid film according to the mass balance law. Liquid film diffusion model is represented by the following simple relation.

$$\ln(-F) = -k_{fd}t \tag{26}$$

where F is the fractional attainment of equilibrium (=q_t/q_e) and k_{fd} (min^{-1}) is the film diffusion rate coefficient. A linear plot of –ln ($1 - F$) versus t with zero intercept suggests that the kinetics of the adsorption process is controlled by diffusion through the liquid film. The film diffusion mass transfer rate equation has been successfully applied to model several liquid/solid adsorption cases, e.g., phenol adsorption by a polymeric adsorbent NDA-100 under different temperature and initial concentration conditions. The effect of transport is eliminated by mechanical mixing with the rapid and constant stirring of the solution.

7 Characterization Technology of Adsorption

7.1 Component Characterization Technology

Chemical characterization of the products and spent adsorbent may be explored to understand the mechanism of the process involved [98].

(1) Chemical composition

Chemical composition of key component determines the properties of the adsorbent, while small quantities of other components will have significant effect on the processing and micro-structural development of the adsorbent materials. Various techniques can be used for quantitative analysis of chemical composition. (1) X-ray fluorescence spectroscopy (XRF) can be used to directly analyze solid adsorbent. However, for elements with atomic number $Z < 9$, it has very low sensitivity. (2) Optical atomic spectroscopy, especially atomic absorption (AA) or atomic emission (AE), has been the most widely used for chemical analysis of adsorbent powders. It can be used to determine the contents of both major and minor elements, as well as trace elements, because of its high precision and low detection limits. Adsorbent powders require complete dissolution because optical atomic spectroscopy is suitable to analyze solutions. Moreover, mass spectrometry, electrochemistry, and nuclear and radioisotope analysis are also used for chemical analysis.

(2) Crystal Structure and Phase Composition

X-ray diffraction (XRD) is the most widely used characterization technique to determine the structure and chemical composition of crystalline mineral adsorbent. It offers qualitative and quantitative characterization of mineral composition of the adsorbents. Moreover, the changes on the phases present in mineral adsorbent during the preparation process could be detected by XRD [99].

(3) Chemical bonds between atoms

Fourier transform infrared spectrometer (FT-IR) is usually used to determine the existence of organic or inorganic functional groups in mineral adsorbents [100]. A broad band source of IR radiation is reflected from the sample (or transmitted, for thin samples). The wavelengths at which absorption occurs are identified by measuring the change in intensity of the light after reflection (transmission) as a function of wavelength. These absorption wavelengths represent excitations of vibrations of the chemical bonds and are specific to the type of bond and the group of atoms involved in the vibration. The sensitivity toward different bonds (chemical groups) is extremely variable, from zero (no coupling of the IR radiation to vibrational excitations because of dipole selection rules) to high enough to detect submonolayer quantities. Intensities and line shapes are sensitive to local solid state effects, such as stress, strain, and defects, so quantification is difficult, but with suitable standards 5–10% accuracy in concentrations are achievable.

Nuclear Magnetic Resonance (NMR) involves transitions between different spin states of the atomic nucleus. It provides some information on the local bonding arrangement around an atom. The local environment of the atoms concerned the coordination number, local symmetries, the nature of neighboring chemical groups, and bond distances can be studied using NMR. Quanyuan Chen used NMR to study the chemical structure of silicate phases of the composite adsorbent [101].

7.2 *Structural and Morphology Characterization Technology*

(1) Imaging techniques

Light microscope is likely to be the first imaging instrument used, which is also the cheapest "modern" instrument. It is capable of handling various types of sample, providing magnification up to 1400×. By utilizing polarizer, many other properties, in addition to size and shape, become accessible (e.g., refractive index, crystal system, melting point, etc.).

The Scanning Electron Microscopy (SEM) where magnified images of up to 300 K× are obtainable, the wavelength of electrons not being nearly so limiting as that of visible light and lateral features down to a few nm become resolvable. The depth from which all this information comes varies from nanometers to micrometers, depending on the primary beam energy used and the particular physical process providing the contrast. A variety of contrast mechanisms exist, in addition to the topological, enabling the production of maps distinguishing high- and low-Z elements, defects, magnetic domains, and even electrically charged regions in semiconductors.

Transmission Electron Microscopy (TEM) has been used in materials science for 30 years. The combination of imaging (with lateral magnification up to 1 M×) with a variety of contrast modes at an atomic resolution for crystalline material, together with small and large area diffraction modes, provide a wealth of characterization information. That is, information on the various aspects of a specimen, including but not limited to materials size, shape, crystallinity, composition, and elemental mapping, can be collected during the interaction of the electron beam with the specimen.

Scanning Tunneling/Scanning Force Microscopy (STM/SFM) are used in imaging single atoms or molecules, and moving them under control on dean surfaces in pristine UHV conditions. Features on the nanometer scale, sometimes not easily seen in SEM, can be observed in STM/SFM.

(2) Size Measurement

Particulate products make up around 80% of all chemical products containing one or more particulate base materials, which may come from a wide variety of sources. Particle size measurement techniques such as Sieves, Laser diffraction, Electrical sensing zone, Optical Microscopy and Nano Sizers are widely used, the choice should be made base on the technique which will provide the most useful information for the particle process or application under consideration.

Size analysis by sieves is perhaps the oldest technique of establishing the size distribution of a particulate material as well as being the most popular and least expensive. Size analysis can be valid in sizes ranging from a few centimetres to 10 μm (dry: 125 mm to 20 μm; wet: 10 μm). The results of the technique are dependent on the apertures, the shape of the particles, the details of the applied vibration and the length of time the stack is shaken.

Laser diffraction is a popular sizing technique, where a collimated beam of mono chromatic light (from a laser) is projected on to a dilute suspension of particles. Particles in the range 2 mm–0.1 μm can be sized with this technique, the suspension medium can be a liquid or a gas.

The electrical sensing zone (ESZ) technique is based on the electrical properties of a conductive fluid (an electrolyte) with suspended particles. The technique was developed by W. H. Coulter for counting blood cells and has now been extended to industrial powders with devices that can cover size range 1200 μm down to about 0.6 μm, where number and size distributions of particles can be determined.

Optical microscopy is typically used with particles in the 0.8–150 μm size range. The technique has the advantage of being able to directly observe, examine and measure individual particles and can often be used where other techniques fail. This is a very time consuming procedure if performed manually; however, image analysis techniques can be used to drastically cut the time taken to obtain particle size data.

Nano Sizers including dynamic light scattering (DLS), particle tracking, electron microscopy (TEM, SEM), atomic force microscopy (AFM), and Multi-Angle Static Light Scattering (MALS) were typically used with particles for a range of less than a nanometre to several micrometres.

(3) Surface Area and Porosity

Surface area of adsorbent is another important parameter. Moreover, it is desirable to quantitatively characterize the porosity and pore size and distribution of the adsorbents.

The surface area is determined in terms of the amount of gas adsorbed by a given mass of solid powder at a given temperature, under different gas pressures. As the standard technique, nitrogen adsorption method is usually used to determine the specific surface area of the solid. Sometimes, Argon gas method is also used as an alternative.

Three types of pores have been classified according to their sizes: micropores (<2 nm), mesopores (2–50 nm), and macropores (>50 nm). For accessible pores, generally, gas condensation is applicable to the measurement in mesopores, whereas mercury porosimetry is more suitable to macropores. For the isolated porosity within particles, Pycnometry is usually used.

7.3 Surface and Interface Properties Characterization Technology

(1) Structure of Surface

The surface of a particle is usually not smooth and homogeneous at microscopic or submicroscopic scale, with the presence of various types of irregularities, such as adatom, step adatom, terrace, terrace vacancy, kink, etc. The techniques for characterization of surface structure of solids include electron diffraction and scanning probe microscopy. Scanning probe microscopy (SPM) is a branch of microscopy that forms images of surfaces using a physical probe that scans the specimen at an atomic level.

(2) Chemistry of Surface

Electron, ion, and photon emissions from the outermost layers of the particle surface can be used to reveal the qualitative or quantitative information on chemical composition of the surface of adsorbent powders. The most widely used techniques include (i) Auger electron spectroscopy (AES), (ii) X-ray photoelectron spectroscopy (XPS), which is also known as electron spectroscopy for chemical analysis (ESCA). X-ray photoelectron spectroscopy (XPS) analysis was used for the study of sorbent and the metal speciation on the sorbent [98]. In addition, XPS could be performed to investigate the oxidation states of different elements [102]. (iii) Secondary ion mass spectrometry (SIMS).

(3) Zeta potential analysis

The zeta potential is a key indicator of the stability of colloidal dispersions. It is also a powerful tool to analyze the adsorption mechanism between adsorbent and adsorbate. Electrokinetic (ζ-potential) measurements are well known as a useful method to characterize adsorbent surfaces. Since the presence of electrical charges often determines the interaction phenomena at interfaces, such as the reciprocation of dissolved ingredients (dyes, ions, enzymes, etc.) with solid surfaces. The variation of the composition of the electrolyte solution used for ζ-potential measurements provides the possibility to determine the surface characteristics, i.e. acidic or basic, polar or nonpolar (hydrophobic). Point of zero charge of mineral adsorbent surfaces can also be determined using zeta potential [103].

7.4 Characterization Technology of Other Property

(1) Magnetic property

At present, many kinds of clay minerals have been used as carrier to prepare magnetic composites to adsorb contaminants from water. The magnetic property parameters such as the saturation magnetization (Ms), remanent magnetization (Mr) and coercivity (Hc) were obtained from the hysteresis loop. The hysteresis loop is usually measurements of the samples were carried out at room temperature using a vibrating sample magnetometer [104].

(2) Cation exchange capacity (CEC) measurement

Cation-exchange capacity is defined as the amount of positive charge that can be exchanged per mass of clay minerals adsorbents, usually measured in $cmol_c/kg$. It indicates the capacity to retain pollutant (e.g. Pb^{2+}). It is measured by displacing all the bound cations with a concentrated solution of another cation, and then measuring either the displaced cations or the amount of added cation that is retained. For example, CEC was determined using sodium acetate procedure (SW-846 Method 9080).

(3) Thermo gravimetric analysis

Thermogravimetric analysis or thermal gravimetric analysis (TGA) is a method of thermal analysis in which the mass of a sample is measured over time as the temperature changes. This method provides information regarding the physical phenomena, such as phase transitions, absorption and desorption; as well as chemical phenomena including chemisorptions [105]. During the preparation of minerals adsorbents, thermogravimetric analysis (TG) was used to probe the microenvironment and packing arrangement of organic surfactant within the organoclays and to calculate the real surfactant loading within organoclays [65].

References

1. Hyun, S.P., Cho, Y.H., Hahn, P.S., Kim, S.J.: Sorption mechanism of U(VI) on a reference montmorillonite: binding to the internal and external surfaces. J. Radioanal. Nucl. Ch. **250**, 55–62 (2001)
2. Uddin, M.K.: A review on the adsorption of heavy metals by clay minerals, with special focus on the past decade. Chem. Eng. J. **308**, 438–462 (2017)
3. Kobayashi, I., Owada, H., Ishii, T., Iizuka, A.: Evaluation of specific surface area of bentonite-engineered barriers for Kozeny-Carman law. Soils Found. **57**, 683–697 (2017)
4. Yener, N., Bicer, C., Onal, M., Sarikaya, Y.: Simultaneous determination of cation exchange capacity and surface area of acid activated bentonite powders by methylene blue sorption. Surf. Sci. **258**, 2534–2539 (2012)
5. Xian, Z., Hao, Y., Zhao, Y., Song, S.: Quantitative determination of isomorphous substitutions on clay mineral surfaces through AFM imaging: a case of mica. Colloid Surf. A **533**, 55–60 (2017)

6. Tournassat, C., Ferrage, E., Poinsignon, C., Charlet, L.: The titration of clay minerals II. Structure-based model and implications for clay reactivity. J. Colloid Interface Sci. **273**, 234–246 (2004)
7. Leroy, P., Tournassat, C., Bernard, O., Devau, N., Azaroual, M.: The electrophoretic mobility of montmorillonite. zeta potential and surface conductivity effects. J. Colloid Interface Sci. **451**, 21–39 (2015)
8. Heuser, M., Spagnoli, G., Leroy, P., Klitzsch, N., Stanjek, H.: Electro-osmotic flow in clays and its potential for reducing clogging in mechanical tunnel driving. B Eng. Geol. Environ. **71**, 721–733 (2012)
9. Zhao, F., Du, Y.-K., Xu, J.-K., Liu, S.-F.: Determination of surfactant molecular volume by atomic force microscopy. Colloid J. **68**, 784–787 (2006)
10. Assemi, S., Nalaskowski, J., Miller, J.D., Johnson, W.P.: Isoelectric point of fluorite by direct force measurements using atomic force microscopy. Langmuir **22**, 1403–1405 (2006)
11. Ijagbemi, C.O., Baek, M.H., Kim, D.S.: Montmorillonite surface properties and sorption characteristics for heavy metal removal from aqueous solutions. J. Hazard. Mater. **166**, 538–546 (2009)
12. Zhao, X., Dong, F.: Applied mineralogy and its future: a review of the 11(th) international congress for applied mineralogy. Acta Petrologica et Mineralogica **33**, 194–198 (2014)
13. Gupta, V., Miller, J.D.: Surface force measurements at the basal planes of ordered kaolinite particles. J. Colloid Interf. Sci. **344**, 362–371 (2010)
14. Ikhsan, J., Wells, J.D., Johnson, B.B., Angove, M.J.: Surface complexation modeling of the sorption of Zn(II) by montmorillonite. Colloid Surf. A **252**, 33–41 (2005)
15. Leroy, P., Revil, A.: A triple-layer model of the surface electrochemical properties of clay minerals. J. Colloid Interf. Sci. **270**, 371–380 (2004)
16. Malikova, N., Cadene, A., Dubois, E., Marry, V., Durand-Vidal, S., et al.: Water diffusion in a synthetic hectorite clay studied by quasi-elastic neutron scattering. J. Phys. Chem. C **111**, 17603–17611 (2007)
17. Gonzalez-Sanchez, F., Juranyi, F., Gimmi, T., Van Loon, L., Seydel, T., et al.: Dynamics of supercooled water in highly compacted clays studied by neutron scattering. J. Phys-Condens. Mat. **20**(41), 9 (2008)
18. Sanchez, F.G., Juranyi, F., Gimmi, T., Van Loon, L., Unruh, T., et al.: Translational diffusion of water and its dependence on temperature in charged and uncharged clays: a neutron scattering study. J. Chem. Phys. **129**(17), 174706–174716 (2008)
19. Rao, W.B., Gao, Z.M., Yang, Z.S., Luo, T.Y.: Geology and geochemistry of the Shangmanggang red clay-type gold deposit in West Yunnan. J. Geochem. Explor. **84**, 105–125 (2004)
20. Swenson, J., Bergman, R., Longeville, S.: A neutron spin-echo study of confined water. J. Chem. Phys. **115**, 11299–11305 (2001)
21. Mamontov, E.: Comment on "Quasielastic neutron scattering of two-dimensional water in a vermiculite clay". J. Chem. Phys. 113, 2873 (2000) and "A neutron spin-echo study of confined water" J. Chem. Phys. 115, 11299 (2001). J. Chem. Phys. **121**, 9193–9194 (2004)
22. Swenson, J., Bergman, R., Howells, W.S., Longeville, S.: Response to Comment on "Quasielastic neutron scattering of two-dimensional water in a vermiculite clay" J. Chem. Phys. 113, 2873 (2000) and "A neutron spin-echo study of confined water" J. Chem. Phys. 115, 11299 (2001). J. Chem. Phys. **121**, 9195–9195 (2004)
23. Charnay, C., Lagerge, S., Partyka, S.: Assessment of the surface heterogeneity of talc materials. J. Colloid Interf. Sci. **233**, 250–258 (2001)
24. Chang, S.J., Kim, S.J., Kwon, K.: Effects of some water repellents on the structure and water sorption of smectite. Clay Clay Miner. **52**, 25–30 (2004)
25. Medhi, H., Bhattacharyya, K.G.: Kinetic and mechanistic studies on adsorption of Cu(ii) in aqueous medium onto montmorillonite K10 and its modified derivative. New J. Chem. **41**, 13533–13552 (2017)

26. Silva, L.C.A., Silva, E.A., Monteiro, M.R., Silva, C., Teleken, J.G., et al.: Effect of the chemical composition of smectites used in KF/Clay catalysts on soybean oil transesterification into methyl esters. Appl. Clay Sci. **102**, 121–127 (2014)
27. Czimerova, A., Iyi, N., Bujdak, J.: Fluorescence resonance energy transfer between two cationic laser dyes in presence of the series of reduced-charge montmorillonites: effect of the layer charge. J. Colloid Interf. Sci. **320**, 140–151 (2008)
28. Aggarwal, V., Li, H., Boyd, S.A., Teppen, B.J.: Enhanced sorption of trichloroethene by smectite clay exchanged with Cs^+. Environ. Sci. Technol. **40**, 894–899 (2006)
29. Xu, J.C., Stucki, J.W., Wu, J., Kostka, J.E., Sims, G.K.: Fate of atrazine and alachlor in redox-treated ferruginous smectite. Environ. Toxicol. Chem. **20**, 2717–2724 (2001)
30. Skipper, N.T., Soper, A.K., McConnell, J.D.C.: The structure of interlayer water in vermiculite. J. Chem. Phys. **94**, 5751–5760 (1991)
31. Pitteloud, C., Powell, D.H., Fischer, H.E.: The hydration structure of the Ni(2 +) ion intercalated in montmorillonite clay: a neutron diffraction with isotopic substitution study. Phys. Chem. Chem. Phys. **3**, 5567–5574 (2001)
32. Tunega, D., Gerzabek, M.H., Lischka, H.: Ab initio molecular dynamics study of a monomolecular water layer on octahedral and tetrahedral kaolinite surfaces. J. Phys. Chem. B **108**, 5930–5936 (2004)
33. Clausen, P., Andreoni, W., Curioni, A., Hughes, E., Plummer, C.J.G.: Adsorption of low-molecular-weight molecules on a dry clay surface: an Ab initio study. J. Phys. Chem. C **113**, 12293–12300 (2009)
34. Sakuma, H., Tsuchiya, T., Kawamura, K., Otsuki, K.: Local behavior of water molecules on brucite, talc, and halite surfaces: a molecular dynamics study. Mol. Simulat. **30**, 861–871 (2004)
35. Sobolev, O., Favre Buivin, F., Kemner, E., Russina, M., Beuneu, B., et al.: Water–clay surface interaction: a neutron scattering study. Chem. Phys. **374**, 55–61 (2010)
36. Malamis, S., Katsou, E.: A review on zinc and nickel adsorption on natural and modified zeolite, bentonite and vermiculite: examination of process parameters, kinetics and isotherms. J. Hazard. Mater. **252**, 428–461 (2013)
37. O'Day, P.A., Vlassopoulos, D.: Mineral-based amendments for remediation. Elements **6**, 375–381 (2010)
38. Waychunas, G.A., Fuller, C.C., Davis, J.A.: Surface complexation and precipitate geometry for aqueous Zn(II) sorption on ferrihydrite I: X-ray absorption extended fine structure spectroscopy analysis. Geochim. Cosmochim. Acta **66**, 1119–1137 (2002)
39. Mahmudov, R., Chen, C., Huang, C.-P.: Functionalized activated carbon for the adsorptive removal of perchlorate from water solutions. Front. Chem. Sci. Eng. **9**, 194–208 (2015)
40. Prietzel, J., Thieme, J., Paterson, D.: Phosphorus speciation of forest-soil organic surface layers using P K-edge XANES spectroscopy. J. Plant. Nutr. Soil. Sci. **173**, 805–807 (2010)
41. Bhattacharyya, K.G., Sen Gupta, S.: Influence of acid activation of kaolinite and montmorillonite on adsorptive removal of Cd(II) from water. Ind. Eng. Chem. Res. **46**, 3734–3742 (2007)
42. Tekin, R., Bac, N.: Antimicrobial behavior of ion-exchanged zeolite X containing fragrance. Micropor. Mesopor. Mat. **234**, 55–60 (2016)
43. Xie, J., Wang, Z., Lu, S., Wu, D., Zhang, Z., et al.: Removal and recovery of phosphate from water by lanthanum hydroxide materials. Chem. Eng. J. **254**, 163–170 (2014)
44. Chiou, C.T., Peters, L.J., Freed, V.H.: A physical concept of soil-water equilibria for nonionic organic compounds. Science (New York, NY) **206**, 831–832 (1979)
45. Chiou, C.T., Peters, L.J., Freed, V.H.: Soil-water equilibria for nonionic organic compounds. Science (New York, NY) **213**, 684–684 (1981)
46. Grafe, M., Nachtegaal, M., Sparks, D.L.: Formation of metal-arsenate precipitates at the goethite-water interface. Environ. Sci. Technol. **38**, 6561–6570 (2004)
47. Ler, A., Stanforth, R.: Evidence for surface precipitation of phosphate on goethite. Environ. Sci. Technol. **37**, 2694–2700 (2003)

48. Scheidegger, A.M., Lamble, G.M., Sparks, D.L.: The kinetics of nickel sorption on pyrophyllite as monitored by X-ray absorption fine structure (XAFS) spectroscopy. J. de Physique IV (Colloque) **7**, 773–775 (1997)
49. Liu, J., Zhu, R., Xu, T., Xu, Y., Ge, F., et al.: Co-adsorption of phosphate and zinc(II) on the surface of ferrihydrite. Chemosphere **144**, 1148–1155 (2016)
50. Schlegel, M.L., Manceau, A.: Zn incorporation in hydroxy-Al- and Keggin Al(13)-intercalated montmorillonite: a powder and polarized EXAFS study. Environ. Sci. Technol. **41**, 1942–1948 (2007)
51. Brown Jr., G.E., Calas, G.: Environmental mineralogy—understanding element behavior in ecosystems. C.R. Geosci. **343**, 90–112 (2011)
52. Elzinga, E.J., Kretzschmar, R.: In situ ATR-FTIR spectroscopic analysis of the co-adsorption of orthophosphate and Cd(II) onto hematite. Geochim. Cosmochim. Acta **117**, 53–64 (2013)
53. Wang, G., Wang, S., Sun, Z., Zheng, S., Xi, Y.: Structures of nonionic surfactant modified montmorillonites and their enhanced adsorption capacities towards a cationic organic dye. Appl. Clay Sci. **148**, 1–10 (2017)
54. Aytas, S., Yurtlu, M., Donat, R.: Adsorption characteristic of U(VI) ion onto thermally activated bentonite. J. Hazard. Mater. **172**, 667–674 (2009)
55. Sen Gupta, S., Bhattacharyya, K.G.: Kinetics of adsorption of metal ions on inorganic materials: a review. Adv. Colloid Interf. **162**, 39–58 (2011)
56. Leodopoulos, C., Doulia, D., Gimouhopoulos, K.: Adsorption of cationic dyes onto bentonite. Sep. Purif. Rev. **44**, 74–107 (2015)
57. Banat, F., Al-Asheh, S., Al-Anbar, S., Al-Refaie, S.: Microwave- and acid-treated bentonite as adsorbents of methylene blue from a simulated dye wastewater. B Eng. Geol. Environ. **66**, 53–58 (2007)
58. Eren, E., Afsin, B.: Removal of basic dye using raw and acid activated bentonite samples. J. Hazard. Mater. **166**, 830–835 (2009)
59. Brezovska, S., Marina, B., Panova, B., Burevski, D., Bosevska, V., et al.: The adsorption characteristics and porous structure of bentonite adsorbents as determined from the adsorption isotherms of benzene vapor. J. Serb. Chem. Soc. **69**, 145–151 (2004)
60. Jaynes, W.F., Boyd, S.A.: Trimethylphenylammonium-smectite as an effective adsorbent of water soluble aromatic hydrocarbons. J. Air Waste Manage. **40**, 1649–1653 (1990)
61. Gamba, M., Flores, F.M., Madejova, J., Torres Sanchez, R.M.: Comparison of imazalil removal onto montmorillonite and nanomontmorillonite and adsorption surface sites involved: an approach for agricultural wastewater treatment. Ind. Eng. Chem. Res. **54**, 1529–1538 (2015)
62. Han, L., Sun, K., Kang, M., Wu, F., Xing, B.: Influence of functional groups and pore characteristics of organic matter on the sorption of hydrophobic organic pollutants. Environ. Chem. **33**, 1811–1820 (2014)
63. Zhu, R., Chen, W., Shapley, T.V., Molinari, M., Ge, F., et al.: Sorptive characteristics of organomontmorillonite toward organic compounds: a combined LFERs and molecular dynamics simulation study. Environ. Sci. Technol. **45**, 6504–6510 (2011)
64. Khenifi, A., Bouberka, Z., Sekrane, F., Kameche, M., Derriche, Z.: Adsorption study of an industrial dye by an organic clay. Adsorption **13**, 149–158 (2007)
65. Xi, Y., Mallavarapu, M., Naidu, R.: Preparation, characterization of surfactants modified clay minerals and nitrate adsorption. Appl. Clay Sci. **48**, 92–96 (2010)
66. Churchman, G.J.: Formation of complexes between bentonite and different cationic polyelectrolytes and their use as sorbents for non-ionic and anionic pollutants. Appl. Clay Sci. **21**, 177–189 (2002)
67. Wu, P., Wu, W., Li, S., Xing, N., Zhu, N., et al.: Removal of Cd^{2+} from aqueous solution by adsorption using Fe-montmorillonite. J. Hazard. Mater. **169**, 824–830 (2009)
68. Kasama, T., Watanabe, Y., Yamada, H., Murakami, T.: Sorption of phosphates on Al-pillared smectites and mica at acidic to neutral pH. Appl. Clay Sci. **25**, 167–177 (2004)

69. Peak, D., Saha, U.K., Huang, P.M.: Selenite adsorption mechanisms on pure and coated montmorillonite: an EXAFS and XANES spectroscopic study. Soil Sci. Soc. Am. J. **70**, 192–203 (2006)
70. Borgnino, L., Giacomelli, C.E., Avena, M.J., De Pauli, C.P.: Phosphate adsorbed on Fe(III) modified montmorillonite: surface complexation studied by ATR-FTIR spectroscopy. Colloid Surf. A **353**, 238–244 (2010)
71. Zhu, R., Zhao, J., Ge, F., Zhu, L., Zhu, J., et al.: Restricting layer collapse enhances the adsorption capacity of reduced-charge organoclays. Appl. Clay Sci. **88–89**, 73–77 (2014)
72. Ma, L., Zhu, J., Xi, Y., Zhu, R., He, H., et al.: Simultaneous adsorption of Cd(II) and phosphate on Al-13 pillared montmorillonite. Rsc. Adv. **5**, 77227–77234 (2015)
73. Shi, L-n, Lin, Y.-M., Zhang, X., Chen, Z-l: Synthesis, characterization and kinetics of bentonite supported nZVI for the removal of Cr(VI) from aqueous solution. Chem. Eng. J. **171**, 612–617 (2011)
74. Shi, L.-N., Zhou, Y., Chen, Z., Megharaj, M., Naidu, R.: Simultaneous adsorption and degradation of Zn^{2+} and Cu^{2+} from wastewaters using nanoscale zero-valent iron impregnated with clays. Environ. Sci. Pollut. R **20**, 3639–3648 (2013)
75. Wu, L., Liao, L., Lv, G., Qin, F.: Stability and pH-independence of nano-zero-valent iron intercalated montmorillonite and its application on Cr(VI) removal. J. Contam. Hydrol. **179**, 1–9 (2015)
76. Wu, P.X., Liao, Z.W., Zhang, H.F., Guo, J.C.: Adsorption of phenol on inorganic-organic pillared montmorillonite in polluted water. Environ. Int. **26**, 401–407 (2001)
77. Li, J., Li, Y., Lu, J.: Adsorption of herbicides 2,4-D and acetochlor on inorganic-organic bentonites. Appl. Clay Sci. **46**, 314–318 (2009)
78. Ren, X., Zhang, Z., Luo, H., Hu, B., Dang, Z., et al.: Adsorption of arsenic on modified montmorillonite. Appl. Clay Sci. **97–98**, 17–23 (2014)
79. Zhu, L., Zhu, R.: Simultaneous sorption of organic compounds and phosphate to inorganic-organic bentonites from water. Sep. Purif. Technol. **54**, 71–76 (2007)
80. Zhu, R., Zhu, L., Zhu, J., Ge, F., Wang, T.: Sorption of naphthalene and phosphate to the CTMAB-Al-13 intercalated bentonites. J. Hazard. Mater. **168**, 1590–1594 (2009)
81. Chowdhury, S., Chakraborty, S., Das, P.: Adsorption of crystal violet from aqueous solution by citric acid modified rice straw: equilibrium, kinetics, and thermodynamics. Sep. Sci. Technol. **48**, 1339–1348 (2013)
82. Huang, W.Y., Chen, J., He, F., Tang, J.P., Li, D., et al.: Effective phosphate adsorption by Zr/Al-pillared montmorillonite: insight into equilibrium, kinetics and thermodynamics. Appl. Clay Sci. **104**, 252–260 (2015)
83. Dotto, G.L., Sellaoui, L., Lima, E.C., Ben Lamine, A.: Physicochemical and thermodynamic investigation of Ni(II) biosorption on various materials using the statistical physics modeling. J. Mol. Liq. **220**, 129–135 (2016)
84. Liu, Y.: Is the free energy change of adsorption correctly calculated? J. Chem. Eng. Data **54**, 1981–1985 (2009)
85. Das, B., Mondal, N.K., Bhaumik, R., Roy, P.: Insight into adsorption equilibrium, kinetics and thermodynamics of lead onto alluvial soil. Int. J. Environ. Sci. Technol. **11**, 1101–1114 (2014)
86. Saha, P., Chowdhury, S., Gupta, S., Kumar, I.: Insight into adsorption equilibrium, kinetics and thermodynamics of malachite green onto clayey soil of Indian origin. Chem. Eng. J. **165**, 874–882 (2010)
87. Islam, M.A., Khan, M.R., Mozumder, S.I.: Adsorption equilibrium and adsorption kinetics. Chem. Eng. Technol. **27**, 1095–1098 (2004)
88. Uslu, H.: Adsorption equilibria of formic acid by weakly basic adsorbent Amberlite IRA-67: equilibrium, kinetics, thermodynamic. Chem. Eng. Technol. **155**, 320–325 (2009)
89. Tan, K.L., Hameed, B.H.: Insight into the adsorption kinetics models for the removal of contaminants from aqueous solutions. J. Taiwan Inst. Chem. Eng. **74**, 25–48 (2017)
90. McKay, G., Ho, Y.S., Ng, J.C.Y.: Biosorption of copper from waste waters: a review. Sep. Purif. Methods **28**, 87–125 (2008)

91. Chmiel, K., Palica, M.: The dynamics of biofiltration of methyl ethyl ketone through a natural hed of pine-tree hark. Przem. Chem. **83**, 391–392 (2004)
92. Plazinski, W., Rudzinski, W., Plazinska, A.: Theoretical models of sorption kinetics including a surface reaction mechanism: a review. Adv. Colloid Interf. **152**, 2–13 (2009)
93. Ho, Y.S.: Review of second-order models for adsorption systems. J. Hazard. Mater. **136**, 681–689 (2006)
94. Verma, N., Sharma, R.: Bioremediation of toxic heavy metals: a patent review. Recent Pat. Biotechnol. **11**, 171–187 (2017)
95. Ho, Y.S., McKay, G.: A comparison of chemisorption kinetic models applied to pollutant removal on various sorbents. Process Saf. Environ. **76**, 332–340 (1998)
96. Acevedo, B., Barriocanal, C., Lupul, I., Gryglewicz, G.: Properties and performance of mesoporous activated carbons from scrap tyres, bituminous wastes and coal. Fuel **151**, 83–90 (2015)
97. Largitte, L., Pasquier, R.: A review of the kinetics adsorption models and their application to the adsorption of lead by an activated carbon. Chem. Eng. Res. Des. **109**, 495–504 (2016)
98. Dubey, S.P., Gopal, K., Bersillon, J.L.: Utility of adsorbents in the purification of drinking water: a review of characterization, efficiency and safety evaluation of various adsorbents. J. Environ. Biol. **30**, 327–332 (2009)
99. Al-Harahsheh, M., Shawabkeh, R., Al-Harahsheh, A., Tarawneh, K., Batiha, M.M.: Surface modification and characterization of Jordanian kaolinite: application for lead removal from aqueous solutions. Appl. Surf. Sci. **255**, 8098–8103 (2009)
100. Phothitontimongkol, T., Siebers, N., Sukpirom, N., Unob, F.: Preparation and characterization of novel organo-clay minerals for Hg(II) ions adsorption from aqueous solution. Appl. Clay Sci. **43**, 343–349 (2009)
101. Chen, Q., Hills, C.D., Yuan, M., Liu, H., Tyrer, M.: Characterization of carbonated tricalcium silicate and its sorption capa city for heavy metals: a micron-scale composite adsorbent of active silicate gel and calcite. J. Hazard. Mater. **153**, 775–783 (2008)
102. Penke, Y.K., Anantharaman, G., Ramkumar, J., Kar, K.K.: Aluminum substituted cobalt ferrite (Co-Al-Fe) nano adsorbent for arsenic adsorption in aqueous systems and detailed redox behavior study with XPS. Acs. Appl. Mater. Inter. **9**, 11587–11598 (2017)
103. Unuabonah, E.I., Guenter, C., Weber, J., Lubahn, S., Taubert, A.: Hybrid clay: a new highly efficient adsorbent for water treatment. Acs. Sustain. Chem. Eng. **1**, 966–973 (2013)
104. Liu, H., Chen, W., Liu, C., Liu, Y., Dong, C.: Magnetic mesoporous clay adsorbent: preparation, characterization and adsorption capacity for atrazine. Micropor. Mesopor. Mat. **194**, 72–78 (2014)
105. Quang-Vu, B., Chen, W.-H.: Pyrolysis characteristics and kinetics of microalgae via thermogravimetric analysis (TGA): a state-of-the-art review. Bioresour. Technol. **246**, 88–100 (2017)

Modification of Mineral Surfaces and Microstructures

Shenxu Bao

Abstract The modification of mineral surfaces and microstructures is generally an effective method to change and/or tailor the adsorption performance of mineral adsorbents. In this chapter, the aims, principle processes and mechanisms of the modification methods, including adsorption, intercalation, acid modification, thermal modification, grafting and coating, etching, etc., are summarized in detail. The characteristics of common modifying agents and their applications in the surface modification are also introduced. Finally, the commonly used measurements and/or evaluation methods for the surface modification of materials are listed.

Keywords Modification · Surface · Microstructure · Adsorption performance · Modifying agent

1 Introduction

The surface properties, pore structure and chemical nature of mineral adsorbents generally determine their adsorption capability. The modification of mineral surfaces and microstructures is commonly conducted to improve or change their adsorption capacity and/or adsorption selectivity. The modification is generally implemented chemically, physically or by the combination of chemical and physical treatment to change the physicochemical properties of minerals purposefully. The main aims of modification of mineral surfaces and microstructures can be summarized as:

S. Bao (✉)
School of Resources and Environmental Engineering, Hubei Key Laboratory of Mineral Resources Processing and Environment, Wuhan University of Technology, Luoshi Road 122, Wuhan 430070, Hubei, China
e-mail: sxbao@whut.edu.cn

S. Song and B. Li (eds.), *Adsorption at Natural Minerals/Water Interfaces*, Engineering Materials,
https://doi.org/10.1007/978-3-030-54451-5_3

(i) Modifying wettability (hydrophobicity or hydrophilicity);
(ii) Modifying surface energy;
(iii) Producing special functional groups at the surface for desired adsorption or reaction;
(iv) Carrying out surface cross-linking for aimed purpose;
(v) Modifying the surface morphology by altering surface crystallinity or roughness;
(vi) Improving the chemical inertness, corrosion-resistant ability for example;
(vii) Modifying physical properties, layer charge, specific surface area for example;
(viii) Modifying the structure properties, such as basal spacing, pore size distribution.

This chapter focuses on the representative modification methods for mineral surface and microstructure, such as adsorption, intercalation, acid modification, thermal modification, grafting and coating, etching. The aims, mechanisms and applications of these methods and commonly used modification agents for minerals surface have been comprehensively summarized.

2 Modification Methods for Mineral Surface and Microstructure

2.1 Adsorption

2.1.1 Aims of Adsorption

Clay minerals have often been used in different industrial and technological processes due to their adsorbent properties [1]. However, natural clay minerals have a drawback in a manner that they contain inorganic cations in aqueous phase, which are strongly hydrated (2:1 phyllosilicates for example in Fig. 1), showing a hydrophilicity surface. Thus, they are good adsorbents for ionic or polar compounds but not for non-ionic or hydrophobic organic compounds [2]. Surface hydrophobization of minerals is commonly conducted by reagent adsorption or activation of long-chain compounds. By using this method, one can control the physicochemical properties of mineral surface and microstructure and thus widely regulate the adsorption capacity and adsorption selectivities for toxic and/or organic contaminants. The synthetic or natural mineral can also be used to inhibite and eliminate adhesive impurities. For example, the stickies in waste paper preparation or paper pulp in the paper industry can be adsorbed by the hydrophobized minerals [3].

In addition, reagent adsorption also plays an important role in industrial application, froth flotation process, for instance. Flotation is a physico-chemical separation process utilizing the difference in surface properties of the valuable minerals and the unwanted gangue minerals [4]. The mechanism of froth flotation is

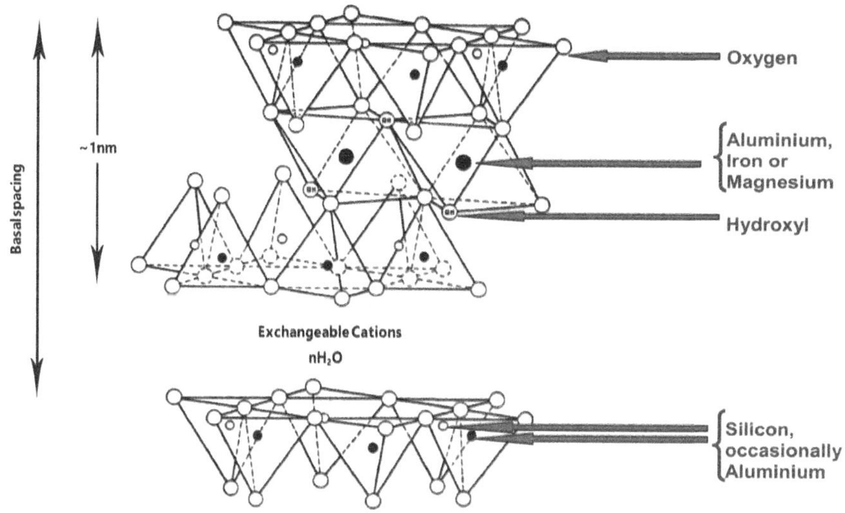

Fig. 1 Idealized structure of 2:1 phyllosilicates

complex, involving three phases (solids, water, and froth) with many subprocesses and interactions, but it is widely recognized that the surface hydrophobicity (or wettability) of minerals is a predominant factor affecting the flotation process.

The hydrophobicity of a solid surface is conventionally quantified by the contact angle, which are formed at the junction between vapor, solid and liquid phases, as shown in Fig. 2. The mineral particles can only attach to the air bubbles if they are somewhat water-repellent, i.e. hydrophobic and rising to the surface of aqueous solution with the bubbles via buoyancy. The attachment of the particles to the bubbles is determined by the interfacial energies between the solid, liquid, and gas phases, which can be expressed by the Yong/Dupre equation.

$$\gamma_{w/a}\cos\theta = \gamma_{s/a} - \gamma_{s/w} \tag{1}$$

where $\gamma_{w/a}$ is the surface energy of the liquid/vapor interface, $\gamma_{s/a}$ is the surface energy of the solid/vapor interface, $\gamma_{s/w}$ is the surface energy of the solid/liquid

Fig. 2 Contact angle between bubble and particle in an aqueous solution

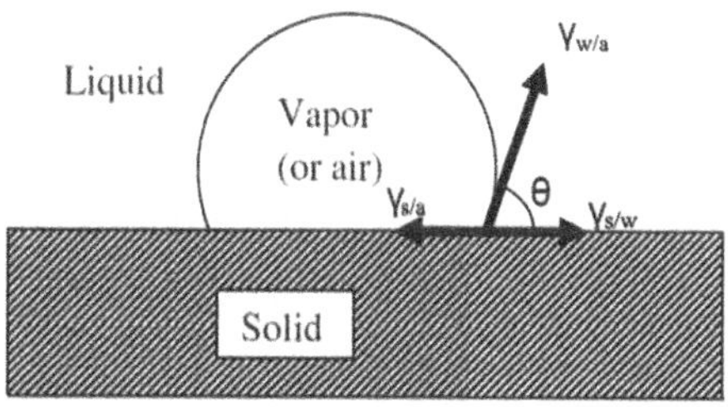

interface. If the contact angle is very small, the particle surface has strong hydrophilia and the particle does not attach to the bubbles. On the contrary, a very large contact indicates the strong hydrophobicity, leading to the high affinity of particle to the bubbles. Generally, a contact angle near 90° is sufficient for effective froth flotation in most cases.

2.1.2 Adsorption of Surfactants

Surfactants can be selectively adsorbed onto the mineral surface and render the particle surface hydrophobic by forming a thin film of non-polar hydrophobic hydrocarbons on the surface of particles. The chemicals are generally classed based on their dissociation in water: cationic, anionic, nonionic, which are summarized in Fig. 3 [5].

The nonionic surfactants are simple hydrocarbon oils, while the hydrophobic mechanism of anionic and cationic surfactants is that their polar part is selectively attaches to the mineral surfaces while a non-polar part projects out into the solution, pressing a hydrophobic surface. Surfactants can be either chemically bonded to the mineral surface by forming chemical bonds or adhere to the surface of minerals by physical forces, as illustrated in Fig. 4. The ionic part or end adheres to the particles surface by physical adsorption and the hydrophobic part or end leads to the surface hydrophobicity.

The chemisorption generally dominates the interaction between the minerals and surfactants when the reagent ions can form stable chemical bonds with the ions or compounds on the minerals surface. For example, the interaction of the sodium oleate on the surface of calcite is generally considered to form the precipitation of

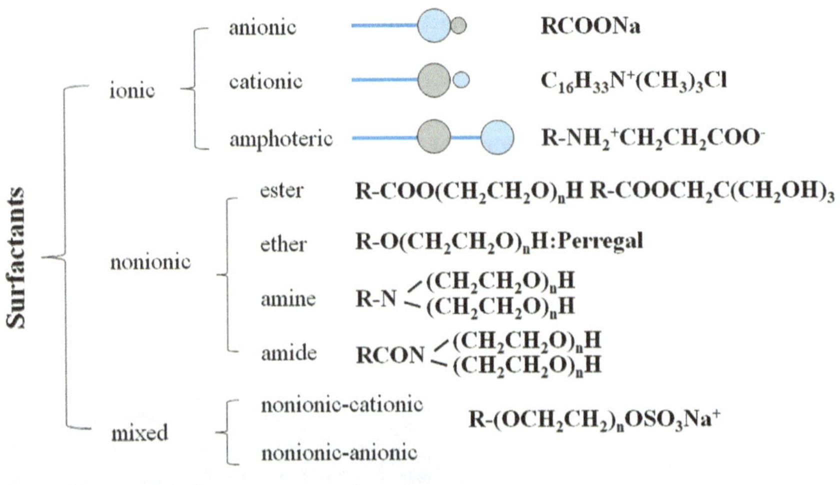

Fig. 3 Commonly used surfactants for surface modification

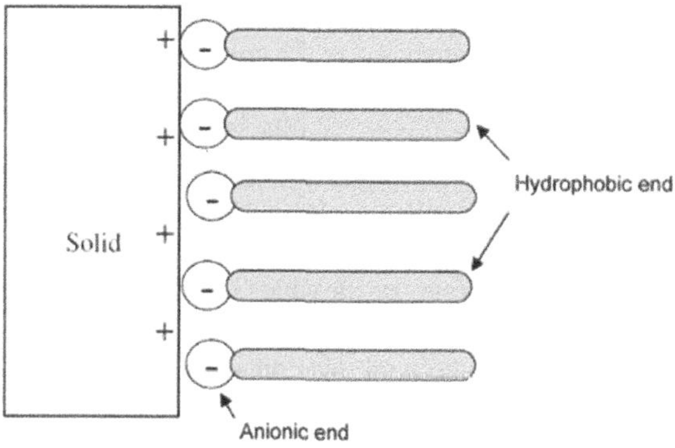

Fig. 4 Physical adsorption of collector onto a solid surface

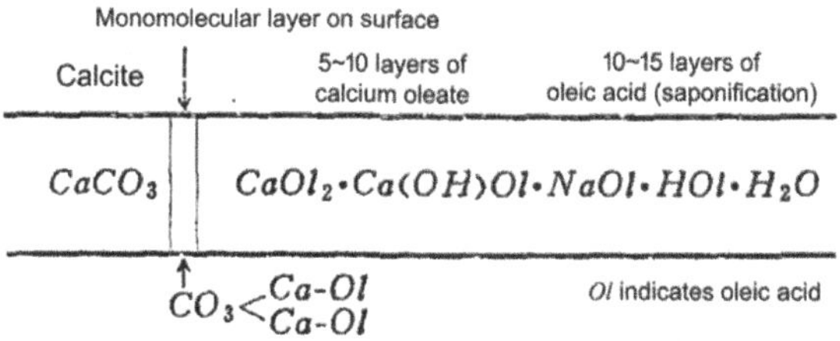

Fig. 5 The adsorption of sodium oleate on the surface of calcite

calcium oleate due to its low solubility [6]. This adsorption is believed to form multilayer adsorption, which enhance the hydrophobicity of the minerals (Fig. 5). Generally, the chemical bonds is strong and irreversible, thus the selective adsorption can permanently changes the nature of the minerals surface. However, the physical adsorption is rendered by electrostatic attraction or van der Waals force. These interactions are relatively weak and the chemicals can be desorbed from the surface in certain case, which may change the wettability of the surface.

Cationic surfactants are still widely employed as modifying agents for minerals modification, especially for clay minerals treatment, due to the electrostatic attraction between the dissociated amphiphilic cation and the negative charged surface of minerals. Anionic surfactants are generally adsorbed much less on layered clay minerals than cationic surfactants because the latter can be adsorbed on the minerals by cation exchange reaction in most cases (see Sect. 2.2 for details) [7]. Thus, the anionic surfactants are seldomly used for clay modification compared to the cationic and nonionic surfactants [8].

Nonionic surfactants display several advantages compared to the cationic or anionic surfactants, such as biodegradable and non-toxic, good thermal and chemical stabilities [9]. Backhaus et al. proved that the Triton-X series (octylphenol, polyethyleneglycol, ethers for example) surfactants can effectively hydrophobize the silica surface other than the montmorillonite surface [10]. The surfactant-covered silica reduces 2, 4-dichlorophenol adsorption drastically with the increasing pH due to Coulomb repulsion and solubility effects. Amirianshoja et al. [11] found that the quantity of a nonionic surfactant (triton X-100) adsorbed by clay minerals increase as the content of minerals in the adsorbents increases and the adsorption power of clay minerals for the nonionic surfactant follows the rank order of montmorillonite ≫ illite > kaolinite.

Recently, gemini surfactants have attracted interests. Gemini surfactants, sometimes also called as bis-surfactants, are a class of amphiphilic molecules containing two head groups and two aliphatic chains that chemically bonded together by a rigid or flexible spacer. Gemini surfactants may present many advantages over the conventional surfactants, such as higher efficiency in reducing the oil/water interfacial tension, better wetting, lower CMC (Critical Micelle Concentration), foaming, and antibacterial activities [12]. Yang et al. modified the wettability of oil-aged mica mineral surface by a gemini cationic surfactant containing double dodecyl chains and two quaternary ammonium headgroups [13]. The contact angle data show that gemini cationic surfactant has great wettability to alter the mica surfaces to mid-wetting even weakly water-wet. The gemini surfactant has stronger ability to modify oil-aged minerals surfaces in comparison with the single chain surfactant. Duval et al. comparatively investigated the adsorbed layer morphologies of cationic gemini surfactants of the type dodecanediyl-alpha, omega-bis (dimethylalkylammonium bromide) and their corresponding monomers, dimethyldodecylalkylammonium bromide, on mica using atomic force microscopy (AFM) soft-contact imaging [14]. Rosen and Li intensively studied the adsorption of two cationic gemini surfactants, ($[C_nH_{2n+1}N^+(CH_3)_2{-}CH_2CH_2]_2 \cdot 2Br^-$, where $n = 12$ and 14) and conventional cationic surfactants with similar single hydrophilic and hydrophobic groups (($C_nH_{2n+1}N^+(CH_3)_3 \cdot Br^-$, where $n = 12$ and 14) on limestone, sand, and clay (Na-montmorillonite) from their aqueous solution [15]. The adsorption amount of the gemini and the conventional surfactants on clay minerals are almost identical and very close to their cation exchange capacities since adsorption is mainly by ion exchange. On sand and limestone, the maximum adsorption amount of gemini adsorbed is about twice that of the comparable conventional surfactant because the former adsorption is caused by both electrostatic interaction and by hydrophobic bonding (Fig. 6). On all three soil minerals, the efficiency of the removal of 2-naphthol is greater for the gemini than that of the conventional surfactants. Besides these cationic surfactants, the adsorption of an anionic gemini surfactant (disodium didecyldiphenylether disulfonate) onto the mineral was also investigated. There is no adsorption from water onto sand for the anionic surfactant. On limestone, the adsorption of the gemini and conventional anionic surfactants is different. The adsorption of gemini surfactant occurs via electrostatic interaction with Ca^{2+} sites on the surface with a single layer and the conventional anionic surfactant adsorbs as a double layer with the hydrophilic groups of the second adsorbed layer oriented toward the aqueous phase (Fig. 7).

The surface hydrophobization not only modifies the adsorption characteristics of the absorbents but also affects the interaction force between the minerals. Sakamoto et al. hydrophobized silicate surface by physical adsorption of trimethylammonium chlorides with alkyl chains of six different lengths and measured the interaction

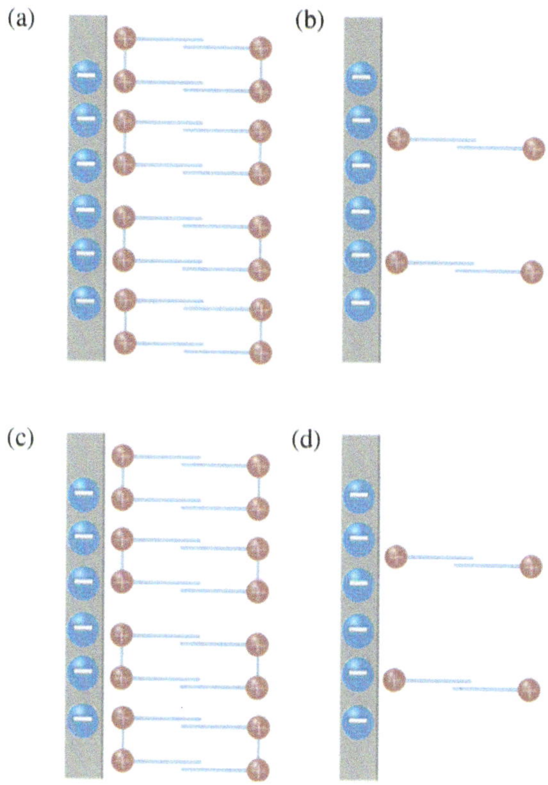

Fig. 6 Schematic illustration of the expected adsorption for both cationic gemini surfactants and conventional surfactants on sand and limestone. **a** Gemini surfactant on sand, **b** conventional surfactant on sand, **c** gemini surfactant on limestone, **d** conventional surfactant on limestone

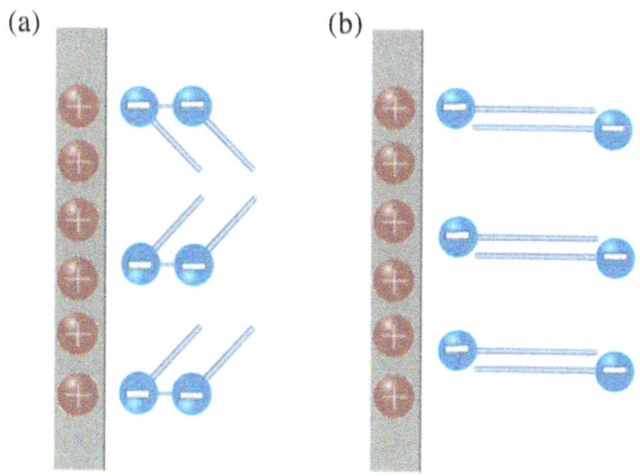

Fig. 7 Schematic illustration of the expected adsorption for both anionic gemini surfactants and conventional surfactants on limestone. **a** Gemini surfactant, **b** conventional surfactant

force between surfaces and observed the interaction force between surfaces in site using a tapping mode AFM [16]. The approaching and separating force curves suggest that long-range attraction is attributable to microscopic bubbles stabilized by surfactants at the interface, as illustrated in Fig. 8.

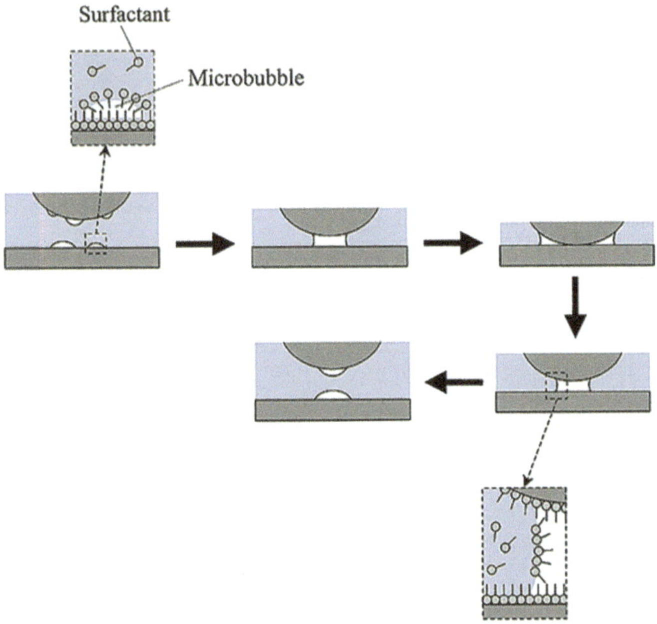

Fig. 8 Illustration of a supposed mechanism for the interaction between surfaces hydrophobized by physical adsorption of surfactants in aqueous solutions. The insets depict the estimated structure at the interface

2.1.3 Adsorption of Polymers

Hogg has proposed a simplified model to provide insight into the distribution of adsorbed polymer among the suspended particles and the nature of flocculation [17]. Hogg's study indicates that polymer adsorption on minerals occurs rapidly and adsorption is essentially complete within a time scale of seconds or less. The polymer-particle collision processes tend to favor adsorption on larger particles (or flocs) and the polymers with high molecular weight are especially ineffective for stable dispersions of very fine particles due to an enhanced preference for adsorption on larger particles and the smaller number of molecules available. Besra et al. [18] established the optimum flocculant concentration for the highest settling rate corresponds to about 50% coverage of the solid surface for untreated as well as surfactant-pretreated kaolin. About 50% coverage of the particle surface by polymer adsorption is necessary for optimum flocculation in which maximum settling rate occurs. This condition is valid for the surfactant-pretreated kaolin and the preadsorbed surfactants on kaolin further improve both settling rates and filtration characteristics on flocculation. Mierczynska-Vasilev and Beattie investigated the effects of impurities and cleavage characteristics of talcs on surface hydrophobicity and polymer adsorption for the minerals from different region [19]. The hydrophobicity and surface composition variation has a significant effect on the adsorption of polymers, which may altere the flotation behaviour for talcs of different impurity content and cleavage/fracture characteristics. Yeap et al. [20] have linked the interfacial chemistry, polyethylene oxide (PEO) flocculant adsorption behaviour, orthokinetic flocculation and dewatering behaviour of talc mineral dispersion. Higher Mg(II) ion concentration and shorter suspension conditioning time, which can enhance the flocculant adsorption, are provend to be less conducive to a better flocculation performance and improved dewaterability. The slower flocculant adsorption conditions promote dewatering behaviour. Chang et al. systematically investigated the equilibrium adsorption rate of adsorption of hydroxyethyl cellulose (HEC), hydroxypropylmethyl cellulose (HPMC), polyvinyl alcohol (PVOH) on clay, mica, talc and limestone [21]. The adsorption amount of PVOH is greatest and that of HEC is least for four minerals. For each polymeric species, the adsorption amount of polymer at equilibrium decreases in the order: clay > talc > mica > limestone. All of the adsorption kinetics can be well correlated by means of a Langmuir adsomtion isotherm expression with the exception of adsorption of HEC and HPMC on mica. The influence of low molecular weight poly(ethylene oxide) (PEO) homopolymer on the adsorption of a representative poly(ethylene oxide)–poly(propylene oxide)–poly(ethylene oxide) (PEO–PPO–PEO) block copolymer (Pluronic P105: EO_{37}–PO_{56}–EO_{37}) at the surface of protonated silica nanoparticles dispersed in aqueous phase was investigated by Bodratti et al. [22]. Pluronic P105 forms hydrophobic domains on the surface of protonated silica at a critical surface micelle concentration (CSMC) in the presence of 0.1 wt% silica nanoparticles in water. The CSMC of PEO–PPO–PEO block copolymer increases with increasing amounts of added PEO homopolymer and the adsorbed layer thickness at the protonated silica surface decreases with increasing PEO homopolymer concentration. The adsorption of the cationic polymer, polyethyleneimine (PEI), on bentonite particles was investigated by Öztekin et al. [23]. The fast adsorption of PEI on bentonite particles is attributed to the electrostatic interaction between the clay particles in ambient basic suspensions and the positively charged high molecular weight polyelectrolyte. The adsorbed polyelectrolyte affects the rheological properties of the bentonite suspensions.

Alemdar et al. investigated the effects of adsorption of a special type of cationic polymer [modified poly(ethylene glycol)] (Fig. 9) on the rheological and colloidal properties of sodium montmorillonite dispersions [24, 25]. The experimental results show that the cationic polymer can be strongly adsorbed onto the sodium montmorillonite surface, and a gradual increase in gelation with the addition of the cationic polymer, which reached a maximum at a cationic polymer concentration of

Fig. 9 Modified poly (ethylene glycol)

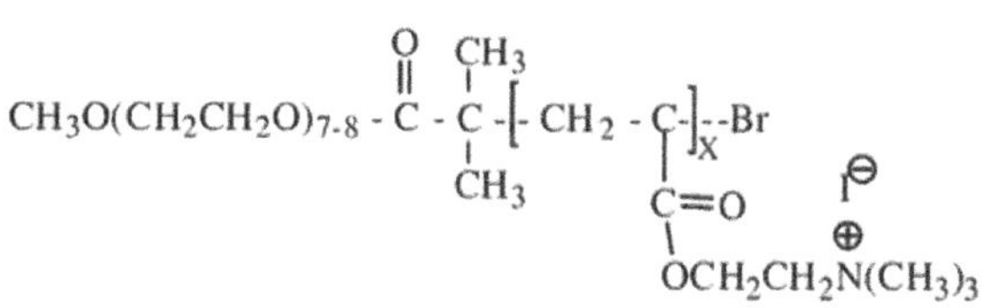

0.4–0.8 g/L. The basal spacing measurements indicate that the cationic polymer strongly adsorbed onto the sodium montmorillonite instead of intercalating the montmorillonite interlayer spaces.

2.1.4 Adsorption of Other Organic Matters

The adsorption of organic matters can also modify the surface characteristics of minerals. The surface of kaolinite can be completely hydrophobized by the adsorption of asphaltene but further adsorption of asphaltene leads to the build up of a polar layer of reverse orientation [26]. The possible interactions between organosiloxanes and active sites of aluminosilicate include heterolytic cleavage of the siloxane bond (Fig. 10a) and/or hydrogen bonding between surface SiOH groups and organosiloxane molecules (Fig. 10b). The hydrophobicity of silica particles can be largely improved by treating with trichloromethylsilane (TMCS) and the adorption capacity of the modified silica particles for ethyl(hydroxyethyl)cellulose (EHEC) also increases with the increasing temperature (Fig. 11) [27].

2.2 *Intercalation*

Intercalation of guest species into layered minerals is widely adopted to modify or fine-tune surface characteristics and structure of inorganic layered minerals, which may modify the adsorption properties of the minerals.

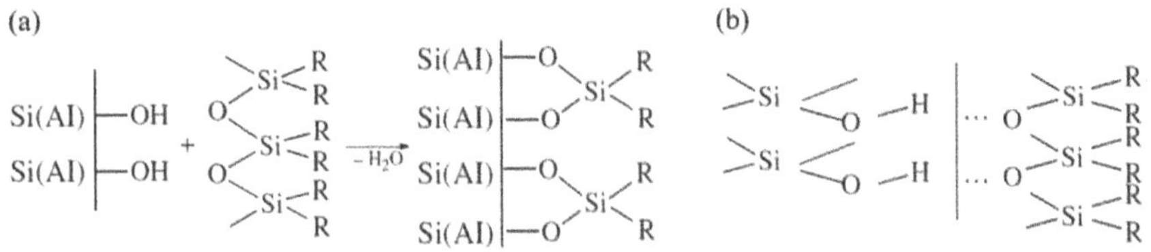

Fig. 10 The interaction between organosiloxanes and active sites of aluminosilicate (R indicates alkyl chain)

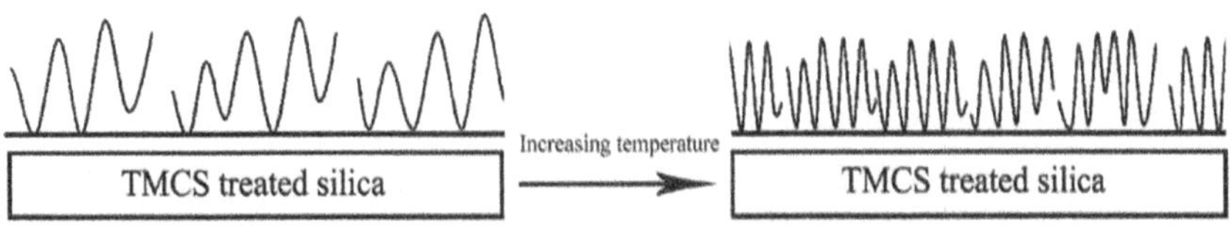

Fig. 11 The adsorbed EHEC conformation on silica particle surfaces at different temperatures

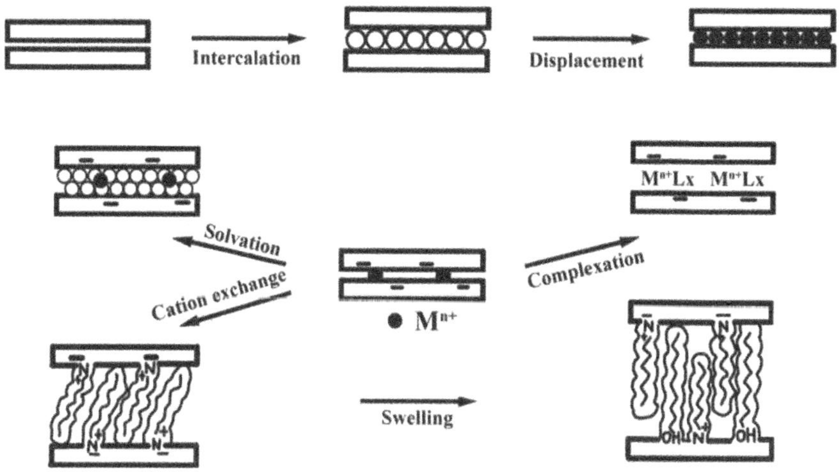

Fig. 12 Schematic illustration of intercalation

Intracrystalline reactivity of layered minerals is the ability to blind guest species between the layers or to exchange interlayer ions, which is schematically illustrated in Fig. 12 [28]. Graphite, kaolinite and the three-layer (or 2:1) clay minerals (Fig. 2), montmorillonite, hectorite and saponite for example, are the typical intracrastalline-reactive clay minerals. Two types of intracrastalline-reactive minerals are usually classified: minerals with neutral layers and minerals consisting of charged layers and gegenions between them [28]. However, this classification is only schematic. For instance, the OH groups pointing into the interlayer space can dissociate and the hydronium ions became the gegenions in the interlayer space when the interlayer space is filled with water molecules. In some literatures, intercalation just indicates the adsorption of neutral guest species between the uncharged layers, intercalation and ion exchange between charged guest species and exchangeable cations in the interlayers are combinedly introduced in this section.

2.2.1 Intercalation of Organocations or Surfactants

Exchanging the interlayer cations of three-layer clay minerals by particular organocations or surfactants provides a simple method to prepare new types of organic–inorganic hybride materials. Besides the three-layer clay minerals, other layered materials containing exchangeable cations are alkali layer silicates, such as kanemite and magadiite, molybdates, niobates, titanates, uranium micas [29]. When the interlayer metal cations of clay minerals are replaced by cationic organic compounds or surfactants, the hydrophilic surface of the minerals are substantially changed to become strongly hydrophobic or organophilic. The combination of the hydrophobic nature of the guest and the stable layered structures of the silicate sheets led unique physiochemical properties [30]. They have been referred as "organoclay" or "organophilic-clay" and present significantly enhanced affinity for organic contaminants. The adsorption performance of the organoclays would be also improved due to their high charge density and special functional groups (e.g. quaternary alkylammonium) in the interlayer spaces [31].

Ion exchange with alkylammonium ions is a well-known method to make these layered minerals hydrophobized and the ion exchange process is generally performed by dispersing the layered minerals in the solutions containing alkylammonium chlorides or bromides [32, 33]. The exchanging process is similar to that on ion exchange resin, i.e., the interlayer space takes up the alkylammonium ions as counterions, and water molecules, including some alkylammonium ions together with chloride or bromide form ion pairs and made the surface of these samples hydrophobic. The molecules of chemicals are not only adsorbed at the external surface but also on the internal surfaces of the layered samples by their cationic groups, and are not displaced when the samples are dispersed in aqueous solution. After washing and drying in vacuum, the basal spacing is determined by the chain length of the alkylammonium ions and the different arrangements of the alkyl chains. Thus, not only the length of the alkyl chains but also their conformation controls the wettability of the surface [28].

Different chain alkylammonium ions present various existing form in the interlayer space. For instance, short chain alkylammonium ions tend to lie flat on the surface in a monolayer, the so-called h1 structure (Fig. 13a). The long chain alkylammonium ions are inclined to be packed in monolayers to form bilayers (h2 strucutre) (Fig. 13b). The chain length determining the form of h1 and h2 structure is depended on the layer charge density and provides the best method of layer charge determination of 2:1 clay minerals [32]. For the highly charged smectites (layer charge >0.4 eq/formula unit) or vermiculites, long alkylammonium ions tend to form the pseudotrimolecular arrangement, i.e. the h3 structure (Fig. 13c). Moreover, for more highly charged vermiculites, micas, etc., the higher layer charges force the alkylammonium ions into the paraffin-type arrangement (v1 structure) (Fig. 13d). Many of the non-silicatic layered materials, such as phosphates, arsenates, niobates, titanates show layer charge densities comparable to vermiculites or even micas, and the alkylammonium ions are often arranged in paraffintype monolayers or bilayers.

Quaternary alkylammonium ions such as trimethylammonium ions can form the structures similar to the primary alkylammonium ions (Fig. 14a, b). The bulkier dialkyl dimethylammonium or ditalloyl dimethylammonium ions also can arrange in paraffin-type structures (Fig. 14c). Moreover, the cation exchange of the interlayer cations with small organic cations such as tetramethylammonium ion, results in the microporous solids where organoammonium ions act as pillar to separate each silicate layers.

The ion exchange reactions of both non-swelling and swelling clay minerals with organocations or surfactants in aqueous solution not only depend on the structure of the layered clay minerals but also on the nature of the exchangeable

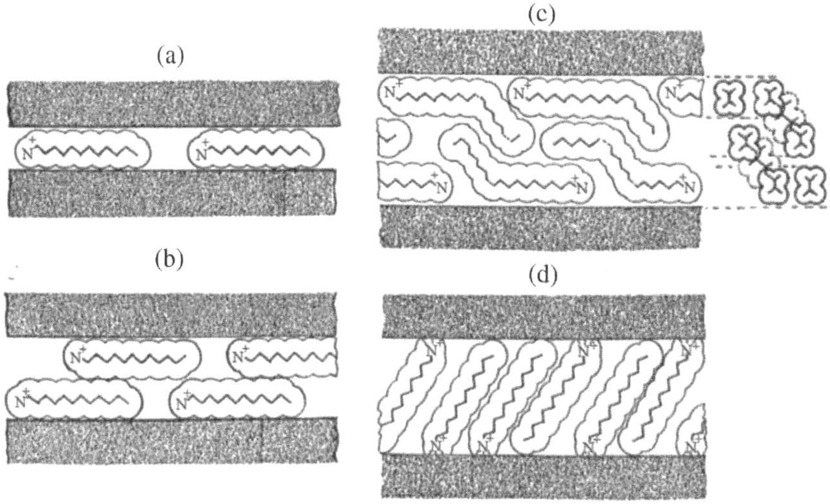

Fig. 13 Different form of alkylammonium ions in the interlayer space of layered materials **a** monolayer of flat-lying chains (h1 structure); **b** bilayers of flat-chains (h2 structure); **c** pseudotrimolecular layers (h3 structure) and **d** paraffin-type monolayers (v1 structure) [34]

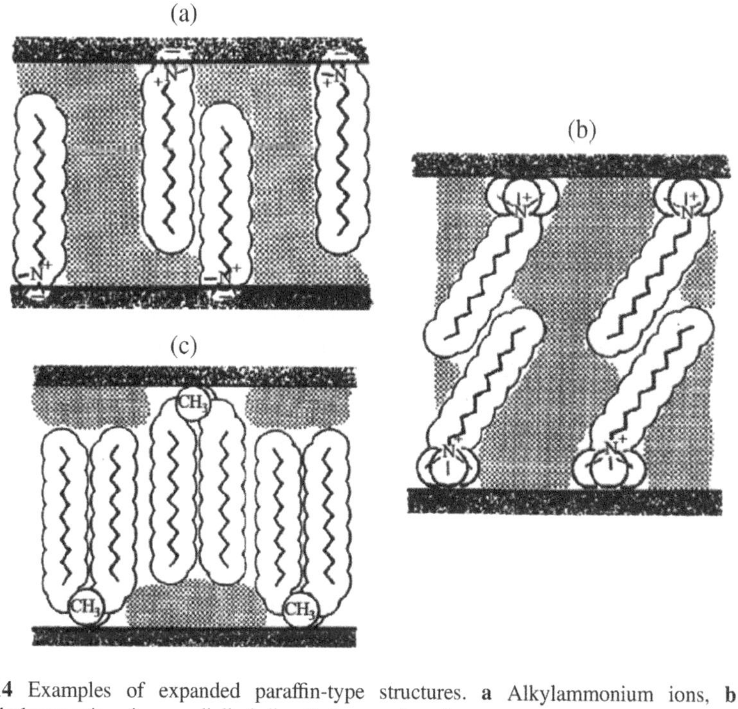

Fig. 14 Examples of expanded paraffin-type structures. **a** Alkylammonium ions, **b** alkyl trimethylammonium ions, **c** dialkyl dimethylammonium ions

inorganic cation. Patzkó and Dékány investigated the adsorption of hexadecylpyridinium chloride (HDPCI) on to the clay minerals kaolinite, attapulgite, vermiculite, allevardite and monocationic montmorillonites in aqueous solution. The values of the cation exchange capacity (CEC) suggest that HDP+ is adsorbed only on the external surface for non-swelling minerals, whereas for swelling minerals adsorption occurs also on the internal surface. The structures of the clay mineral also affect the extent of molecular adsorption (Ion exchange and molecular adsorption of a cationic surfactant on clay minerals). The study of Celis et al. [35] indicates that modification of montmorillonites (SAz-1 and SWy-2) with different organic cations (l-carnitine, spermine, hexadimethrine, tyramine, phenyltrimethylammonium, and hexadecyltrimethylammonium) renders organoclays with excellent affinities for terbuthylazine, diuron, and 4-chloro-2-methylphenoxyacetic acid (MCPA). All three herbicides display very strong affinities for SAz-1 exchanged with hexadecyltrimethylammonium cations, particularly when hexadecyltrimethylammonium is incorporated at 100% of the CEC of the clay mineral. Terbuthylazine and diuron also displayed very strong affinities for SWy-2 exchanged with l-carnitine and spermine, respectively. Shen et al. investigated the removal of phenol from the contaminated water by benzyltrimethylammonium bromide (BTMA) modified bentonite. The organomodified bentonite agglomerates and bond phenol to create flocs in aqueous solution and thus the phenol can be effectively cleaned by removing the flocs. BTMA-bentonite displays a high affinity for phenol, possibly because phenol molecules interact favorably with the benzene ring in BTMA ion through increased π–π type interactions [36]. Sanchez-Martin et al. [37] compared the efficiency of a series of clay minerals (montmorillonite, illite, muscovite, sepiolite and palygorskite) modified with the cationic surfactant, octadecyltrimetylammonium bromide (ODTMA), in the adsorption of the pesticides penconazole, linuron, alachlor, atrazine and metalaxyl. It is concluded that the efficiency of the ODTMA-clays for the adsorption of the pesticides depends on the degree of organic cation saturation; i.e., on the ODTMA content of the clay mineral, but the adsorption capacity of ODTMA is determined by the charge density of the clay mineral, i.e., by the density of the organic cation. The aromatic compounds can also be effectively adsorbed by tetrahedra tetramethylammonium (TMA) modified smectite, due possibly to a tilted orientation in a face-to-face arrangement with the TMA tetrahedra. In the presence of water, TMA-smectite shows high selectivity based on molecular size/shape, resulting in high uptake of benzene and progressively lower uptake of larger aromatic molecules [38]. Miyamoto et al. [39] intercalated a cationic cyanine dye, 1,1′-diethyl-2,2′-cyanine (PIC), into the layer silicate magadiite by ion-exchange between an aqueous PICBr solution and Na-magadiite or odecyltrimethylammonium-exchanged magadiite. The intercalated PIC ions formed aggregates in both systems. Moreover, surface properties of MgAl layered double hydroxides (LDHs) can be modified from hydrophilic to hydrophobic by intercalating two sulfonated surfactants, dodecylsulfonate (DSO) and dodecylbenzenesulfonate (DBS) using the co-precipitation method [40].

The intercalation of cationic surfactants not only changes the surface properties from hydrophilic to hydrophobic but also may greatly increase the anions adsorption capacity especially when surfactant loading exceeds the CEC of clay minerals. The resulting adsorption of surfactant molecules via hydrophobic bonding and the positive charge of ammonium will attract anions [41]. Thus, the clay minerals modified by organic surfactants may be potential adsorbens for the removal of anions, such as chromate [42], nitrate and arsenate [43], chromium(VI) [44] and iodide [45].

The non-ionic surfactant can also be used to modify the clay minerals. Guegan et al. [46] synthesized an organo-clay by intercalating a non-ionic surfactant triethylene glycol monodecyl ether ($C_{10}E_3$) into Ca-montmorillonite at pH of ~6.5 ± 0.2. The BDTAC is ion exchanged in stoichiometric proportions with the Ca^{2+} counter ions present in the interlayer space. For $C_{10}E_3$, the adsorption process between the adsorbed $C_{10}E_3$ and Ca-smectite does not depend on ion exchange but on several possible mechanisms (ion-dipole, H-bonding, van der Waals forces, and hydrophobic interactions).

2.2.2 Intercalation of Polymers

Recently, polymer/layered silicate (PLS) nanocomposites also have attracted great attention because they can exhibit different physicochemical properties when compared with virgin polymer or pristine layered silicates [47, 48]. Depending on the strength of interactions between the polymer matrix and layered minerals (modified or not), PLS nanocomposites can be classified as three different types (Fig. 15).

(1) Intercalated nanocomposites: In these nanocomposites, the insertion of a polymer matrix into the layered clay minerals occurs in a crystallographically regular fashion, regardless of the clay to polymer ratio. Intercalated nanocomposites are normally interlayered by a few molecular layers of polymer (Fig. 15a) [49].
(2) Flocculated nanocomposites: This type is same as the intercalated nanocomposites. However, the clay layers are sometimes flocculated due to hydroxylated edge–edge interaction of the silicate layers (Fig. 15b).

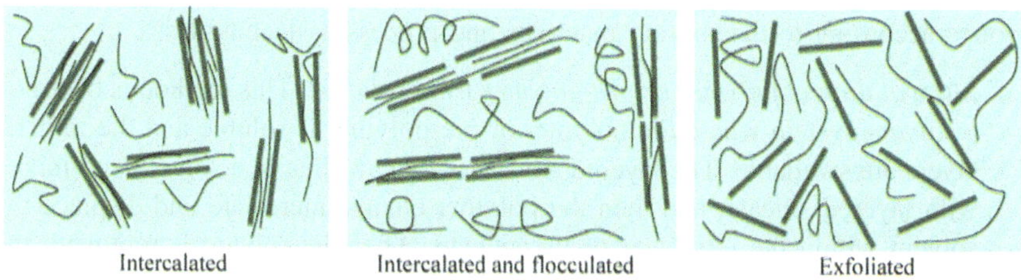

Fig. 15 Schematically illustration of three different types of thermodynamically achievable PLS nanocomposites

Table 1 The polymers used in PLC nanocomposites preparation

Types	Representative reagents
Vinyl polymers	Acrylamide, acrylic acid, acrylonitrile, ethyl vinyl alcohol copolymer, methyl methacrylate, methyl methacrylate copolymers, poly (N-isopropylacrylamide), poly(N-vinyl pyrrolidone), poly(vinyl pyrrolidinone), poly(vinyl pyridine), poly(ethylene glycol), poly (ethylene vinyl alcohol), poly(vinylidene fluoride), poly (p-phenylenevinylene), polybenzoxazole, poly (styrene-co-acrylonitrile), polystyrene–polyisoprene diblock copolymer, PVA, styrene, tetrafluoro ethylene, 4-vinylpyridine
Condensation (step) polymers	Butadiene copolymers, epoxidized natural rubber, epoxy polymer resins (EPR), fluoropoly (ether-imide), nylon-6, poly(1-caprolactone) (PCL), poly(ethylene terephthalate), poly(trimethylene terephthalate), poly(butylene terephthalate), polycarbonate (PC), PEO, ethylene oxide copolymers, poly(ethylene imine), poly(dimethyl siloxane), polybutadiene, phenolic resins, polyurethanes (PU), polyurethane urea, polyimides, poly(amic acid), polysulfone, polyetherimide
Polyolefins	Ethylene propylene diene methylene linkage rubber (EPDM), polypropylene (PP), polyethylene (PE), polyethylene oligomers, copolymers such as poly(ethylene-covinyl acetate) (EVA) and poly (1-butene)
Specialty polymers	Aryl-ethanyl-terminated coPoss imide oligomers, cyanate ester, Nafion®, polypyrrole (PPY), poly(N-vinylcarbazole) (PNVC), polyaniline (PANI), poly(p-phenylene vinylene)
Biodegradable polymers	Aliphatic polyester, polylactide (PLA), poly(butylene succinate) (PBS), PCL, unsaturated polyester, polyhydroxy butyrate

(3) Exfoliated nanocomposites: the individual clay layers are separated in a continuous polymer matrix by an average distances that depends on clay loading in this type of nanocomposite. Commonly, the clay content of an exfoliated nanocomposite is much lower than that of an intercalated nanocomposite (Fig. 15c).

Intercalation of polymers in layered hosts, such as layered clay minerals, has proven to be a successful method to synthesize PLS nanocomposites. The large variety of polymers used in nanocomposites preparation with layered silicates can be conventionally classified as below (Table 1) [50].

The preparation methods for PLS nanocomposites can be divided into three main groups according to the starting materials and processing techniques:

(a) *Intercalation of polymer or pre-polymer from solution.* This method is based on a solvent system where the polymer or pre-polymer is soluble and the mineral layers are swellable. The layered silicate is first swollen in a solvent and mixed with layered silicate, and then the polymer chains intercalate and displace the solvent within the interlayer of the silicate. After the solvent is removed, the intercalated structure remains and PLS nanocomposite can be obtanied. The water-soluble polymers, such as PEO, PVA, PVP, and PEVA are commonly intercalated into the layered clay minerals by using this method [51, 52].

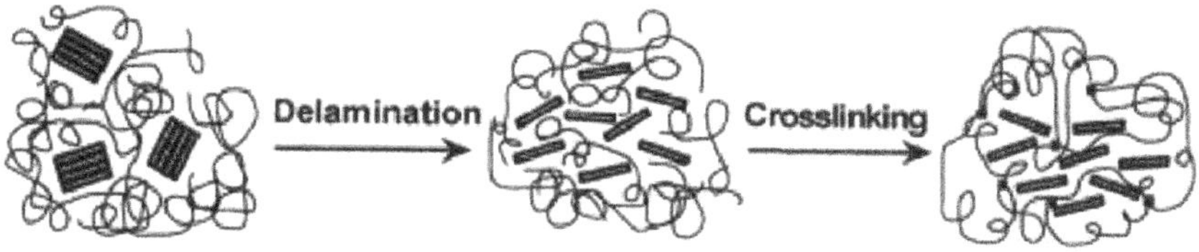

Fig. 16 Schematic illustration of nanocomposite synthesis

Burnside and Giannelis [53] used this method to prepare Poly(dimethylsiloxane) (PDMS)/MMT nanocomposites, which is illustrated in Fig. 16. The processes involve silicate delamination in the polymer matrix followed by cross-linking. Delamination of the layered silicate in the PDMS matrix was accomplished by suspending and sonicating the organosilicate in PDMS at room temperature.

(b) In situ *intercalative polymerization method.* The layered clay mineral is swollen within the liquid monomer or a monomer solution so the polymer formation can occur between the intercalated layers in this method. Polymerization can be ignited by heat or radiation, by the diffusion of a suitable initiator, or by an organic initiator or catalyst fixed through cation exchange inside the interlayer before the swelling process. Okada et al. [54] firstly reported the ability of α ω-amino acids modified Na^+-MMT to be swollen by the ε-caprolactam monomer at 100 °C and subsequently initiate its ring opening polymerization to obtain Nylon-6/montmorillonite nanocomposites, which can be illustrated in Fig. 17.

(c) *Melt intercalation method.* This method generally includes annealing, statically or under shear, a mixture of the polymer and organically modified layered clay mineral above the softening point of the polymer (Fig. 18) [55]. This method has become the standard for the preparation of PLS nanocomposites because it has great advantages such as environmentally friendliness and industrial compatibility.

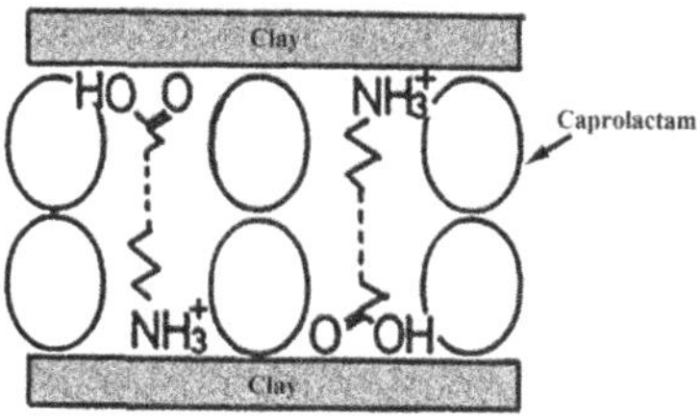

Fig. 17 Swelling behavior of ω-amino acid modified montmorillonite by ε-caprolactam

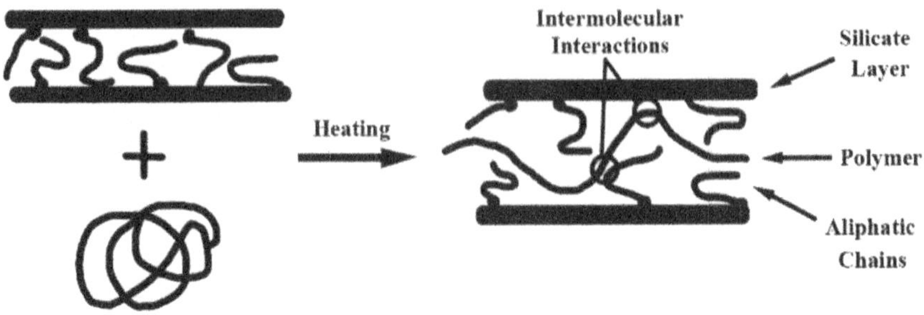

Fig. 18 Schematic illustration of the intercalation process between a polymer melt and organically modified layered clay mineral

2.2.3 Intercalation of Metals

As mentioned above, the ion exchange between the exchangeable cations in the interlayer of natural minerals with organocations such as tetramethylammonium, dodecyl-trimethylammonium, hexadecyl-trimethylammonium is one principal modification for clay minerals. However, the second method commonly used to modify clay minerals can be carried out by exchanging the interlayer cations with highly charged polymeric metal species, while the metal oxide pillared clays (PILCs) with a bi-dimensional porous network can be obtained by subsequent calcination [56]. Various polyoxocations (Al, Ni, Zr, Fe, Cr, Mg, Si, Bi, Be, B, Nb, Ta, Mo, Ti and Cu for example) and multimetallic pillars have been reported [57–59]. However, among all of these, only the Al-polyoxocation, i.e., $[Al_{13}O_4(OH)_{24}(H_2O)_{12}]$ or Al_{13}, has been well defined for its chemical composition, structure and charge [60].

The preparation processes for PILCs generally include a controlled hydrolysis reaction, which may be conducted in solution or in the interlamerllar space of the layered clay minerals. When the hydrolysis reaction is performed in aqueous phase different methods may be carried out: (a) addition of metal carbonates (Na, Mg, Zn, etc.) to a solution of $AlCl_3$ (or $FeCl_3$); (b) addition of metal hydroxides (Na, K, etc.); (c) addition of metallic aluminium in HCl and/or $AlCl_3$; and (d) electrolysis of $AlCl_3$ solution (or other metal chloride). Similar to the preparation of the pre-polymerized metals coagulants, the amount of polycation species depends not only on the preparation method, but also on other factors, such as the nature of the reactants and their initial concentration, the degree of hydrolysis (the ratio of OH to M^{n+}), the reaction temperature, the rate of addition of the reactants, and the aging conditions of the hydrolyzed solutions [57, 61].

Jiang and Zeng [62] investigated the effects of two types of montmorillonites (K10 and KSF) on the adsorption behavior. The modifiers and preparation processes are shown in Fig. 19. The basal spacing of the modified clays varied after modification, and it is highly related to the types of raw clay and modification conditions. The inorganic contaminant in solution (e.g., Cu) tends to be adsorbed by

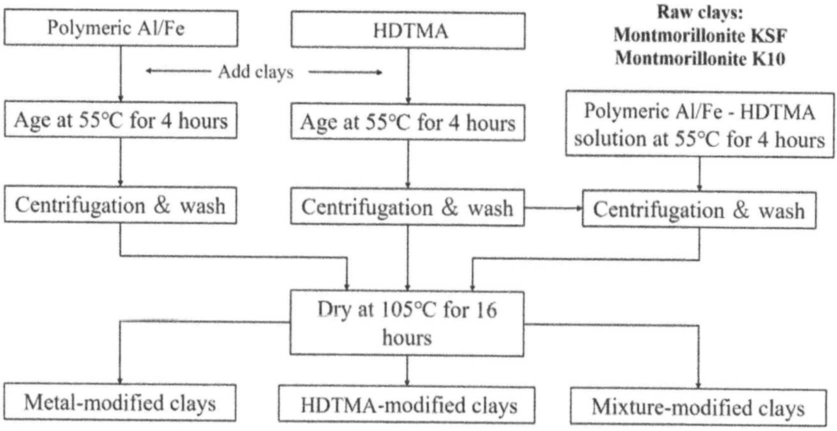

Fig. 19 The preparation processes for modified clay (HDTMA: hexadecyl-trimethylammonium)

the polymeric Al/Fe modified clay and the organic impurities (e.g., phenol) will be preferably removed by the surfactant modified clay; both due to the specific surface properties resulting from introducing the modifiers. The complex modified clays possess the ability of adsorbing both inorganic and organic contaminants.

Matthes et al. [63] investigated the adsorption of various heavy metals on Na-rich bentonite, Al and Zr-pillared Na-rich bentonite, the uncalcined hydroxy-intercalated precursors, and commercial Al-pillared bentonite, respectively. The adsorption of Cd, Cu, Pb, and Zn by A1 and Zr-hydroxy-intercalated and pillared bentonites is governed by cation exchange but the higher and partially non-exchangeable quantities of Zn sorbed by these materials suggest the dominance of surface complexation of Zn-ions with hydroxyl groups of the A1 and Zr-polyhydroxo cations and pillars, indicating Al-hydroxy-interlayered and Al-pillared bentonites are potential sorbents for the removal of Zn from aqueous phase.

Yoda et al. [64] reported a new method for preparation of PILCs by supercritical intercalation performed using a flow type supercritical carbon dioxide ($scCO_2$) extraction/impregnation system. This method consists of two steps. Firstly, hydrophobitization of interlayer space by ion exchange of interlayer cations into organic cations in hydrous solution, and secondly, the intercalation of organic metal compounds followed by hydrolysis with interlayer adsorbed water in $scCO_2$ and calcination. The basal spacing can be controlled by intercalation time on silica pillared clay, indicating that the potential of seemless control of pore size and pillar size using this method.

2.3 Acid Modification

Treatment of minerals with inorganic acids, such as sulfuric acid, hydrochloric acid and nitric acid, referred to as "acid activation" or "acid modification", is also a commonly used method for production of sorbents or catalysts used in industry or environmental protection measures [65]. Acid treatment is a useful and effective way to improve the adsorption capacity of minerals by increasing the specific surface area and altering the pore volume and chemical composition of the minerals, which is accomplished by ion exchange between H^+ ions and the exchangeable ions in the layered minerals and the partical dissolution of Al^{3+}, Mg^{2+} and Fe^{3+} from the silicate sheets during acid treatment [66]. Moreover, acid activation can also change the swelling ability, CEC, layer charge and surface acid sites of minerals and thus affect the adsorption properties of the minerals [67, 68].

The acid treatment on bentonite has been intensively studied and the acid-leached, also known as bleaching earth, has been used for decolorization of vegetable oils, removal of contaminant from aqueous solution and the catalyst support in industrial scale. Sarma et al. [68] proved that the montmorillonite treated with 0.25 and 0.50 M sulfuric acid would be suitable adsorbent for removing crystal violet from aqueous solution. Akpomie and Dawodu [69] investigated the adsorption of heavy metals on the acid modified montorillonite. Their results revealed that the acid treatment can improve the adsorption capacity of the montmorillonite and the removal of heavy metals from the effluent followed the order Zn > Cu > Mn > Cd > Pb > Ni, which is closely related to the concentration of these ions in the automobile effluents. The adsorption of Acid Blue 80 (AB80) on the chemical modified bentonite from aqueous solutions also can be largely improved by treating the mineral with sulfuric acid [70]. Sakizci et al. investigated the adsorption SO_2 on two different bentonites (BK and BY) treated with 1 and 2 M HCl [66]. Acid-treated bentonites have larger specific surface areas, greater SiO_2 contents and smaller contents of Al_2O_3, Fe_2O_3, MgO and CaO than the raw clays. The adsorption capcity of bentonites for SO_2 can be largely improved by acid treatment. The adsorption capacity for SO_2 depends on the clay type, the cation content in the channels, the amount and content of impurities and the conditions of acid treatment process, etc. The study of Alver and Sakizci also proves that acid treatment leds to an increase in both the specific surface areas and adsorption capacities towards ethylene for five clay minerals (sepiolite, two kaolinites and two bentonites) [71]. Both kaolinite samples exhibit the lowest adsorption capacities due to their structural characteristics. Vuković et al. [72] investigated the effect of the concentration of HCl on the chemical composition and structure of bentonite. The increase in SiO_2 content with the increasing HCl content can be fitted by a second degree polynomial, while all other oxides display an exponential decrease with the increasing concentration, which is consistent with the previous study conducted by Falaras et al. [73] although the acid activation was performed with another type of bentonite (Texas smectite) and a different mineral acid (H_2SO_4). The acid activation reduces the size of bentonite particles and increases the specific

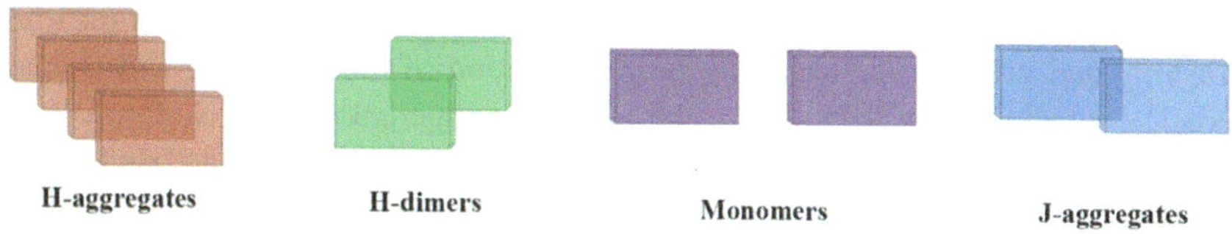

Fig. 20 The adsorption species of MB on clay minerals

surface area and pore volume of the minerals. These effects are improved with the increasing acid concentration up to 4.5 M HCl and further increase in acid concentration does not led to the development of new porous structure [74].

Panda et al. [75] refluxed natural kaolin with sulphuric acid at different concentrations. The acid treatment under reflux conditions leads to the removal of the octahedral Al^{3+} cations along with other impurities and the leaching of Al^{3+} ions increases progressively with severity of the acid treatment. Thus, acid leaching method is useful to manufacture the surface activity, porous and high surface area material which can be used for catalyst as well as an adsorbent. Wambu et al. [76] found that the diatomaceous mineral treated by diluted HCl can present extremely strong selective adsroption for F ions because of high affinity of the acid pretreated mineral for the ion, suggesting that the acid treated diatomaceous earth may be used as a plausible, cheap and safe F adsorbent for defluoridating fluoride polluted aqueous streams.

Pentrák et al. [77] characterize the adsorption properties of methylene blue (MB) on the acid modified clay minerals (smectites, illite–smectites and kaolinite). Second derivative absorption spectra (SDS) were used to distinguish the absorption of different dye species, such as the monomers, dimers, H-aggregates and J-aggregates (Fig. 20). Decomposition of the clay minerals structure due to HCl attack results in the reduction of the layer charge of all minerals as confirmed their decreasing CEC values. The presence of non-expandable layers in smectite/illite and illites reduces their dissolution rate in HCl solution compared to smectites. The shape of the SDS of kaolinitic sand indicates the adsorption of MB as H-aggregates on the accessory clay minerals (e.g. illite or muscovite) and monomers, most probably on the basal surfaces of middle-ordered kaolinite, present as a main mineral in this sample.

2.4 *Thermal Modification*

Thermal modification has been widely applied into modification of the structures and characteristics of many minerals, for instance, goethite and siderite, alunite, zeolite, diatomite, ulexite, feldspar, sepiolite and bentonites for using as adsorbents. Thermal modification to prepare adsorbents with abundant pore structures and large surface area has drawn increasing attention since 1970s. Generally, the natural minerals are calcinated in a conventional or microwave oven for a certain time after crushing and grinding, resulting in the modification of the structures and characteristics, which greatly depends on the roasting temperature [78]. Thermal modification has been found to be good for preparing homogenous microporous mineral adsorbents [79].

2.4.1 Conventional Heating

Conventional thermal treatment is commonly carried out in electrical ovens. In this process, the natural minerals are usually calcinated at different temperature for certain time to get desired product. Calcination is one thermal method to modify the surface properties, such as morphology, wettability, catalytic and/or photocatalytic activity, electrochemical capability, for minerals, synthesized catalysts, inorganic nanoparticles, etc. The calcination is used to modify the structure and surface properties of components at relative high temperature in most cases.

Zhang et al. reported that the crystal structure of trapezoid-like MgO microparticles evolved from mesocrystal to polycrystal, then to pseudomorph, and finally to cubic single crystal with the increase of calcination temperature ranging from 400 to 1000 °C [80] (Fig. 21). Along with the process, its surface changed from smooth to fractured structures followed by the structure composed of uniform nanoparticles resulting from the sintering of MgO at a higher temperature, which cause the fluctuation of the electrochemical performance. Their catalytic activities were highly dependent on the textural properties (e.g., surface area and porosity) although the reaction selectivity was still closely related with calcination temperature.

Bojemueller et al. [81] investigated the enhancement of pesticide adsorption by thermal modified bentonites. They found that the adsorption capacity for metolachlor from aqueous was enhanced by thermally modified bentonites due to the increase of the specific mesopore surface area with the calcination (Fig. 22).

Aivalioti et al. [82] studied the adsorption of BTEX and MTBE onto raw and thermally modified diatomite. It was found that adsorption capacity is closely related to the calcination temperature due to the different specific surface, pore volume distribution and porosity changes in the modified diatomite. Pelte et al. [83]

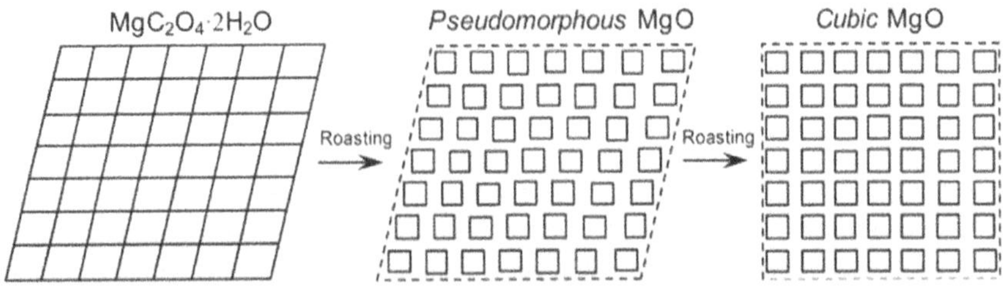

Fig. 21 Schematic illustration of the evolution of the crystal plane of MgO

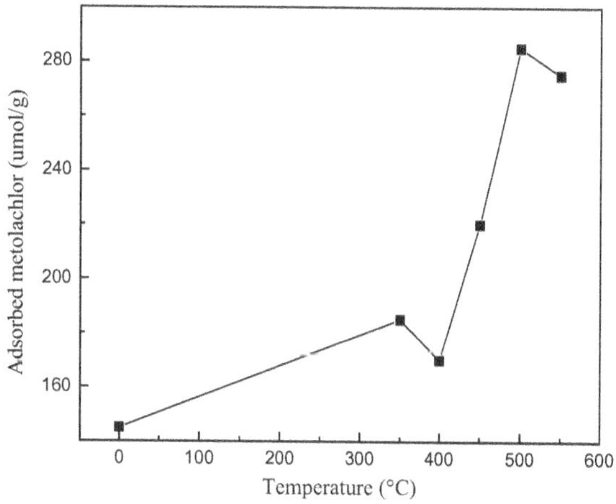

Fig. 22 Adsorption of metolachlor on bentonite and bentonite calcinated at 350–550 °C

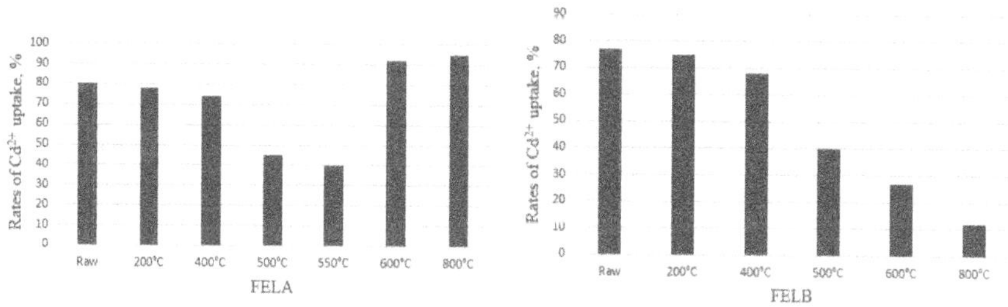

Fig. 23 Cd^{2+} sorption rates at room temperature of FELA and FELB, untreated or heated at different temperatures

investigated the effects of thermal treatment on feldspar sorptive properties. They have demonstrated that cadium removal efficiency of treated minerals is strongly dependent on heating temperature because of the different migration of Na and K cations caused by thermal treatment at various temperature (Fig. 23).

Carbonization is usually a thermal modification method to produce carbon layer on the surface of minerals, which can change the physicochemical properties and surface chemistry. Wang et al. [84] investigated the formation of spinel $Li_4Ti_5O_{12}$ nano-particles by heat treatment of polyaniline coated TiO_2 particles at 800 °C under an argon atmosphere containing 5% H_2 as shown in Fig. 24. In this process, the carbonization of polyaniline not only effectively restricts the particle-size growth of $Li_4Ti_5O_{12}$, but also reduces the surface Ti(IV) into Ti(III). The surface modification combined with tailored particle size can improve the surface conductivity, shorten the Li-ion diffusion path and increase the equilibrium solid solution that is beneficial for Li-ion mobility. Janus et al. carbonized a commercial

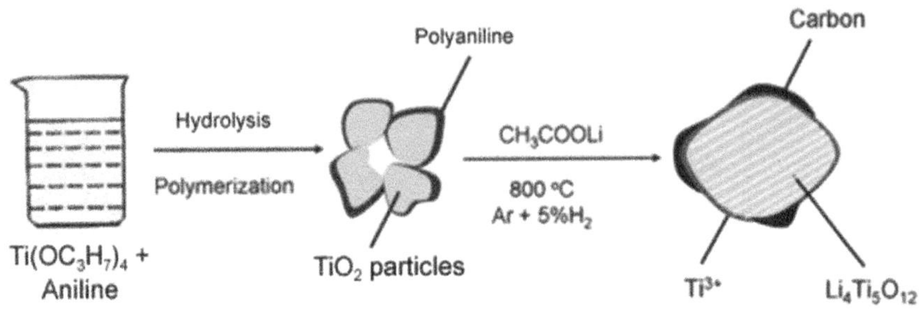

Fig. 24 Schematic of the formation of $Li_4Ti_5O_{12}$

anatase-TiO_2 in a pressure reactor in an ethanol atmosphere at 120 °C for 24 h and compared the adsorption of azo dyes onto surfaces of the pristine TiO_2, commercial P25 and carbon-modified TiO_2 [85]. Modification of TiO_2 by carbon leads to the change from the Freundlich model to the Langmuir model of adsorption isotherm and the increase of photocatalytic activity of carbon-modified TiO_2 photocatalyst was also observed.

The surface area and pore size distribution of the minerals containing hydroxyl, C, N or S could be modified by calcination in aerobic conditions. The release of H_2O, CO_2, NO_x, SO_2 creates pores with different size inside and transforms the mineral species. For instance, both goethite and siderite could be transformed into porous hematite after thermal treatment [86]. The reactions are as follows:

$$\alpha - FeOOH \xrightarrow{300^{\circ}C} \alpha - Fe_2O_3 + H_2O \tag{2}$$

$$FeCO_3 + O_2 \xrightarrow{500^{\circ}C} Fe_2O_3 + CO_2 \tag{3}$$

Considering the different molecular size of H_2O and CO_2, the corresponding pores produced in hematite are different. It is reported the average pore size of hematite calcined from siderite is 9.09 nm, which is twice of that calcined from goethite [86]. Moreover, the formation of pores would increase the surface area of the minerals. According to the reports, the surface area of porous hematite produced from goethite reach to 85 and 140 m^2/g [87, 88]. Further study shows that the calcination temperature could also affect the surface area and pore size of the produced minerals. Jia et al. [89] reported the major micropores of porous hematite derived from goethite were formed around 300 °C. Then the micropores were merged into mesopores with temperature increasing. The surface area of the products decreased accordingly, from 121 m^2/g at 300 °C to 42 m^2/g at 600 °C.

In sum, conventional heating has been proved to be an effective method to modify the structures and properties of the natural minerals, especially the pore structure and surface area, which strongly influence the adsorption properties of the minerals when using as adsorbents.

2.4.2 Microwave Heating

Conventional thermal modification is featured by high energy consumption, long time and nonuniform heating. Recently, microwave irradiation has obtained great attention to replace for conventional oven in modification of natural minerals through thermal treatment. The microwave heating is an extremely efficient process in selective heating of minerals as no energy is wasted in "bulk heating" the sample, which is an obvious advantage that microwave heating has over conventional heating methods [90]. Thus, the advantages of microwave heating over conventional heating are energy consumption savings [90], uniform heating [91] and time savings [92].

Microwave heating process needs less time than conventional heating process for synthesis of zeolite T [93]. It was reported that microwave heating yielded zeolite T within shorter synthesis time, producing an appropriate ratio of SiO_4 to AlO_4 tetrahedra in the reaction gel to form stable phase phillipsite. The minerals prepared by microwave heating modification feature large specific area and high adsorption capacity due to the instantaneous and fast heating. For instance, a specific area of 310 $m^2\ g^{-1}$ was obtained at 100 °C for 48 h [93].

2.5 Grafting and Coating

The notable advantage of reagent adsorption (Sect. 2.1) is that the structure of mineral is not altered. However, the main potential disadvantage is that the forces between the adsorbed molecules and the mineral may be weak. Grafting or coating of functional polymers to mineral surface can significantly change the surface properties of minerals and thus control the adsorption characteristics on the minerals [94].

Generally, two methods are widely used to graft polymer chains on the surface of inorganic particles. One is the "grafting to" method, where the end-functionalized polymers react with an appropriate surface. The other is the "grafting from (also called surface-initiated polymerizations)" method in which polymer chains are grown from an initiator-terminated self-assembled monolayer. The process of the latter method can be simply described as below. First, the substrate of mineral particles or compound is modified with initiator-bearing self-assembled monolayers. These monolayers can be formed on almost any surface. Then, the initiator surfaces are exposed to solution containing catalyst and monomer (solvent are added if necessary). Ideally, the polymerization is not only surface-initiated but also surface-confined, i.e. no polymerization in the solution. The two methods can be illustrated in Fig. 25 [95].

A higher percentage of successful grafts can be achieved in polymer-grafted inorganic particles by initiating the graft polymerization from initiating groups located on the particles' surfaces. The polymerization processes, which may include

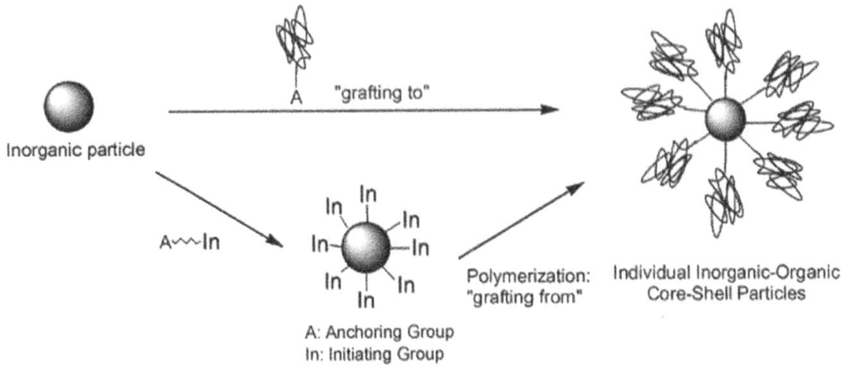

Fig. 25 The illustration of "grafting in" and "grafting from" methods

radical, anionic and cationic polymerization methods, involves propagation of the grafted polymers from the surface of the particles [96].

Mesnage et al. [97] firstly investigated the polymer grafting of TiO_2 nanoparticles by a "green process" based on diazonium salts initiating radical polymerization. They used The Graftfast™ process, which is based on the chemical reduction of diazonium salts by reducing agents in presence of a vinylic monomer [i.e. 2-hydroxyethyl methacrylate (HEMA)], to synthesize stable, homogeneous and covalently grafted polymer films by a short one-step reaction occurring at atmospheric pressure, room temperature, in water and requiring no external energy source. The mechanism of the Graftfast™ process for the grafting of methacrylate polymer onto TiO_2 nanoparticles is illustrated in Fig. 26, which is originally proposed by Mévellec et al. [98]. This process relies on the chemical reduction of diazonium salt, in the presence of a vinylic monomer, into aryl radicals (Fig. 26a) which are then not only able to form a grafted polyphenylene-like film on the TiO_2 particles (Fig. 26b) but also to initiate the radical polymerization of the monomers (Fig. 26c). Then, the growing radical oligomers eventually graft on the aryl rings present on the surface (Fig. 26d) to form a grafted polymer shell around the TiO_2 core. Rong et al. [99] successfully grafted polystyrene and polyacrylamide onto nanosized alumina particles to modify the surface characteristics of alumina nanoparticles by firstly treating the particles with silane, followed by radical grafting polymerization in aqueous or non-aqueous systems.

Shirai et al. [100] modified the dispersibility of polymethylsiloxane-coated titanium dioxide with alcoholic hydroxyl groups (Ti/Si–R–OH) in solvents, the radical graft polymerization of vinyl monomers initiated by azo groups introduced onto Ti/Si–R–OH was investigated (Fig. 27a, b). In addition, the graft polymerization of vinyl monomers initiated by the system consisting of trichloroacetyl groups on titanium dioxide surface and $Mo(CO)_6$ was also examined (Fig. 27c). The surface of polystyrene-grafted titanium dioxide shows extremely hydrophobic nature, but that of polyDEAM-grafted and polyNVF-graft titanium dioxide show hydrophilic

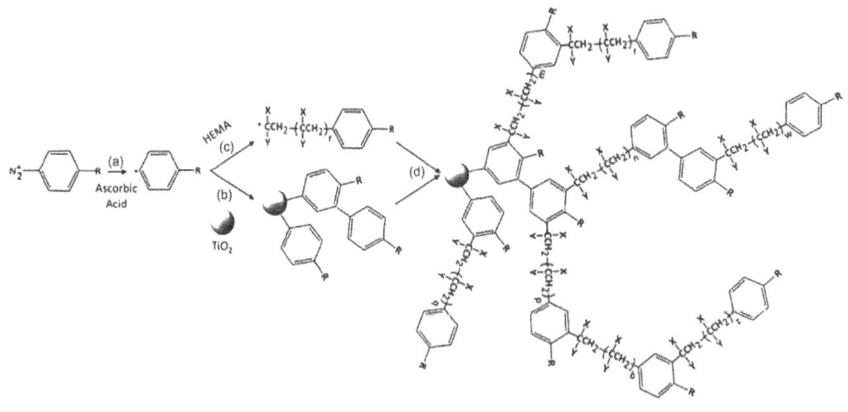

Fig. 26 Mechanism of the Graftfast^{TM} process for the grafting of methacrylate polymer onto TiO_2 nanoparticles

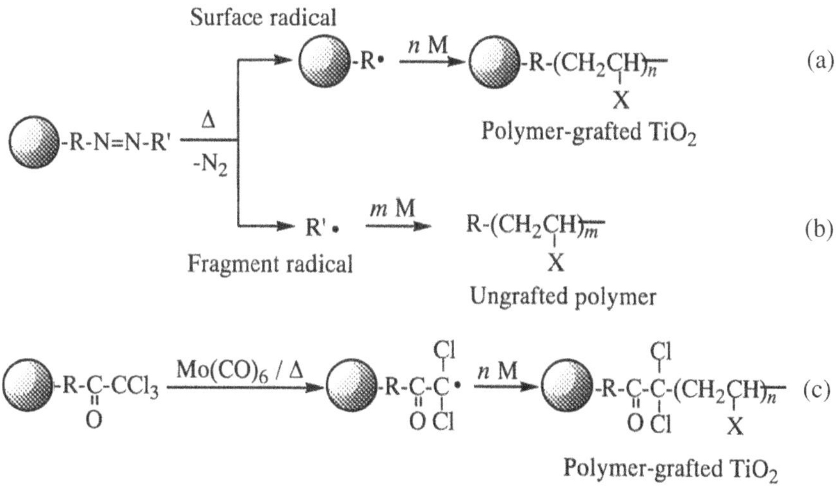

Fig. 27 Radical graft polymerization onto Ti/Si-R-OH (**a**, **b**) and by the system consisting of trichloroacetyl groups on titanium dioxide surface and $Mo(CO)_6$ (**c**)

nature, revealing that the wettability of titanium dioxide surface is readily controlled by grafting of hydrophobic or hydrophilic polymer onto the surface.

Figure 28 shows the anionic polymerization and cationic polymerization for mineral surface and microstructure modification. Jordan et al. [101] applied a self-assembled monolayer of biphenyllithium moieties to initiate the anionic polymerization of styrene on gold substrates. A self-assembled monolayer containing bromobiphenyl groups was formed initially, and converted to the initiating species by reaction with sec-butyllithium. The quite uniform films (about 18 nm)

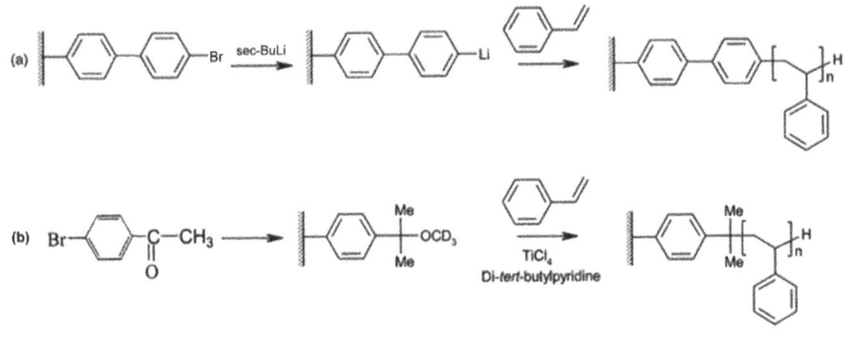

Fig. 28 Typical surface modification by **a** anionic polymerization and **b** cationic polymerization

can be formed by adding styrene. Zhao and Brittain [102] described the synthesis of polystyrene (PS) on silicate wafer substrates via carbocationic polymerization from self-assembled monolayers. Activation with $TiCl_4$ in the presence of styrene and a proton scavenger (di-tert-butylpyridine) leads to the growth of tethered PS films to 30 nm thickness (Fig. 28b). This method is also applicable to other cationic monomers, such as isobutylene and styrene derivatives.

In addition, coating of silane on minerals is a commonly used method to produce superhydrophobic surface for substrate. Superhydrophobicity is defined as water contact angle greater than 150° [103]. This remarkable characteristics of extreme liquid wettability has been widely applied in both fundamental research and practical applications, for example in the surface modification of nano-particles and polymer membranes for water purification, desalination, heavy metal removal, food industry and purification of pharmaceutical products [104]. Research on superhydrophobic surfaces has been motivated by mimicking nature [105]. For example, superhydrophobicity can be imparted to boehmite and silica films by the sublimation of aluminium acetylacetonate (AACA) during calcination, and transparent superhydrophobic films of these substrates are prepared by subsequent coating with fluoroalkylsilane [106].

2.6 Etching

Etching is a conventional technology for surface modification. Etching is the removal of a portion of the film that is not masked by the resist, by means of chemical or physical methods (or both). Etching technology is divided into dry etching and wet etching. Dry etching mainly uses reactive gas or plasma while wet etching mainly utilizes chemical reagent and etched material to undergo chemical reaction. Etching is also used as surface modification method to enhance the surface properties of minerals.

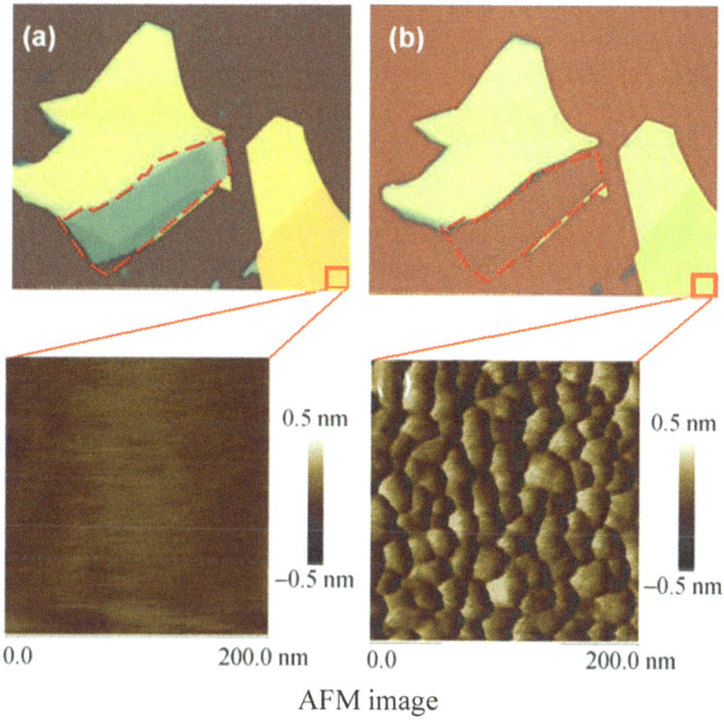

Fig. 29 The optical images of a MoS_2 flake before (**a**) and after ething (**b**)

Huang et al. studied the etching of MoS_2 crystals by using XeF_2 as a gaseous reactant because XeF_2 has high oxidizing properties, leading to easy reaction with MoS_2, and the reaction by-products can be efficiently removed as gases, such as Xe, SF_6, F_2, and MoF_3, et al. (see Reaction 4 and 5) [107]. The strong oxidation-reduction reaction produces a large amount of heat and the increased temperature then enhance the reaction rate.

$$MoS_2 + XeF_2 \rightarrow Xe + MoF_4 + SF_6 \quad (4)$$

$$MoF_4 \rightarrow MoF_3 + F_2 \quad (5)$$

Figure 29 presents the multilayered MoS_2 flake on Si/SiO_2 substrate before and after etching. The surface morphology of this flake was changed considerably after etching, and the root-mean-square (rms) surface roughness has been largely improved. The roughness may be caused by the ununiform surface and the different reaction rates for S layers and Mo layers.

Yamamoto et al. investigated the anisotropic etching of atomically thin molybdenum disulfide (MoS_2) [108]. They found that triangular etch pits with uniform orientation on the surfaces of MoS_2 can be obtained when exposed to

Fig. 30 Schematic drawing of hexagonal lattice of the MoS_2 structure with riangular pits

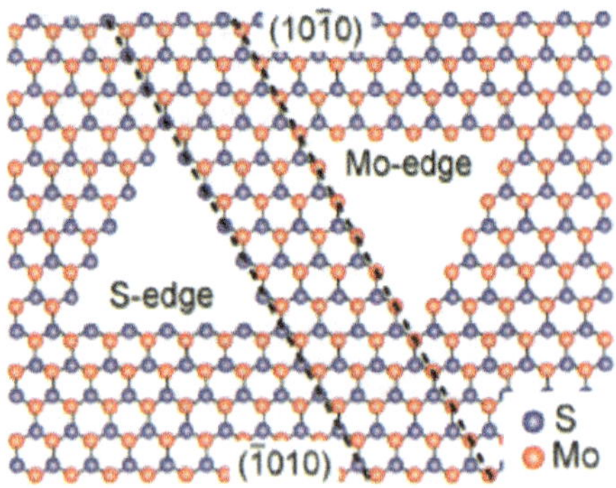

oxygen at 300–340 °C, indicating anisotropic etching terminating on lattice planes. The triangular shapes of the pits reflect the lattice of MoS_2 basal plane surface while the edges of the pits are along the zigzag directions with only a single chemical termination, that is, terminated on either the Mo-edge $(10\bar{1}0)$ or S-edge $(\bar{1}010)$ (see Fig. 30).

Zhu et al. [109] studied remote plasma oxidation and atomic layer etching of MoS_2. After 5 min or longer remote O_2 plasma exposure at 200 °C, there was a uniform coverage of monolayer amorphous MoO_3. The MoO_3 can be completely removed at 500 °C, leaving a clean ordered MoS_2. Qian and Shen [110] developed a simple chemical etching method for the surface roughening to fabricate superhydrophobic surfaces on three polycrystalline metals (aluminum, copper, and zinc). The key to this etching technique is the use of a dislocation etchant that preferentially dissolves the dislocation sites in the grains and it may offer a simple and convenient strategy for fabricating superhydrophobic surfaces. Zhang et al. [111] studied MoS_2 sheet exposed to water using AFM to obtain a better understanding for the preparation and application of two-dimensional MoS_2.

H_2O and O_2 can react with the surface of the sample in aqueous solution (Eq. 6), resulting in the etching of the surface layer.

$$MoS_2 + O_2 + H_2O \rightarrow MoO_3 \cdot H_2O + Mo(VI)_xO_y^{a-} + SO_4^{2-} + H^+ \quad (6)$$

where $Mo(VI)_xO_y^{a-}$ indicates the molybdate species containing Mo(VI), such as ${Mo_7O_{24}}^{6-}$ and ${HMoO_4}^{-}$.

According to Eq. 6, MoS_2 was oxidized into $MoO_3 \cdot H_2O$ and then remains on the surface of MoS_2, which renders the hydrophobic surface of MoS_2 to be hydrophilic due to the formation of hydrophilic $MoO_3 \cdot H_2O$ (Fig. 31).

Spychalski et al. [112] studied the microscale characteristic of oxidation of single MoS_2 crystals in air. Accroding to AFM topographs, they speculated that a surface diffusion of the initially physisorbed oxygen could control diffusion limited aggregation (DLA) process.

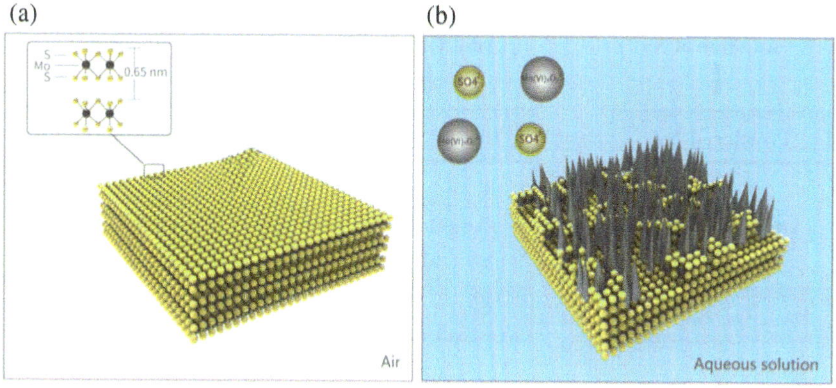

Fig. 31 Schematic representation of the oxidation of MoS_2 in water: **a** before oxidation, **b** after oxidation

2.7 *Other Methods*

Chemical conditioning is another technique to chemically tailoring the surface properties of clay minerals. In this process, the exchangeable cations in the interlayer of clay minerals can be exchanged with certain ions to enhance the adspriton capability for specific matters. Oliveira and Rubio pre-treated the natural powdered zeolites (NZ) (from Chile) with NaCl solution (Na-AZ) and then conditioned Na-AZ with barium chloride solution (Ba-Z). It can be found that the natural powdered zeolites does not adsorb sulphate ions, but after the treatment with Ba^{2+} ions, the zeolite is able to uptake sulphate significantly [113]. The adsorption capacity is not influenced by the medium pH showing that the mechanism involved is chemisorption between barium and sulphate ions. The sulphate loaded (saturated) zeolites can also be used as an adsorbent for the barium ions, and these barium loaded zeolites (flocculated or not) for the sulphates in series, which is shown in Fig. 32.

Chansuvarn's study indicates that the surface area of natural clay (NC) can be enlarged by about 10-times after the clay is treated with manganese chloride [114]. By comparing with NC, MnO modified NC provides higher adsorptive capability than NC by about 30-times for Cd(II). Lihareva et al. [115] found that the adsorption capacity of clinoptilolite for Ag ions can be increased by conditioning with sodium chloride solution compared with the natural mineral.

Besides acid treatment, the clay minerals can be modified with bases to change their structure and adsorption capability. Jozefaciuk and Bowanko [116] compared the effects of alkali and acid treatments on surface areas and adsorption energies of bentonite, biotite, illite, kaolin, vermiculite and zeolite. The surface areas of most minerals increase with acid and alkali treatments. Dissolution of Al prevailes over Si in acid treatments, and the opposite was observed in alkali treatments. The research conducted by Owabor et al. [117] also indicates that sodium hydroxide

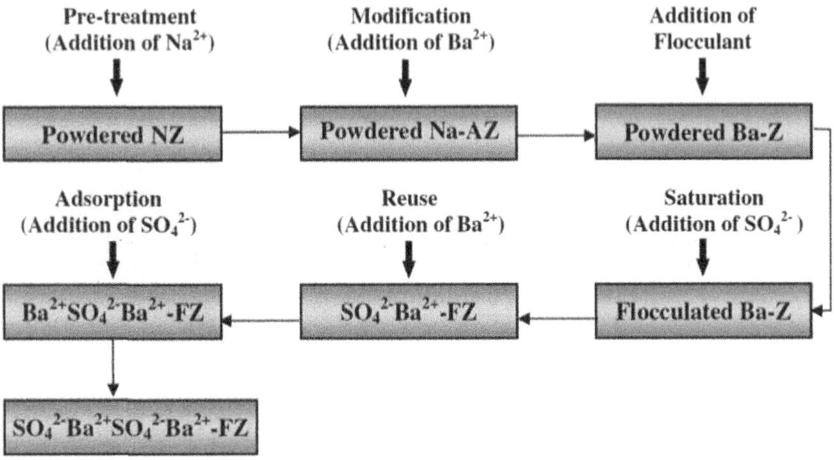

Fig. 32 Staged reuse studies of the modified zeolites as adsorbent (FZ: flocculated zeolite)

modified clay showed relatively good adsorption performance on the removal of naphthalene from the bulk solution, while ammonium hydroxide modified clay showed a slow rate of adsorption of naphthalene from the bulk solution.

Corrosion of minerals in certain environment is also a feasible method to modify the surface properties of the minerals and accelerate the interaction between the minerals and aqueous phases. Pasek and Lauretta [118] investigated the phosphide corrosion in five different solutions. Different corrosion products can be obtained in various aqueous solutions, indicating the distinct surface corrosion of the minerals.

3 Modifying Agents and Applications

The properties of modifying agents determine the performance and function of the modified minerals [119]. There are many types of surface modifying agents for minerals, but the commonly used are introduced below.

3.1 Coupling Reagent

Coupling agent is a chemical substance with amphoteric structure. According to its chemical structure and composition, it can generally be divided into titanate, silane, aluminate, zirconium aluminate and their organic complex. A part of coupling agent can react with various functional groups on the surface of mineral powder to form a strong chemical bond, and the other part of the agent can react with base material by establishing a special "molecular bridge" [120].

3.1.1 Titanate Coupling Agent

Titanate coupling agents are widely used for inorganic fillers (pigments) and other surface modification. The molecular structure of titanate coupling agent can be presented as below and each part has different functions.

$$(RO)_M{-}Ti{-}(OX{-}R'{-}Y)_N \tag{5}$$

where $1 \leq M \leq 4, M + N \leq 6$, R indicates alkane with short chain, R′ indicates alkanc with long chain, X indicatcs C, N, P, S etc., and Y indicates functional groups, such as hydroxyl, amino, double bond etc.

$(RO)_M$ is the group coupled with the surface of inorganic minerals. The titanate coupling agent is chemically adsorbed or reacted with the trace hydroxyl and proton on the surface of inorganic component forming a monomolecular layer. The Ti–O can crosslink ester or carboxyl in organic polymers, resulting in cross-linking between titanate, inorganic minerals powder and organic polymers. The groups linked with X determine the special properties, antioxidant for example, of the coupling agent. The R′ group can react with organic base materials to improve their compatibility and enhance the extension, tear and impact strength of the composite material. Y is the fixing reaction groups and N is non-hydrolyzable groups.

The titanate coupling agent can be divided into three categories: monoalkoxy type, chelate type and coordination type according to the chemical structure.

A. *Monoalkoxy type*

Most varieties are poor in water resistance except for monoalkoxy groups containing ethanol and pyruvate groups [121]. The reaction mechanism between this type of reagent with mineral powder can be illustrated as Fig. 33.

B. *Chelate type*

The chelate type of titanate has good water resistance and is suitable for surface treatment of high water content inorganic powder. The representative mechanism of coupling is shown in Fig. 34.

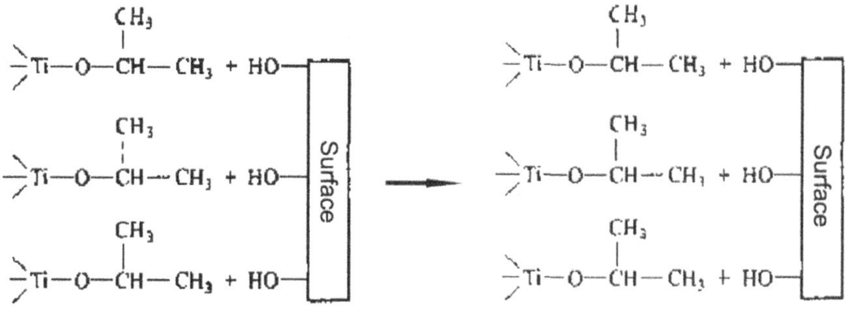

Fig. 33 The reaction mechanism between monoalkoxy titanate and mineral powder

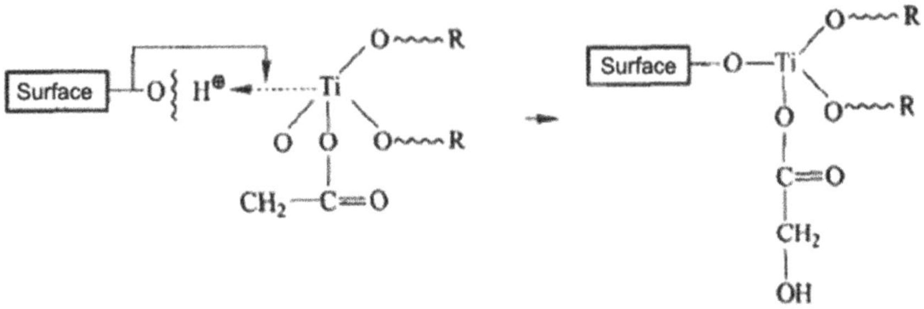

Fig. 34 The representative mechanism of coupling for chelate type of titanate

C. *Coordination type*

KR-46 is one typical coordination type of titanate coupling agent, which is shown as Fig. 35.

It can be seen from the above formula that the titanium atom is transform from the normal 4-valent to a 6-valent, reducing its reactivity and improving water resistance property. The coupling mechanism of the coordination titanate is shown in Fig. 36.

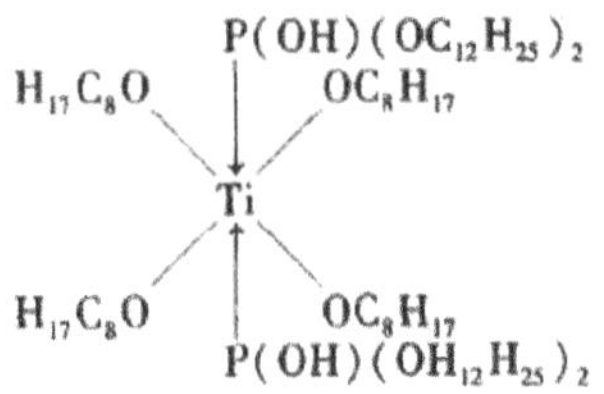

Fig. 35 The formula titanate coupling agent

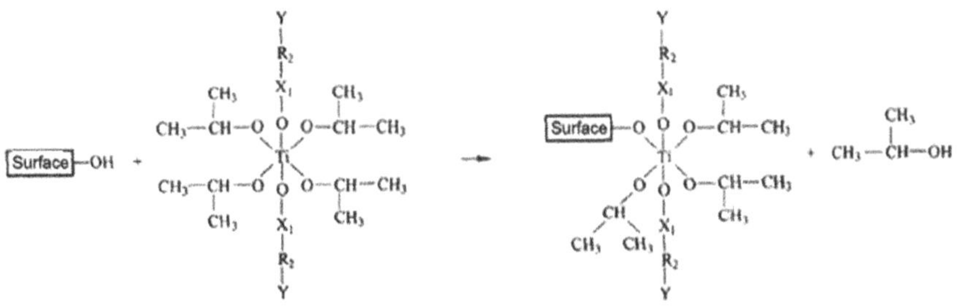

Fig. 36 The coupling mechanism of the coordination titanate

3.1.2 Silane Coupling Agent

Silane coupling agent is one type of organosilicon compounds with low molecular weight and specific structure. The general chemical formula is $RSiX_3$ where R represents an active functional group having an affinity or reactivity with polymers, such as oxy group, mercapto group, vinyl group, epoxy group, amide group etc. X represents the hydrolysable alkoxy group, such as halogen, alkoxy groupacyloxy group etc. [106]. When the coupling reaction is carried out, the X group is initially hydrolyzed to form a silanol, then reacting with the hydroxyl group on the surface of mineral particles to form hydrogen bond, and finally condensing into –SiO–M covalent bond (M represents the surface of mineral particles). At the same time, the silanol molecules of silane are associated with each other to form a network structure film covering the surface of mineral particles. The brief process of the chemical reactions is shown as follows:

Hydrolysis: $$RSiX_3 + 3H_2O \xrightarrow[\text{Catalyst}]{\text{pH}} RSi(OH)_3 + 3HX \quad (6)$$

(HX generally is alcohol or acid)

Condensation:

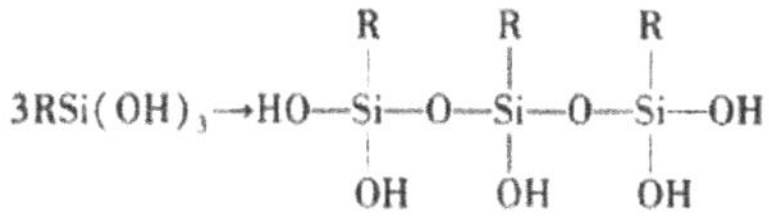

(7)

Formation of hydrogen bond:

(8)

Formation of covalent bond:

(9)

According to the R group in the molecular structure, silane coupling agent can be divided into aminosilane, epoxysilane, thiol silane, methacryloxy silane, vinylsilane, ureidosilane, isocyanatesilane etc. [122].

Silane coupling agent is good for quartz powder and carbon black which contain abundant silicic acid component and is also suitable to treat kaolin, hydrated alumina, magnesium oxide etc.

Cheng et al. investigated the graft modification of Titanium nitride (TiN) nano-particles by silane coupling agent (KH-570) via a direct blending method [123]. The hydroxyl groups on the surface of TiN nano-particles can interact with silanol groups [–Si–OCH_3] of KH-570 forming an organic coating layer, which could improve the dispersibility of nano-TiN particles in ethyl acetate.

3.1.3 Aluminate Coupling Agent

The chemical formula of aluminate coupling agent can be illustrated as Fig. 37

where D_n represents the coordinating groups, such as N, O etc., RO is the group reacting with the active proton or functional group on the mineral surface, COR′ is the group reacting with the polymer base.

The molecular space structure of aluminate coupling agent is shown in Fig. 38.

The modification mechanism of aluminate coupling agent for mineral surface is schematically illustrated as Fig. 39.

3.1.4 Other Coupling Agents

(1) Zircoaluminate coupling agent

Zircoaluminate coupling agent is widely used as surface modifying reagent or adhesion additives in many applications, such as plastics, rubbers, coatings, and pigments. It can be synthesized by zirconium chloride hydrate ($ZrOCl_2 \cdot 8H_2O$), aluminum chlorohydrate ($Al_2(OH)_5Cl$), propylene alcohol, carboxylic acid etc. Li et al. [124] synthesized a range of zircoaluminates with different functional groups by using polyaluminum chloride (PAC), zirconium oxychloride, 1,2-propanediol, and others. The molar ratio of reactants and selected organofunctional ligands can affect not only the properties but also the thermal stabilities of final products with the highest decomposition temperature of 374 °C.

The molecular structure of common zircoaluminate coupling agent is shown in Fig. 40, where X is the organic functional group.

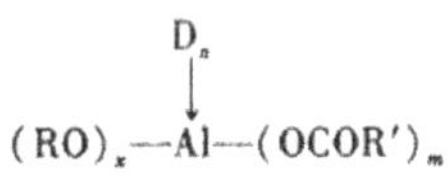

Fig. 37 The chemical formula of aluminate coupling agent

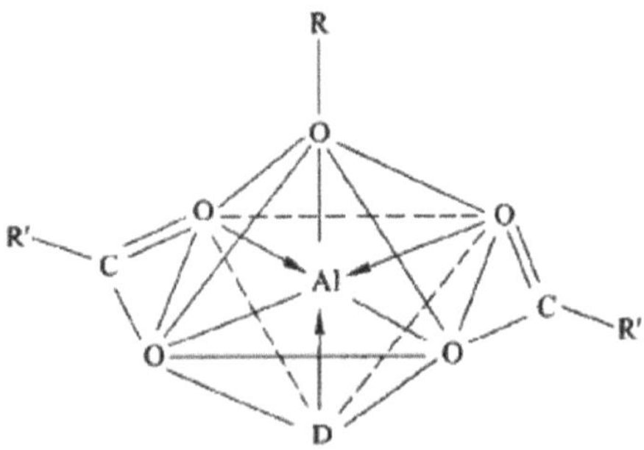

Fig. 38 The molecular space structure of aluminate coupling agent

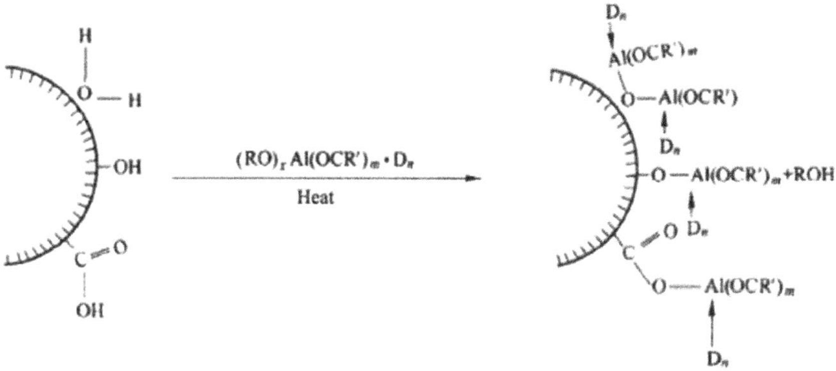

Fig. 39 The mechanism of interaction between aluminate coupling agent and mineral powder

Fig. 40 The molecular structure of zirconium aluminate coupling agent

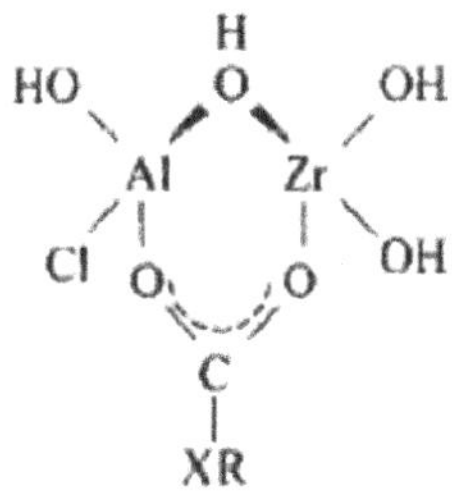

Zircoaluminate coupling agent contains two inorganic parts (zirconium and aluminum) and an organic functional ligand. Therefore, compared with silane and other coupling agents, it has stronger inorganic property. Thus, it has more abundant inorganic reaction sites than silane coupling agents to enhance the surface modification effects for inorganic minerals.

Zircoaluminate coupling agent can be covalently linked to the hydroxylated surface by condensating of zirconium hydroxide and aluminum hydroxide groups. Furthermore, another important feature is that this coupling agent has the ability to form the complex oxygen bridge on the metal surface by forming hydroxyl groups, as shown in Fig. 41.

(2) Organic chromium coupling agent

Organic chromium coupling agent is a coordination metal complex formed from unsaturated organic acid and chromium atom. The mostly used types are chromium methacrylate-chromium chloride and fumaric acid chromium nitrate complexes, in which one end of the molecular contains active unsaturated groups that can react with the polymer matrix and the other end is binded to silicon-oxygen bond on the surface of the materials. The mechanism of the coupling reaction can be shown in Fig. 42.

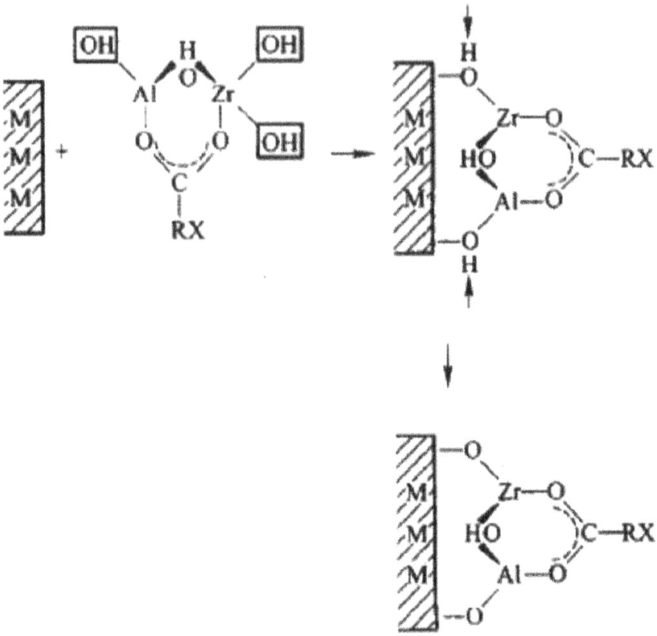

Fig. 41 The reaction of zirconium aluminate coupling agent and metal surface

Fig. 42 The modification mechanism of organic chromium coupling agent

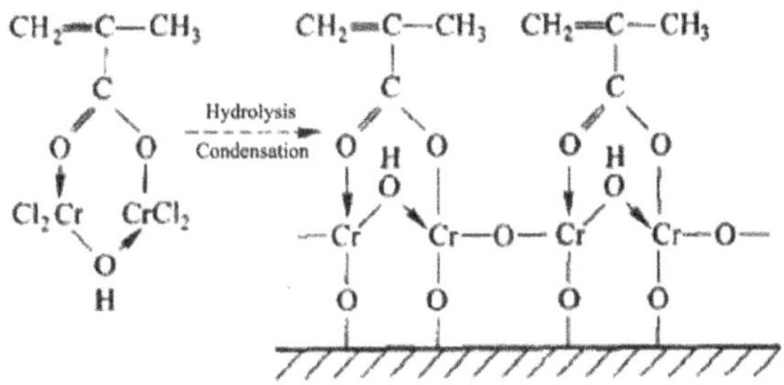

3.2 *Surfactant*

3.2.1 Overview

Surfactant molecule [125] consists of two distinct parts, i.e., one is lipophilic (hydrophobic) groups with strong affinity for oil or organic matter, and the other is hydrophilic groups with strong affinity water or inorganic matter (oleophobic). This structural feature allows surfactant to be used for surface modification. The hydrophilic groups can adsorb on the surface of minerals by physical or chemical reaction. Because hydrophobic groups are outward, the surface of the mineral can be converted from hydrophilicity to hydrophobicity to improve the affinity of the mineral with organic materials.

Surfactant can be classified as anionic, cationic and nonionic surfactant depending on ion type. The former can dissociate in water but the latter does not dissociate in aqueous phase.

3.2.2 Anionic Surfactant

The commonly used anionic surfactants for the surface modification of mineral powder include (1) fatty acids and their salts. The general formula is RCOOH(Me), where Me indicates the metals, such as Na^+, K^+, while R represents long-chain alkyl (2) sulfonates and their esters. Their general formula is RSO_3Me and (3) phosphatic ester with the general formula of $ROPO_3Me$.

Song et al. used two kinds of anionic surfactants, the linear alkylbenzene sulfonic acid (LABSA) and branched alkylbenzene sulfonic acid (BABSA), to modify $CaCO_3$ nanoparticles. Contact angle measurement has shown that $CaCO_3$ surface becomes more hydrophobic as surfactant concentration increases but the use of excess surfactant has led to the reverse change in surface property of $CaCO_3$ particles from hydrophobic to hydrophilic due to the bilayer adsorption of the surfactants on $CaCO_3$ surface, which can be illustrated in Fig. 43 [126].

3.2.3 Cationic Surfactant

The widely used cationic surfactants are primary amine, secondary amine, tertiary amine and quaternary amine salts etc. where there are at least 1–2 long-chain alkyl (C_{12}–C_{22}).

Özcan et al. [127] modified bentonite using a cationic surfactant (dodecyltrimethylammonium (DTMA) bromide) to adsorb textile dye (Reactive Blue 19 (RB19)) as the anionic dyes are negatively charged. They found that the adsorption of RB19 onto DTMA–bentonite was favored with the negative Gibbs free energy values and the adsorption kinetics fits the pseudo-second-order rate equation.

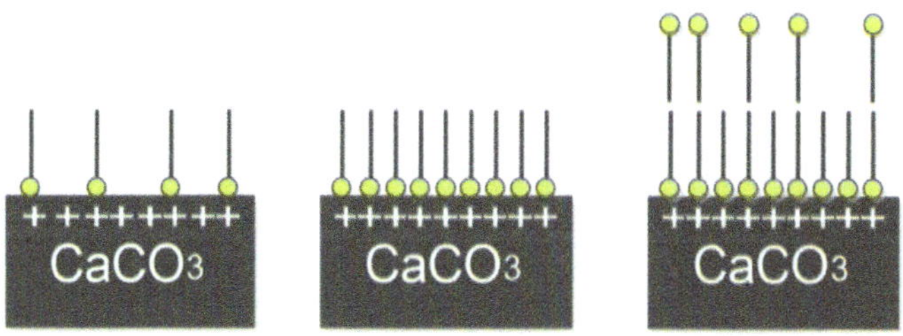

Fig. 43 The illustration of adsorption of anionic surfactants on $CaCO_3$ surface

3.2.4 Nonionic Surfactant

Nonionic surfactants do not present ionic state in water thus they have strong stability and are not easily affected by strong electrolytes and alkaline or acid. Shen prepared organobentonite by modifying natural bentonite with different nonionic surfactants and one cationic surfactant [128]. The nonionic surfactants can intercalate into the interlamellar space of bentonite and may demonstrate higher sorption capacity than cationic surfactant because ethylene oxide can form hydrogen bonds with SiO_2 on the surface of bentonite. In addition, the nonionic surfactant modified bentonites are more chemically stable than cationic surfactant derived organobentonites.

3.3 Organosilicon

Organosilicon is a special kind of surfactant, where the hydrophobic group is siloxane chain, and the hydrophilic groups are polyoxyethylene chain, carboxyl group, ketone group or other polar groups. The common organosilicon series of surfactant include polydimethylsiloxane, organic modified polysiloxane, copolymer of organosilicon and organic compound.

3.3.1 Polydimethylsiloxane

The molecular structure of polydimethylsiloxane is shown in Fig. 44.

Where Me stands for methyl group. Because the main molecule is composed of methyl, it is insoluble in water, low alcohol, acetone, ethylene glycol, etc., but can dissolve in most organic solvents such as aliphatic hydrocarbons, aromatic hydrocarbons, higher alcohols, ethers, lipids, chlorinated hydrocarbons.

Polydimethylsiloxane can be grafted with acrylic acid, acrylamide, dimethylacrylamide, 2-hydroxylethyl acrylate, and poly(ethylene glycol)monomethoxyl acrylate to yield hydrophilic surfaces.

3.3.2 Organomodified Polysiloxane

The commonly used Organomodified polysiloxane can be illustrated as Fig. 45.

Fig. 44 The molecular structure of polydimethylsiloxane

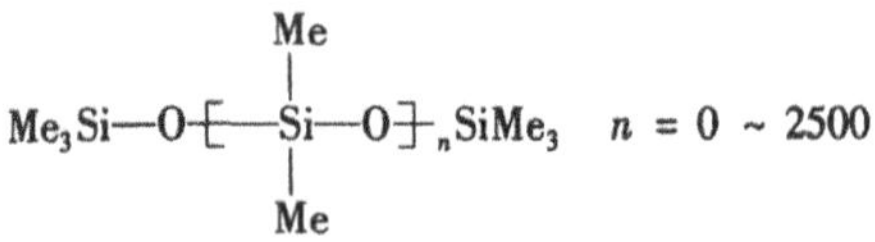

Fig. 45 The molecular structure of Organomodified polysiloxane

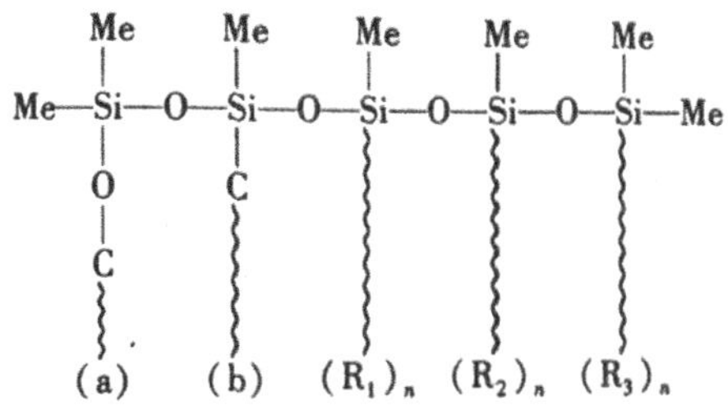

The performance of organomodified polysiloxane is determined by (1) the connection bond between the modifier and the main chain, (a) –Si–O–C or (b) –Si–C, (2) the types of modifier R1, R2, R3 and the number and location of the n.

3.4 Unsaturated Organic Acids

Some unsaturated organic acids, such as acrylic acid, methacrylic acid, maleic acid, cinnamic acid, sorbic acid, butenoic acid, alpha-chloroacrylic acid, itaconic acid, vinyl acetate, can be used as surface modifier. These organic acids have one or more unsaturated double bond(s) or several hydroxyls and the number or carbon atoms in these acids are generally less than 10. Generally, the stronger the acid, the easier it is to form an ionic bond. Therefore, acrylic acid, methacrylic acid and butenoic acid are the most commonly used [129, 130].

The surface modifiers commonly have two functional groups, unsaturated double bond and carboxyl group. During the process of modification, carboxyl can react with active metals on inorganic particles to form strong chemical bonds and coats the particles as monolayer film [131].

3.5 Polyolefin Oligomer

Polyolefin oligomer is a kind of non-polar oligomers with average molecular weight of 1500–5000, mainly including irregular polypropylene and polyethylene wax.

Polyolefin oligomer has high adhesion properties and can be well infiltrated, adhered and coated with inorganic powder. As a result, it can be used as a surface coating modifier for particles.

3.6 Water-Soluble Macromolecule

Water-soluble macromolecule, also called water-soluble polymer, is a hydrophilic macromolecule material. The hydrophilicity of the water-soluble polymer is derived from the hydrophilic group, carboxyl, hydroxy, amide, amine, ether for example, in the molecule. Water-soluble polymer can be divided into three categories: natural water-soluble polymer, semisynthetic water-soluble polymer and synthetic water-soluble polymer. Currently, synthetic water-soluble polymers are used for surface modification of mineral powder, such as polyacrylic acid and its salts, polyacrylamide, polyethylene glycol, maleic anhydride and maleic acid-acrylic acid copolymer.

3.7 Inorganic Surface Modifier

Some metal salts, such as titanium tetrachloride, titanyl sulfate, ferrous sulfate, chromium salt, aluminium salt, are often used as a surface modifier for mineral particles. For example, titanium tetrachloride and titanyl sulfate are commonly used as modifier for mica [132]. Aluminum salt can be used as surface coating reagent for titanium dioxide [133]. Metal oxides, alkali or alkaline earth metals, rare earth oxides, inorganic acids and salts, and metals such as Cu, Pt, Pd, Ag, Au, Mo, Co, Ni are commonly used as surface modifier for adsorbent and catalytic powder materials, for instance, diatomite, molecular sieves, zeolites, silica, sepiolite, bentonite and so on.

4 Evaluation of Modification

The chemical and physical properties of modified minerals should be fully determined or analysed not only for the evaluation of modification processes but also for the improvements in modification techniques. In addition, the in-depth understanding of the properties of modified minerals is necessary for probing various modification mechanisms for mineral surface and microstructure. The commonly used measurements for mineral modification are briefly introduced below.

4.1 Chemical and Morphology Analysis

Techniques and methods commonly used to probe bulk properties may not be suitable because only the chemical and physical properties within the first few nanometers of the modified surface are important for success [134]. In recent years,

Table 2 Surface-sensitive techniques used to analyse surface characteristics

AES	Auger electron spectroscopy
AFM	Atomic force microscopy
ATIR	Attenuated total internal-reflection spectroscopy
ATR-FTIR	Attenuated total reflection Fourier transform infrared spectroscopy
DRIFT	Diffuse reflectance Fourier transform infrared spectroscopy
DSIMS	Dynamic secondary ion mass spectrometry
EDX	Energy dispersive X-ray spectroscopy
ESCA	Electron spectroscopy for chemical analysis
FABMS	Fast atom bombardment mass spectrometry
FRS	Forward-recoil spectroscopy
FTIR	Fourier transform infrared spectroscopy
IRAS	Infrared reflection absorption spectroscopy
ISS	Ion-scattering spectroscopy
LEED	Low-energy electron diffraction
LIMA	Laser ionization mass analyzer
MS	Mossbauer spectroscopy
NEXAFS	Near-edge X-ray absorption fine structure
PMIM	Phase measurement interference microscopy
RBS	Rutherford backscattering
SEM	Scanning electron microscopy
SIMS	Secondary ion mass spectrometry
SSIMS	Static secondary mass spectrometry
STEM	Scanning transmission electron microscopy
STM	Scanning tunneling microscopy
TDS	Thermal desorption Spectroscopy
TEM	Transmission electron microscopy
UPS	Ultraviolet photoelectron spectroscopy
XAES	X-ray-induced Auger electron spectroscopy
XPS	X-ray photoelectron spectroscopy

many surface-sensitive techniques, including atomic force microscopy (AFM), scanning tunnelling microscopy (STM), high resolution electron energy loss spectroscopy (HREELS), secondary ion mass spectrometry (SIMS), X-ray photoelectron spectroscopy (XPS), scanning electron microscopy (SEM), have been developed to analyse surface characteristics. The common used techniques are listed in Table 2.

These techniques analyse the surface with different sampling depths to give the morphology, chemical and structural information of the surface. It is important to choose suitable surface analysis techniques according to sampling depth, surface information, analysis environment, and sample suitability. The high-resolution, 3-D images of surface can be obtained by AFM, STM and SEM. SSIMS and XPS, providing detailed chemical information of surface. If the information of grafted or crosslinked functional components are needed, FTIR and ATR-FTIR may be good choices.

4.2 Specific Surface Area and Pore Structure Analysis

Modification of mineral surface and microstructure commonly changes the specific surface area and pore structure of minerals and then improve their adsorption capability. Thus specific surface area and pore structure analysis are generally necessary to characterize the modified minerals, especially for layered and/or porous minerals.

Brunauer–Emmett–Teller (BET) theory aims to explain the physical adsorption of gas molecules on a solid surface for analyzing the specific surface area of samples. Nitrogen is the most commonly used gaseous adsorbate at the boiling temperature (77 K) for surface probing by BET methods. Adsorption of nitrogen at a temperature of 77 K leads to a so-called adsorption isotherm, sometimes referred to as BET isotherm, which is mostly measured over porous materials. Further probing adsorbates, including argon, carbon dioxide, and water, are also utilized, albeit with lower frequency, allowing the measurements of surface area at different temperatures and measurement scales. The pore size distribution of samples is commonly calculated by applying Horvath-Kawazoe method and Barrett-Joyner-Halenda method for micropores and mesopores, respectively, to the N_2 adsorption-desorption isotherms at 77 K [135].

The layer structure, especially the basal spacing, is commonly characterized by XRD analysis.

4.3 Contact Angle Analysis

The wettability of surface can be directly reflected by contact angle. The definition and implication of the contact angle have been introduced in Sect. 2.1.1. The smaller a liquid's contact angle on the surface of a material, the greater is that liquid's ability to wet that material, and thus, these characteristic contact angles can be used to rank the ability of a liquid to wet, or cover, a substrate and to calculate the capillary pressure. Various methods have been developed for measuring the contact angle. Generally, the contact angle may be easy to measure directly as one drop on a flat surface, but for the contact angle between liquid and powder, direct measuring may not work and the dynamic method, also called capillary rise should be adopted [136].

References

1. Yariv, S., Cross, H.: In: Organo-Clay Complexes and Interactions, p. 688. Marcel Dekker, New York (2002)
2. Cox, L., Hermosín, M.C., Celis, R., et al.: Sorption of two polar herbicides in soils and soil clays suspensions. Water Res. **31**(6), 1309–1316 (1997)
3. Pohlen A (2001) Method for eliminating stickies from treatment of waste paper using hydrophobized minerals: US, US6210526
4. Kenneth, J.I.: In: The Scientific of Flotation, p. 429. Martinus Nijhoff, Hague (1984)
5. Glembotskii, A.V. (ed.): Flotation Methods for Extracting Valuable Components from Solutions and for Purifying Effluent. Russian, Nauka Moscow (1972)
6. Pugh, R., Stenius, P.: Solution chemistry studies and flotation behaviour of apatite, calcite and fluorite minerals with sodium oleate collector. Int. J. Miner. Process. **15**(3), 193–218 (1985)
7. Sánchezmartín, M.J., Dorado, M.C., Del, H.C., et al.: Influence of clay mineral structure and surfactant nature on the adsorption capacity of surfactants by clays. J. Hazard. Mater. **150**(1), 115 (2008)
8. Schoonheydt, R.A., Johnston, C.T.: Chapter 3 surface and interface chemistry of clay minerals. In: Developments in Clay Science, pp. 87–113. Elsevier Ltd. (2006)
9. Guégan, R.: Self-assembly of a non-ionic surfactant onto a clay mineral for the preparation of hybrid layered materials. Soft Matter **9**(45), 10913–10920 (2013)
10. Backhaus, W.K., Klumpp, E., Narres, H.D., et al.: Adsorption of 2,4-dichlorophenol on montmorillonite and silica: Influence of nonionic surfactants. J. Colloid Interf. Sci. **242**(1), 6–13 (2001)
11. Amirianshoja, T., Junin, R., Idris, A.K., et al.: A comparative study of surfactant adsorption by clay minerals. J. Petrol. Sci. Eng. **101**(2), 21–27 (2013)
12. Yang, S.F., Gao, M., Luo, Z.: Adsorption of 2-naphthol on the organo-montmorillonites modified by Gemini surfactants with different spacers. Chem. Eng. J. **256**, 39–50 (2014)
13. Zhang, R., Zhou, J., Peng, L., et al.: The wettability alteration behavior of cationic gemini surfactants on oil-aged mica mineral surfaces. Tenside Surfact. Det. **50**(3), 214–218 (2013)
14. Franck, P.D., Raoul, Z., Gregory, G.W.: Adsorbed layer structure of cationic gemini and corresponding monomeric surfactants on mica. Langmuir **22**(3), 1143–1149 (2006)
15. Rosen, M.J., Fang, L.: The adsorption of gemini and conventional surfactants onto some soil solids and the removal of 2-naphthol by the soil surfaces. J. Colloid Interf. Sci. **234**(2), 418–424 (2001)
16. Sakamoto, M., Kanda, Y., Minoru Miyahara, A., et al.: Origin of long-range attractive force between surfaces hydrophobized by surfactant adsorption. Langmuir **18**(15), 5713–5719 (2002)
17. Hogg, R.: The role of polymer adsorption kinetics in flocculation. Colloid Surf. A **146**(1–3), 253–263 (1999)
18. Besra, L., Sengupta, D.K., Roy, S.K., et al.: Polymer adsorption: its correlation with flocculation and dewatering of kaolin suspension in the presence and absence of surfactants. Int. J. Miner. Process. **66**(1–4), 183–202 (2002)
19. Mierczynska-Vasilev, A., Beattie, D.A.: The effect of impurities and cleavage characteristics on talc hydrophobicity and polymer adsorption. Int. J. Miner. Process. **118**(1), 34–42 (2013)
20. Yeap, K.Y., Addai-Mensah, J., Beattie, D.A.: The pivotal role of polymer adsorption and flocculation conditions on dewaterability of talcaceous dispersions. Chem. Eng. J. **162**(2), 457–465 (2010)
21. Chang, S.H., Ryan, M.E., Gupta, R.K., et al.: The adsorption of water-soluble polymers on mica, talc limestone, various clay minerals. Colloid Surf. **59**(9), 59–70 (1991)
22. Bodratti, A.M., Sarkar, B., Song, D.D., et al.: Competitive adsorption between PEO-containing block copolymers and homopolymers at silica. J. Disper. Sci. Technol. **36**(1), 1–9 (2015)

23. Öztekin, N., Alemdar, A., Güngör, N., et al.: Adsorption of polyethyleneimine from aqueous solutions on bentonite clays. Mater. Lett. **55**(1–2), 73–76 (2002)
24. Alemdar, A., Butun, V.: Interaction between a tertiary aminemethacrylate based polyelectrolyte and a sodium momtmorillonite dispersion and its rheological and colloidal properties. J. Appl. Polym. Sci. **95**, 300–306 (2005)
25. Alemdar, A., Oztekin, N., Gungor, N., Ece, O.I., Erim, F.B.: Effects of polypropyleneimine adsorption on the rheological properties of purified bentonite suspensions. Colloid Surf. **25**, 95–98 (1998)
26. Pernyeszi, T., Patzkó, Ágnes, Berkesi, O., et al.: Asphaltene adsorption on clays and crude oil reservoir rocks. Colloid Surf. **137**(1–3), 373–384 (1998)
27. Kapsabelis, S., Prestidge, C.A.: Adsorption of ethyl(hydroxyethyl)cellulose onto silica particles: the role of surface chemistry and temperature. J. Colloid Interf. Sci. **228**(2), 297–305 (2000)
28. Lagaly, G., Beneke, K.: ChemInform abstract: intercalation and exchange reactions of clay minerals and non-clay layer compounds. Colloid Polym. Sci. **23**(18), 1198–1211 (1992)
29. Lagaly, G.: Interaction of alkylamines with different types of layered compounds. Solid State Ionics **22**(1), 43–51 (1986)
30. Jones, T.R.: The properties and uses of clays which swell in organic solvents. Clay Miner. **18** (4), 399–410 (1983)
31. Socías-Viciana, M.M., Hermosin, M.C., Cornejo, J.: Removing prometrone from water by clays and organic clays. Chemosphere **37**(2), 289–300 (1998)
32. Lagaly, G.: Charge characteristics of 2:1 clay minerals. In: Mermut, A. (ed.) CMS Workshop Lectures, vol. 6, pp. 1–46. The Clay Minerals Society, Boulder, Co. (1994)
33. Mullassery, M.D., Fernandez, N.B., Anirudhan, T.S.: Adsorptive removal of acid red from aqueous solutions by cationic surfactant-modified bentonite clay. J. Colloid Interf. Sci. **299** (1), 116–124 (2006)
34. Lagaly, G., Dékany, I.: Adsorption on hydrophobized surfaces: clusters and self-organization. Adv. Colloid Interf. **114–115**, 189–204 (2005)
35. Celis, R., Trigo, C., Facenda, G., et al.: Selective modification of clay minerals for the adsorption of herbicides widely used in olive groves. J. Agr. Food Chem. **55**(16), 6650–6658 (2007)
36. Shen, Y.H.: Removal of phenol from water by adsorption–flocculation using organobentonite. Water Res. **36**(5), 1107–1114 (2002)
37. Sanchez-Martin, M.J., Rodriguez-Cruz, M.S., Andrades, M.S., et al.: Efficiency of different clay minerals modified with a cationic surfactant in the adsorption of pesticides: influence of clay type and pesticide hydrophobicity. Appl. Clay Sci. **31**(3), 216–228 (2006)
38. Lee, J.F., Mortland, M.M., Boyd, S.A., et al.: Shape-selective adsorption of aromatic molecules from water by tetramethylammonium–smectite. J. Chem. Soc. Faraday Trans. **85** (85), 2953–2962 (1989)
39. Miyamoto, N., Kawai, R., Kuroda, K., et al.: Intercalation of a cationic cyanine dye into the layer silicate magadiite. Appl. Clay Sci. **19**(1–6), 39–46 (2001)
40. Wang, B., Hui, Z., Evans, D.G., et al.: Surface modification of layered double hydroxides and incorporation of hydrophobic organic compounds. Mater. Chem. Phys. **92**(1), 190–196 (2005)
41. Lee, S.Y., Kim, S.J.: Expansion of smectite by hexadecyltrimethylammonium. Clay Clay Miner. **50**(4), 435–445 (2002)
42. Li, Z.H.: Sorption of Chromate and PCE by surfactant-modified clay minerals. Environ. Eng. Sci. **15**(3), 237–245 (1998)
43. Li, Z., Bowman, R.S.: Retention of inorganic oxyanions by organo-kaolinite. Water Res. **35** (16), 3771–3776 (2001)
44. Krishna, B.S., Murty, D.S., Jai Prakash, B.S.: Thermodynamics of chromium(vi) anionic species sorption onto surfactant-modified montmorillonite clay. J. Colloid Interf. Sci. **229**(1), 230 (2000)

45. Kaufhold, S., Pohlmann-Lortz, M., Dohrmann, R., et al.: About the possible upgrade of bentonite with respect to iodide retention capacity. Appl. Clay Sci. **35**(1), 39–46 (2007)
46. Guegan, R., Gautier, M., Beny, J.M., et al.: Adsorption of a C10E3 non-ionic surfactant on a Ca-smectite. Clay Clay Miner. **57**(4), 502–509 (2012)
47. Biswas, M., Ray, S.S.: Recent progress in synthesis and evaluation of polymer-montmorillonite nanocomposites. Adv. Polym. Sci. (2001)
48. Bharadwaj, R.K.: Modeling the barrier properties of polymer-layered silicate nanocomposites. Macromolecules **34**(26), 9189–9192 (2001)
49. Ray, S.S., Okamoto, K., Okamoto, M.: Structure-property relationship in biodegradable poly (butylene succinate)/layered silicate nanocomposites. Macromolecules 36(7):2355–2367 (2003)
50. Ray, S.S., Okamoto, M.: Polymer/layered silicate nanocomposites: a review from preparation to processing. Prog. Polym. Sci. **28**(11), 1539–1641 (2003)
51. Aranda, P., Ruiz-Hitzky, E.: Poly(ethylene oxide)-silicate intercalation materials. Chem. Mater. **4**, 1395–1403 (1992)
52. Zhao, X., Urano, K., Ogasawara, S.: Adsorption of poly(ethylene vinyl alcohol) from aqueous solution on montmorillonite clays. Colloid Polym. Sci. **267**, 899–906 (1989)
53. Burnside, S.D., Giannelis, E.P.: Synthesis and properties of new poly(dimethylsiloxane) nanocomposites. Chem. Mater. **7**(9), 1597–1600 (1995)
54. Okada, A., Kawasumi, M., Usuki, A., Kojima, Y., Kurauchi, T., Kamigaito, O.: Synthesis and properties of nylon-6/clay hybrids. In: Schaefer, D.W., Mark, J.E. (eds.) Polymer Based Molecular Composites. MRS Symposium Proceedings, Pittsburgh, vol. 171, pp. 45–50 (1990)
55. Richard, A.V., Giannelis, E.P.: Lattice model of polymer melt intercalation in organically-modified layered silicates. Macromolecules **30**(25), 7990–7999 (1997)
56. Lee, W.Y., Raythatha, R.H., Tatarchuk, B.J., et al.: Pillared-clay catalysts containing mixed-metal complexes: I. Preparation and characterization. J. Catal. **115**(1), 159–179 (1989)
57. Jiang, J.Q., Cooper, C., Ouki, S.: Comparison of modified montmorillonite adsorbents: part I: preparation, characterization and phenol adsorption. Chemosphere **47**(7), 711–716 (2002)
58. Bergaoui, L., Mra, I., Lambert, J.F., Ghorbel, A., et al.: A comparative study of the acidity toward the aqueous phase and adsorptive properties of Al13-pillared montmorillonite and Al13-pillared saponite. J. Phys. Chem. B **103**(15), 2897–2902 (1999)
59. Cañizares, P., Valverde, J.L., Kou, M.R.S., et al.: Synthesis and characterization of PILCs with single and mixed oxide pillars prepared from two different bentonites. A comparative study. Micropor. Mesopor. Mat. **29**(3), 267–281 (1999)
60. Michot, L.J., Pinnavaia, T.J.: Improved synthesis of alumina-pillared montmorillonite by surfactant modification. Chem. Mater. **4**(6), 1433–1437 (1992)
61. Jiang, J.Q., Graham, N.J.D.: Preparation and characterisation of an optimal polyferric sulphate (PFS) as a coagulant for water treatment. J. Chem. Technol. Biot. **73**(4), 351–358 (2010)
62. Jiang, J.Q., Zeng, Z.: Comparison of modified montmorillonite adsorbents. Part II: The effects of the type of raw clays and modification conditions on the adsorption performance. Chemosphere **53**(1), 53–62 (2003)
63. Winnie, M., Fritz, T.M., Guenther, K.: Sorption of heavy-metal cations by Al and Zr-hydroxy-intercalated and pillared bentonite. Clay Clay Miner. **47**(5), 617–629 (1999)
64. Yoda, S., Nagashima, Y., Endo, A., Yanagishita, H., Otake, K.: Preparation of metal-oxide-pillared clay using supercritical intercalation. In: 6th International Symposium on Supercritical Fluids PRm4 (2003)
65. Ravichandran, J.: Properties and catalytic activity of acid-modified montmorillonite and vermiculite. Clay Clay Miner. **45**(6), 854–858 (1997)
66. Sakizci, A.M., Yörükoğullari, B.E.: SO_2 adsorption on acid-treated bentonites from Turkey. Clay Miner. **46**(1), 73–83 (2011)

67. Madejová, J., Andrejkovičová, S., Bujdák, J., et al.: Characterization of products obtained by acid leaching of Fe-bentonite. Clay Miner. **42**(4), 527–540 (2007)
68. Sarma, G.K., Sen, G.S., Bhattacharyya, K.G.: Adsorption of crystal violet on raw and acid-treated montmorillonite, K10, in aqueous suspension. J. Environ. Manag. **171**, 1–10 (2016)
69. Akpomie, K.G., Dawodu, F.A.: Acid-modified montmorillonite for sorption of heavy metals from automobile effluent. Beni-Suef Univ. J. Basic Appl. Sci. **5**(1), 1–12 (2016)
70. Gomri, F., Boutahala, M., Zaghouaneboudiaf, H., et al.: Removal of acid blue 80 from aqueous solutions by adsorption on chemical modified bentonites. Desalin. Water Treat. **30** (3), 26240–26249 (2016)
71. Burcu, E.A., Meryem, S.: Ethylene adsorption on acid-treated clay minerals. Adsorpt. Sci. Technol. **30**(3), 265–273 (2012)
72. Vuković, Z., Milutonović, A., Rožić, L., et al.: The influence of acid treatment on the composition of bentonite. Clay Clay Miner. **54**(6), 697–702 (2006)
73. Falaras, Kovanis P., Lezou, I., et al.: Cottonseed oil bleaching by acid-activated montmorillonite. Clay Miner. **34**(2), 221–232 (1999)
74. Vuković, Z., MilutinovićNikolić, A., Krstić, J., et al.: The influence of acid treatment on the nanostructure and textural properties of bentonite clays. Mater. Sci. Forum **494**(3), 339–344 (2005)
75. Panda, A.K., Mishra, B.G., Mishra, D.K., et al.: Effect of sulphuric acid treatment on the physico-chemical characteristics of kaolin clay. Colloid Surf. A **363**(1–3), 98–104 (2010)
76. Wambu, E.W., Onindo, C.O., Ambusso, W.J., et al.: Fluoride adsorption onto acid-treated diatomaceous mineral from Kenya. Mater. Sci. Appl. **2**(11), 1654–1660 (2015)
77. Pentrák, M., Czímerová, A., Madejová, J., et al.: Changes in layer charge of clay minerals upon acid treatment as obtained from their interactions with methylene blue. Appl. Clay Sci. **55**(7), 100–107 (2012)
78. Gan, F., Zhou, J., Wang, H., et al.: Removal of phosphate from aqueous solution by thermally treated natural palygorskite. Water Res. **43**(11), 2907–2915 (2009)
79. Meng, L.Y., Park, S.J.: Effect of heat treatment on CO_2 adsorption of KOH-activated graphite nanofibers. J. Colloid Interf. Sci. **352**(2), 498–503 (2010)
80. Zhang, X., Zheng, Y., Feng, X., et al.: Calcination temperature-dependent surface structure and physicochemical properties of magnesium oxide. Rsc. Adv. **5**(105), 86102–86112 (2015)
81. Bojemueller, E., Nennemann, A., Lagaly, G.: Enhanced pesticide adsorption by thermally modified bentonites. Appl. Clay Sci. **18**(5–6), 277–284 (2001)
82. Aivalioti, M., Vamvasakis, I., Gidarakos, E.: BTEX and MTBE adsorption onto raw and thermally modified diatomite. J. Hazard. Mater. **178**(1–3), 136 (2010)
83. Pelte, S., Flamant, G., Flamand, R., et al.: Effects of thermal treatment on feldspar sorptive properties: Identification of uptake mechanisms. Miner. Eng. **13**(6), 609–622 (2000)
84. Wang, Y., Liu, H., Wang, K., et al.: Synthesis and electrochemical performance of nano-sized Li4Ti5O12 with double surface modification of Ti(III) and carbon. J. Mater. Chem. **19**(37), 6789–6795 (2009)
85. Janus, M., Kusiak, E., Choina, J., et al.: Enhanced adsorption of two azo dyes produced by carbon modification of TiO_2. Desalination **249**(1), 359–363 (2009)
86. Dai, M., Xia, L., Song, S., et al.: Adsorption of As (V) inside the pores of porous hematite in water. J. Hazard. Mater. **307**, 312–317 (2016)
87. Perez-Maqueda, L.A., Criado, J.M., Real, C., et al.: Study of thermal evolution of porous hematite by emanation thermal analysis. J. Eur. Ceram. Soc. **22**, 2277–2281 (2002)
88. Landers, M., Gilkes, R.J.: Dehydroxylation and dissolution of nickeliferous goethite in new caledonian lateritic Ni ore. Appl. Clay Sci. **35**, 162–172 (2007)
89. Jia, F., Ramirez-Muñiz, K., Song, S.: Preparation and characterization of porous hematite through thermal decomposition of a goethite concentrate. Miner. Process. Extr. Metall. Rev. **35**(3), 193–201 (2014)

90. Jones, D.A., Lelyveld, T.P., Mavrofidis, S.D., et al.: Microwave heating applications in environmental engineering—a review. Resour. Conserv. Recy. **34**(2), 75–90 (2002)
91. Clark, D.E., Folz, D.C., West, J.K.: Processing materials with microwave energy. Maters Sci. Eng. A **287**(2), 153–158 (2000)
92. Anuwattana, R., Khummongkol Jr., P., Balkus, K.J., et al.: Conventional and microwave hydrothermal synthesis of zeolite ZSM-5 from the cupola slag. Micropor. Mesopor. Mater. **111**(1), 260–266 (2008)
93. Zhou, R., Zhong, S., Lin, X., et al.: Synthesis of zeolite T by microwave and conventional heating. Micropor. Mesopor. Mater. **124**(1–3), 117–122 (2009)
94. Liu, P.: Polymer modified clay minerals: a review. Appl. Clay Sci. **38**(1), 64–76 (2007)
95. Kickelbick, G.: Concepts for the incorporation of inorganic building blocks into organic polymers on a nanoscale. Prog. Polym. Sci. **28**(1), 83–114 (2003)
96. Tsubokawa, N., Kogure, A., Sone, Y.: Grafting of polyesters from ultrafine inorganic particles: copolymerization of epoxides with cyclic acid anhydrides initiated by COOK groups introduced onto the surface. J. Polym. Sci. Pol. Chem. **28**, 1923–1933 (1995)
97. Mesnage, A., Magied, M.A., Simon, P., et al.: Grafting polymers to titania nanoparticles by radical polymerization initiated by diazonium salt. J. Mater. Sci. **46**(19), 6332–6338 (2011)
98. Mévellec, V., Roussel, S., Tessier, L., et al.: Grafting polymers on surfaces: a new powerful and versatile diazonium salt-based one-step process in aqueous media. Chem. Mater. **19**(19), 6323–6330 (2007)
99. Rong, M.Z., Ji, Q.L., Zhang, M.Q., et al.: Graft polymerization of vinyl monomers onto nanosized alumina particles. Eur. Polym. J. **38**(8), 1573–1582 (2002)
100. Shirai, Y., Kawatsura, K., Tsubokawa, N.: Graft polymerization of vinyl monomers from initiating groups introduced onto polymethylsiloxane-coated titanium dioxide modified with alcoholic hydroxyl groups. Prog. Org. Coat. **36**(4), 217–224 (1999)
101. Jordan, R., Ulman, A., Kang, J.F., et al.: Surface-initiated anionic polymerization of styrene by means of self-assembled monolayers. J. Am. Chem. Soc. **121**(5), 1016–1022 (2013)
102. Zhao, B., Brittain, W.J.: Synthesis of polystyrene brushes on silicate substrates via carbocationic polymerization from self-assembled monolayers. Macromolecules **33**(2), 342–348 (2000)
103. Mchale, G., Shirtcliffe, N.J., Newton, M.I.: Super-hydrophobic and super-wetting surfaces: analytical potential? Analyst **129**(4), 284–287 (2004)
104. El-Bourawi, M.S., Ding, Z., Ma, R., et al.: A framework for better understanding membrane distillation separation process. J. Membr. Sci. **285**(1), 4–29 (2006)
105. Zhang, X., Shi, F., Niu, J., et al.: Superhydrophobic surfaces: from structural control to functional application. J. Mater. Chem. **18**(6), 621–633 (2008)
106. Nakajima, A., Fujishima, A., Hashimoto, K., et al.: Preparation of transparent superhydrophobic boehmite and silica films by sublimation of aluminum acetylacetonate. Adv. Master. **11**(16), 1365–1368 (1999)
107. Huang, Y., Wu, J., Xu, X., et al.: An innovative way of etching MoS_2: characterization and mechanistic investigation. Nano. Res. **6**(3), 200–207 (2013)
108. Yamamoto, M., Einstein, T.L., Fuhrer, M.S., et al.: Anisotropic etching of atomically thin MoS_2. J. Phys. Chem. C **117**(48), 25643–25649 (2013)
109. Zhu, H., Qin, X., Cheng, L., et al.: Remote plasma oxidation and atomic layer etching of MoS_2. Acs. Appl. Mater. Interf. **8**(29), 19119 (2016)
110. Qian, B., Shen, Z.: Fabrication of superhydrophobic surfaces by dislocation-selective chemical etching on aluminum, copper, and zinc substrates. Langmuir **21**(20), 9007–9009 (2017)
111. Zhang, X., Jia, J., Yang, B., Song, S.: Oxidation of molybdenum disulfide sheet in water under in situ atomic force microscopy observation. J. Phys. Chem. C **121**(18), 9938–9943 (2017)
112. Spychalski, W.L., Pisarek, M., Szoszkiewicz, R.: Microscale insight into oxidation of single MoS_2 crystals in air. J. Phys. Chem. C **121**, 26027–26033 (2017)

113. Oliveira, C.R., Rubio, J.: New basis for adsorption of ionic pollutants onto modified zeolites. Miner. Eng. **20**(6), 552–558 (2007)
114. Chansuvarn, W.: Modification of natural clay surface with manganese-oxide for removal of cadmium(ii) ion from aqueous solution. Appl. Mech. Mater. **866**, 124–127 (2017)
115. Lihareva, N., Dimova, L., Petrov, O., et al.: Ag^+ sorption on natural and Na-exchanged clinoptilolite from Eastern Rhodopes. Bulgaria. Micropor. Mesopor. Mater. **130**(1–3), 32–37 (2010)
116. Jozefaciuk, G., Bowanko, G.: Effect of acid and alkali treatments on surface areas and adsorption energies of selected minerals. Clay Clay Miner. **50**(5), 647–656 (2002)
117. Owabor, C.N., Ono, U.M., Isuekevbo, A.: Enhanced Sorption of naphthalene onto a modified clay adsorbent: effect of acid, base and salt modifications of clay on sorption kinetics. Adv. Chem. Eng. Sci. **2**(3), 330–335 (2012)
118. Pasek, M.A., Lauretta, D.S.: Aqueous corrosion of phosphide minerals from iron meteorites: a highly reactive source of prebiotic phosphorus on the surface of the early Earth. Astrobiology **5**(4), 515 (2005)
119. Agnihotri, R., Mahuli, S.K., Chauk, S.S., et al.: Influence of surface modifiers on the structure of precipitated calcium carbonate. Ind. Eng. Chem. Res. **38**(38), 2283–2291 (1999)
120. Nachtigall, S.M.B., Cerveira, G.S., Rosa, S.M.L.: New polymeric-coupling agent for polypropylene/wood-flour composites. Polym. Test **26**(5), 619–628 (2007)
121. Jiang, J., Chen, X., Jia, X.: Application of mono-alkoxy titanate coupling agents in composite materials. China Coat **21**(6), 35 (2006)
122. Yao, Z., Silane Coupling Agent: China Plast. Ind. **6**(20), 18 (1991)
123. Cheng, G., Luo, J., Qian, J., et al.: Surface modification of nano-TiN by using silane coupling agent. Mater Sci. Pol. **32**(2), 214–219 (2014)
124. Li, B., Mou, H., Li, Y., et al.: Synthesis and thermal decomposition behavior of zircoaluminate coupling agents. Ind. Eng. Chem. Res. **52**(34), 11980–11987 (2013)
125. Gao, P., Ying, C., Wang, S., et al.: Low temperature hydrothermal synthesis of ZnO nanodisk arrays utilizing self-assembly of surfactant molecules at solid-liquid interfaces. J. Nanopart. Res. **8**(1), 131–136 (2006)
126. Song, E.M., Kim, D.W., Kim, B.J., et al.: Surface modification of $CaCO_3$, nanoparticles by alkylbenzene sulfonic acid surfactant. Colloid Surf. A **461**(1), 1–10 (2014)
127. Özcan, A., Ömeroğlu, Çiğdem, Erdoğan, Y., et al.: Modification of bentonite with a cationic surfactant: an adsorption study of textile dye Reactive Blue 19. J. Hazard. Mater. **140**(1–2), 173 (2007)
128. Shen, Y.H.: Preparations of organobentonite using nonionic surfactants. Chemosphere **44**(5), 989–995 (2001)
129. Zengin, H., Hu, B., Siddiqui, J.A., et al.: Surface modification of glass beads with poly (acrylic acid). Polym. Adv. Technol. **17**(5), 372–378 (2010)
130. Makra, Z., Szőllősi, G.: Hydrogenation of (E)-2-methyl-2-butenoic acid over cinchona-modified Pd catalyst in the presence of achiral amines: solvent and modifier effect. Catal. Commun. **46**(5), 113–117 (2014)
131. Dearmitt, C., Rothon, R.: In: Surface Modifiers for Use with Particulate Fillers. Springer International Publishing (2017)
132. Ren, M., Yin, H., Wang, A., et al.: Mica coated by direct deposition of rutile TiO_2 nanoparticles and the optical properties. Mater. Chem. Phys. **103**(2), 230–234 (2007)
133. Fujino, T., Matzuda, T.: Synthetic process of titanium dioxide coating on aluminum by chemical conversion method. Mater. Trans. **47**(3), 849–853 (2006)
134. Chan, C.M.: In: Polymer Surface Modification and Characterization, p. 264. Carl Hanser, GmbH & Co., Munich (1993)
135. Gregg, S.J., Sing, K.S.W.: Adsorption, Surface Area and Porosity. Academic Press, London (1982)
136. Siebold, A., Nardin, M., Schultz, J., et al.: Effect of dynamic contact angle on capillary rise phenomena. Colloid Surf. A **161**(1), 81–87 (2000)

137. Kadir, E., Erdogan, T.: The adsorption of anionic and cationic surfactants by the different clay minerals: ftir spectroscopic study. Spectrosc. Lett. **34**(4), 443–451 (2001)
138. Bel'chinskaya, L.I., Strel'nikova, O.Y., Novikova, L.A., et al.: Enhancement of the adsorption selectivity of nanoporous clinoptilolite by hydrophobization with organosiloxanes. Prot. Met. **44**(4), 390–393 (2008)
139. Xavier, K.C.M., Santos, M.R.M.C., Osajima, J.A., et al.: Effects of acid treatment on the clay palygorskite: XRD, surface area, morphological and chemical composition. Mater. Res. **17**(1), 3–8 (2014)

Adsorption of Anions on Minerals

Feifei Jia, Min Dai, and Bingqiao Yang

Abstract Natural minerals stand out from conventional materials as promising adsorbents for the anions removal from aqueous solution due to their efficient adsorption performance, cost-effectiveness and easy-availability. This chapter summarizes the advances of natural minerals as adsorbents for the removal of anions (arsenic, phosphorus, fluoride, nitrate) in water. The common natural minerals, as well as their adsorption capacity, are introduced for each anion. The affecting factors and adsorption mechanism are also highlighted.

Keywords Natural minerals · Arsenic · Phosphorus · Fluoride · Nitrate · Adsorption

1 Introduction

Anions, such as arsenic, phosphorus, fluoride et al., are the major water pollutants, which would pose a threat to both humans and ecosystem. Therefore, it is of great importance to develop a simple, cost-effective and highly efficient process for the anions removal from aqueous solution. Minerals in nature are responsible for the

F. Jia (✉)
Hubei Key Laboratory of Mineral Resources Processing and Environment, Wuhan University of Technology, Luoshi Road 122, Wuhan 430070, Hubei, China
e-mail: jff_520@126.com

F. Jia
School of Resources and Environmental Engineering, Wuhan University of Technology, Luoshi Road 122, Wuhan 430070, Hubei, China

M. Dai
School of Environmental and Chemical Engineering, Zhaoqing University, Zhaoqing 526061, China

B. Yang
School of Xingfa Mining Engineering, Wuhan Institute of Technology, Xiongchu Avenue 693, Wuhan 430073, Hubei, China

S. Song and B. Li (eds.), *Adsorption at Natural Minerals/Water Interfaces*, Engineering Materials,
https://doi.org/10.1007/978-3-030-54451-5_4

immobilization and attenuation of anions in water, sediment and soil. And the adsorption on minerals is one of the most promising approaches to remove anions because it meets the above requirements. This chapter focus on the removal of anions from water with mineral adsorbents. A summary of relevant published data with some of the latest findings, and a source of up-to-date literature are presented and the results have been discussed.

Sections 2, 3, 4 and 5 will present the adsorption of arsenic, phosphorus, fluoride, and nitrogen on minerals, respectively. In each section, a discussion on the anion contaminants, the existence forms of anions, mineral adsorbents (types, structure, etc.), adsorption behaviours (adsorption isotherm, adsorption kinetics), factors affecting the adsorption, and mechanisms will be discussed in detail. Generally, the iron, aluminum, calcium based minerals are the major adsorbents responsible for the removal of these anions. The zeolites, clays are also discussed and compared. The factors, such as pH, temperature and competitive ions, exhibit great influences in the removal efficiency. Hence, a good understanding of the effects would facilitate the removal. Adsorption mechanism is essential in the adsorption, which unveils the inner relation between the adsorbents and adsorbates. In summary, the anion adsorption is mainly based on the ligand exchange, ion exchange, cooperating with electrostatic attraction and surface precipitation.

2 Arsenic Adsorption

2.1 Arsenic Species, Toxicity and Distribution

Arsenic is a silver-grey brittle crystalline solid with symbol As. It is a ubiquitous metalloid element, occurring in the earth crusts, soil, sediments, water, air, even in living organisms. The abundance ranks 20th in the earth crust, 14th in the seawater, and 12th in the human body [1]. The natural occurrence of arsenic is usually found in combination with sulfur, oxygen and several metals, including iron, copper, silver and so on [2]. The arsenic mobility is mainly caused by the natural process, including weathering reactions, biological activity and volcanic emissions. But recent years, the elevated discharge from a range of anthropogenic activities worsens the situation. The combustion of fossil fuels, the use of arsenical pesticides, herbicides and wood preserving, the disposal of tailing waste and the discharge from the industry increase the arsenic concentration locally [2, 3].

The Arsenic exists in four oxidation states: arsenate (As(V)), arsenite (As(III)), arsenic (As(0)), and arsine (As(–III)). Among them, the inorganic form as oxyanions of As(V) and As(III) are predominant in the natural water [4]. The speciation are highly controlled by the redox potential (Eh) and pH, shown in Fig. 1. Arsenate is stable in the aerobic conditions, commonly presenting as $HAsO_4^{2-}$ and $H_2AsO_4^-$. The species AsO_4^{3-} and H_3AsO_4 exit in extremely alkaline and strong acid conditions, respectively. Arsenite is predominant in the anaerobic environment, presenting as H_3AsO_3 under pH 9.2. The respective deprotonation dissociation constants (pKa) of arsenate are 2.20 ($H_3AsO_4/H_2AsO_4^-$), 6.97 ($H_2AsO_4^-/HAsO_4^{2-}$) and 11.53 ($HAsO_4^{2-}/AsO_4^{3-}$). For arsenite, the pKa value are 9.22 ($H_3AsO_3/H_2AsO_3^-$), 12.13 ($H_2AsO_3^-/HAsO_3^{2-}$), and 13.40 ($HAsO_3^{2-}/AsO_3^{3-}$) [5]. The inorganic arsenic compounds can be methylated into monomethylarsonic acid (MMA: $CH_3AsO(OH)_2$), dimethylarsinic acid (DMA: $(CH_3)_2AsOOH$) and trimethylarsine oxide (TMAsO: $(CH_3)_3AsO$) by biotransformation under oxidizing conditions [6, 7]. They are found in the soil and the marine environment as the minor components [2, 8]. The microcrystalline As(0) is rare in nature, and the As(–III) can be detected in atmosphere [9].

Arsenic toxicity is the major concern in the environment, following the order of MMA(III) > As(III) > As(V) > DMA(V) > MMA(V) [2, 10]. Because of the toxicity, the arsenic compounds have long been a preferred agent of homicide and suicide. Although it has been found useful in some diseases, most of the contact to human being was harmful. Numerous studies have demonstrated the toxicity of arsenic on human health. melanosis, leuco-melanosis, keratosis, hyperkeratosis, dorsum, non-petting edema, gangrene and skin cancer are related to arsenic exposure [11]. The long term intake of arsenic (above 0.05 mg/L) may lead to cancers (skin, bladder, kidney, lung), diseases of the blood vessels of the legs and feet, and possibly diabetes, high blood pressure and re-productive disorders [12]. The main source of arsenic to human is the consumption of drinking water and food accumulating arsenic [11]. Therefore, the World Health Organisation recommends the maximum concentration of arsenic in drinking water is 10 μg/L.

However, it has been reported that about 88,750 km^2 area in West Bengal is identified as As contaminated zone among which 43.8% area is highly affected [13]. In Argentina, Bangladesh, Cambodia, Chile, China, India, Hungary, Mexico, many parts of the USA Romania, Vietnam, and Nepal, the concentration of arsenic in water is above 50 μg/L [2, 14]. In inner Mongolia, China, As(III) accounts for 60–90% of total arsenic concentration 1860 μg/L [15]. In Tinto and Odiel Rivers,

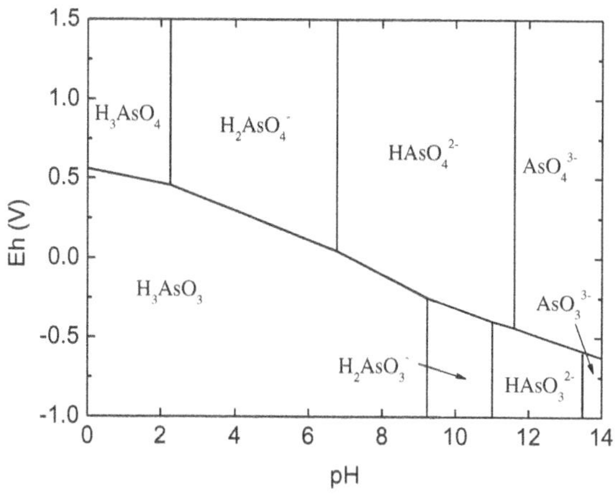

Fig. 1 Eh-pH diagram of arsenite and arsenate

Spain, arsenic concentration range from 0.5 to 7900 μg/L [16]. The groundwater in Calcutta, India contain arsenic with the concentration of 50–23,080 μg/L [17]. Statistically, more than 150 million people are suffering from chronic As poisoning risk all around the world [18]. Thus, remediation of arsenic contamination from water is essential to reduce the healthy threat to human beings.

2.2 Adsorption

Arsenic remediation mainly refers the removal of oxyanions of As(V) and As(III) in water. The removal technologies involve coagulation and coprecipitation, adsorption, ion-exchange, membrane technologies, bioremediation and capacitive deionization (CDI), listed in Table 1.

Adsorption is a process that uses solids to remove substances from either gaseous or liquid solutions, which is also the most common methods to control the distribution of arsenic due to its simple operation, low cost, good regeneration capability, as well as no chemical addition. The adsorbents for arsenic consist of activated carbon, clay, zeolite, biomass, iron oxide, activated alumina and lots of by products from industry [4]. It is commonly accepted that the iron and aluminum based materials are among the most effective adsorbents for their strong affinity to arsenic. Minerals play an important part in the adsorbents.

Table 1 Comparison of arsenic removal technologies

Technologies	Advantages	Disadvantages
Coagulation Eletrocoagulation Coprecipitation	Simple to operate, low capital cost	Filtration is needed; instability of solid; sludge disposal is a potential source of contamination; pre-oxidation of As(III)
Adsorption	Simple without chemical addition	Need replacement after several regeneration
Ion exchange	Arsenic selectivity; pH independent	High cost medium, operation and maintenance; difficult to remove As (III)
Nanofiltration Reverse osmosis Electrodialysis	High removal efficiency	High capital and running cost; high tech operation and maintenance; high water rejection; low As level can be treated
Bioremediation	Environmental compatible	Low efficiency; hard to disposal the plants after remediation
Capacitive deionization (CDI)	Low voltage <2 V, can be operated by solar energy; low energy consumption; low water rejection	As needs to be charged; precipitation of salts and biofouling have a negative effect

2.2.1 Adsorbents

(i) Iron-based minerals

According to the reports, there are sixteen pure phases of iron oxides, hydroxides or oxy-hydroxides [19]. Almost all of them are investigated in arsenic removal. Among them, ferrihydrite (Fh), greenrust, goethite (α–FeOOH), hematite (α–Fe_2O_3), sidetite ($FeCO_3$), magnetite (Fe_3O_4) are the most studied species.

Fh is ubiquitous in soil and sediments, including 2 and 6-line Fh, which is classified according to the two or six identifiable broad reflections in a diffraction pattern. The thermodynamically unstable material has poorly crystal structure and high surface area, ranging from 100 to 700 m^2/g, making it effective in arsenic removal [20, 21]. The arsenic retention is above 100 mg/g with a higher adsorption of arsenite at the appropriate pH [22]. The result was in agreement with that found by Maria Martin. The arsenate and arsenite adsorption capacity was 0.21 and 0.40 mg/m^2 at pH 6.5, respectively [23]. Jiang also reported that the maximum adsorption capacity of arsenate on Fh was 142.86 mg/g at pH 3 [24].

Ferric oxyhydroxides and oxides mainly include goethite (α–FeOOH), akaganeite (β–FeOOH), lepidocrocite (γ–FeOOH) and hematite. The adsorption capacity varied with different reports. But the hematite usually had a higher arsenic retention than goethite [25, 26]. As reported by Silva et al. [26] the adsorption capacity of hematite and goethite were 0.193 and 0.101 mmol/g, respectively, much lower as compared with that on Fh [23, 26].

The siderite, green rust and magnetite are the minerals which are composed of Fe(II) and Fe(III). The adsorption capacity was a little higher than that of hematite and goethite [27]. The magnetite showed a much higher arsenic adsorption in nanoparticles [28]. The comparison of arsenic adsorption between goethite, hematite and magnetite was conducted by Giménez et al. It was reported the adsorption rate was faster on the goethite, magnetite than that on the hematite. The adsorption on three minerals fitted well with Langmuir isotherm [25]. Meanwhile, the ^2C inner-sphere complex was formed on the surface of the minerals, in agreement with ferric oxides. In most of the reports, the existence of Fe(II) could neither reduce As(V) nor oxidize As(III) in anoxic condition. But Zheng et al. [29] found the reduction of arsenate on the zirconium-based magnetic sorbent ($ZrO(OH)_2 \cdot 1.6Fe_3O_4 \cdot 2.5H_2O$), they assumed the divalent iron in the sorbent may provide electrons for the reduction. With the dissolved oxygen, a rapid oxidation of As(III) upon sorption was observed on the Fe(II) containing minerals. This process was mainly caused by a Fenton-type reaction as follows:

$$\mathrm{Fe(II)} + \mathrm{O_2} \rightarrow \mathrm{Fe(III)} + \mathrm{O_2^{\cdot -}} \quad (1)$$

$$\mathrm{Fe(II)} + \mathrm{O_2^{\cdot -}} + 2\mathrm{H^+} \rightarrow \mathrm{Fe(III)} + \mathrm{H_2O_2} \quad (2)$$

$$\mathrm{Fe(II)} + \mathrm{H_2O_2} + \mathrm{H_3AsO_3} + \mathrm{O_2} \rightarrow \mathrm{Fe(III)} + \mathrm{H_3AsO_3} + \mathrm{O_2^{\cdot -}} + \mathrm{H_2O} \quad (3)$$

The dissolved Fe(II) reacted with O_2 and generated H_2O_2 and $O_2^{\cdot -}$ which was responsible for the oxidation of As(III) at acidic or neutral pH range [30, 31]. Same results were observed on the ferric oxides surface with the addition of Fe(II) in oxic environment [32]. In addition, the magnetite is preferred in arsenic removal for the advantage in the separation. It can be easily removed by applying a magnetic field [33–35].

(ii) Aluminum-based minerals

Aluminum, similar with iron, has a strong affinity to arsenic. Activated alumina (AA, Al_2O_3), which is prepared by thermal dehydration of aluminum hydroxide, is classified among the best available technologies for As removal from water by the United Nations Environmental Program agency (UNEP) [4]. It has been used in fixed beds to remove contaminants, including As, fluoride, NOM, selenium and silica [11]. The mesoporous alumina with a high surface area (307 m^2/g) and uniform pore size (3.5 nm) was prepared by Kim. The material showed 7 times higher adsorption capacity and a faster adsorption rate than the conventional AA. The adsorption capacity was 121 mg/g and 47 mg/g for As(V) and As(III), respectively [36]. Both Langmuir and Freundlich isotherm equations could describe the adsorption behavior well [37]. The surface complex was detected by FTIR as monodentate complexes on the nano-AA [38].

Other aluminum-based minerals were also studied. The adsorption capacity on gibbsite was 4.6 and 3.3 mg/g for As(V) and As(III), respectively [39]. The bidentate binuclear surface complex was formed when adsorbed As(V) [40]. Same complexes were identified on the γ–Al_2O_3 for As(V) and As(III) by Arai et al. [41] Meanwhile, the outer-sphere complex was formed by As(III) EANES spectra (pH 5.5–8). They also investigated the arsenate adsorption at amorphous aluminosilicate mineral–water interfaces. As(V) formed bidentate binuclear surface species on aluminum octahedral structures. Except for the hydroxyl groups, the adsorption mechanism contributed to the surface-coordinated water molecules and silicate ions [42]. When precipitating Al, Fe hydoxides at different ratio, As(III) adsorption decreased with increasing Al:Fe ratio [43].

Red mud is a by-product produced when bauxite ore is subjected to caustic leaching [4]. It is mainly composed of oxides of iron, aluminum, silicon and titanium. Production for a ton of alumina generate about 1–2 tons of the by-products [44]. The adsorption capacity was 4.31 and 5.07 μmol/g for As(III) and As(V), respectively. The activation of red mud by heat and acid treatment increased the adsorptive capacity [45]. The neutralization of red mud leachate by gypsum resulted in 86% Aland 81% As removal. But no evidence showed the adsorption or incorporation of arsenic in the calcite structure [46].

(iii) Manganese-based minerals

Manganese oxides in arsenic removal have not been studied as much as Fe-oxides, but Mn-oxides have also shown their efficiency as arsenic adsorbents.

The oxidation of As(III) and adsorption of As(V) are considered as an important pathway to control As species in natural systems [47]. The oxidation of As(III) to As(V) follow two steps after adsorbed on the surface, reduction of Mn(IV) to Mn (III) and subsequent to Mn(II), showing as follows [48, 49]:

$$2MnO_2 + H_3AsO_3 + H_2O \rightarrow 2MnOOH + H_3AsO_4 \quad (4)$$

$$2MnOOH + H_3AsO_3 + 4H^+ \rightarrow 2Mn^{2+} + H_3AsO_4 + 3H_2O \quad (5)$$

$$2Mn{-}OH + H_3AsO_4 \rightarrow (MnO)_2AsOOH + 2H_2O \quad (6)$$

Then As(V) oxyanions were adsorbed on the MnO_2 surface [49, 50]. The synthesized birnessite (MnO_2) could represent many natural manganese oxide minerals. As a result, it was used in arsenic adsorption [49]. The adsorption capacity was ~110 mg/g at an arsenic concentration of 700 mg/L [51]. The EXAFS analysis showed that the As(V)-Mn interatomic distance was 0.322 nm. It was accepted that a bidentate binuclear corner sharing (bridged) complex occurring at MnO_2. However, the complexes would limit further oxidation for the formation of surface layer and prevent further reduction of Mn(IV) in the interior of crystal [49]. According to Eq. (5), the oxidation of As(III) released Mn^{2+} into solution. The surface alteration would create fresh reaction sites and promote further adsorption of arsenic ions. The phenomenon was further confirmed by the comparison of the adsorption capacity of As(III) and As(V) on the MnO_2. The results showed that the arsenic retention with As(III) as initial ions was higher than that of As(V) [49]. But the comparison made by Saleh et al. exhibited an opposite results. This might be caused by the shorter time as they only conducted the experiment for 1 h [52].

(iv) Ti-based minerals

TiO_2 is classified as anatase, rutile and brookite in terms of the crystal structure. The photocatalytic performance of TiO_2 makes it as a good candidate for oxidation. The adsorption of light on the material react with oxygen and generate free radicals such as $O_2^{\bullet-}$, $HO_2^{\bullet}$, $OH^{\bullet}$ which are responsible for oxidation. Many researches have investigated the oxidation of As(III). 2 mg/L As(III) was completely converted to As(V) within 25 min by TiO_2 suspension [53]. In addition, TiO_2 showed a high arsenic retention according to the literatures. The adsorption capacity for As(V) and As(III) were 41.4 and 32.4 mg/g at pH 7. The adsorption data fitted well with the Langmuir model [54]. The EXAFS study by Pena et al. indicated that both As(V) and As(III) formed bidentate binuclear surface complexes. The average bond distance were 0.33 and 0.335 nm for Ti–As(V) and Ti–As(III) [55].

(v) Copper, zirconium, lanthanum, cerium based minerals

The metals, such as copper, zirconium, lanthanum, cerium also show arsenic retention. But the reports are limited when compared with metals above, considering the cost [56]. Early investigations are based on these metal oxides, such as cupric oxide, zirconium oxide and hydrated yttrium oxide.

The adsorption process by cupric oxide(CuO) followed Langmuir isotherm with an adsorption capacity of 1.08 mg/g [57]. Martinson et al. found the adsorption capacity of As(V) on CuO nanoparticles was slightly higher than that of As(III), 26.9 and 22.6 mg/g, respecitively. Besides, in this study, the X-ray photoelectron spectroscopy (XPS) suggested that As(III) was oxidized on the surface of CuO. But no further discussion about the oxidation [58]. Yttrium, lanthanum and cerium are the trivalent rare earth elements and they behave similarly, exhibiting a good arsenic retention. It was reported the arsenate adsorption capacity was 480.2 mg/g at pH 5 on the hydrated yttrium oxide. The adsorption fitted well with pseudo-second-order adsorption kinetic model. Both Langmuir and Freundlich isotherm could describe the adsorption behavior. The ligand exchange between arsenate ions and hydroxyl groups played a key role during the adsorption, forming the linkage of As–O–Y [59]. The arsenic adsorption on the yttrium carbonate followed the same mechanism. But this adsorbent can only be used in neutral and alkaline condition. The dissolution happened when pH was below 6.5 [60]. The hydrous cerium oxide had the adsorption capacity over 170 mg/g g for As(III) and 107 mg/g for As(V) at neutral pH. Different from other adsorbents, the hydrous cerium oxide showed a stable As(III) over a wide pH range 3–10. Similarly, the adsorption mechanism is the formation of inner-sphere complexion [61]. Lanthanum compounds, including, lanthanum hydroxide, lanthanum carbonate and basic lanthanum carbonate, dissolve below pH 4. When adsorbing arsenic, ligand exchange dominated in the neutral to alkaline pH range. The zirconium oxide is another candidate for arsenic removal. The amorphous zirconium oxide (am–ZrO_2) synthesized by a simple hydrothermal process had a adsorption capacity 83 mg/g for As(III) and 32.4 mg/g for As(V), respectively [62].

(vi) Clay and Zeolite

The clay minerals are often used as adsorbents due to its high surface area, excellent performance in ion exchange and low cost. They are mainly composed of silica tetrahedral sheet and alumina of magnesium octahedral sheet in different layer structures [63]. The octahedrally coordinated aluminium is expected to have a strong selectivity of arsenic. The clay minerals used for As adsorption include kaolinite, montmorillonite, illite and et al. The negatively charged surface makes it effective in cation adsorption. But to achieve a higher arsenic adsorption on clay minerals, the solution need to be adjusted to acid to keep the solid surface positively charged. The maximum adsorptions of As(V) on kaolinite, montmorillonite, and illite were 0.86, 0.64, and 0.52 mg/g, respectively. The adsorption decreased with increasing temperature. Thus, the process was exothermic [64]. Same phenomenon was observed by Eloussaief et al. But the As(III) adsorption capacity was much higher, i.e. 75.4 mg/g [65]. Bentahar et al. found no significant adsorption on some clays and concluded that the iron content was essential to the adsorption. Both the Langmuir and Freundlich models could describe the adsorption behavior [66]. Interestingly, the point of zero charge (PZC) shifted positively after As(V) adsorption and it was concluded that arsenic formed inner-sphere complex and the specific ion adsorption happened on the surface [64]. However, insufficient reports focused on the arsenic adsorption mechanism on the clay.

Zeolites are microporous crystalline aluminosilicates, composed of tetrahedral silica and alumina at different ratio. Zeolites with infinite three-dimensional structures are well known for ion exchange. In arsenic adsorption, the arsenic oxyanions were exchanged with terminal aluminal or silanol hydroxyl groups. The adsorption capacity of arsenate was 35.8 mg/g with an initial concentration of 10 mg/L [67]. The ion exchange between hydroxyl groups on aluminum and arsenate were also identified as surface complexation. But studies have demonstrated that the efficiency of As removal by zeolite is greatly affected by the iron and aluminum contents [68]. Considering the flourishing pore structure, zeolites were used as substrate for modification in the adsorption which would be discussed below.

(vii) Sulfide minerals

Arsenic adsorption on the sulfide minerals has received less attention. The adsorption of As(III) on the minerals followed the Langmuir isotherm at low surface coverages [69, 70]. Arsenite bonded on the surface of FeS and FeS_2 as arsenopyrite (FeAsS) with a distance of As–S 0.235 nm and As–Fe 0.240 nm, suggesting an exchange reaction with surface sulfhydryl groups except for the ligang exchange with hydroxyl. The calculated site density was near 2.6 and 3.7 sites/nm^2 for FeS and FeS_2, respectively [69]. But another study reported that the arsenite adsorption on mackinawite fitted with the Freundlich model. And the outer-sphere complexes were observed for arsenite and arsenate. A particular adsorption mechanism was proposed by Bostick et al. when arsenite was adsorbed on the ZnS and PbS. A polynuclear arsenic sulfide similar to the $As_3S_3(SH)_3$ trimer was formed, resulted in the As–S coordination instead of As–O [70].

(viii) Modified minerals

After exploring the natural metal oxides for arsenic removal, modification of these materials has been applied to combine the different advantages together. Except for the intrinsic character of metal oxides, the physical performance, such as surface area and developed pore structures are important during adsorption. Thus, the adsorbents with porous structure or nanoparticle-size are synthesized. The carbon materials, clay and zeolite are used as matrix in the modification.

Apparently, the porous microstructure contributes to the higher surface area, leading to higher adsorption capacity. Therefore, the modification of natural non-porous minerals is explored. As discussed in Chap. 3, minerals including the elements, such as C, N and S, could be porous after releasing CO_2, NO_x and SO_2 through thermal modification. In addition, dehydration and dihydroxylation also work. After thermal treatment, the porous products are used for adsorption. The porous hematite prepared by calcination of siderite-goethite concentrate was used in arsenic adsorption. After calcination, the surface area increased from 5 to 93 m^2/g. The adsorption capacity was increased 7 times than that of the original one [71]. Similar results were shown in the adsorption of As(III) on porous hematite calcined from goethite concentration. The adsorption capacity was increased from 2.965 to 14.46 mg/g [72]. Except for the increase in surface area, thermal modification also changes the pore size distribution inside the minerals, which would affect the adsorption efficiency accordingly. For instance, the porous hematite derived from goethite and siderite possessed different average pore size. Although the samples produced from goethite showed a higher surface area (58.2 m^2/g), the adsorption capacity of the one produced from siderite with the surface area of 39.7 m^2/g was higher [73]. It was deduced that parts of the micropore surface area didn't work in the adsorption for the pore blockage as shown in Fig. 2.

The carbon materials have high surface area, microporous structure and high degree of surface selectivity, making them versatile in adsorption. But the functional groups on the surface have no selectivity to arsenic. Hence, they are not chosen as the adsorbent in arsenic adsorption. Eguez et al. studied the isosteric heat of adsorption of As(III) and As(V). The values ranged from 4–0.75 kcal/mol and 4–2 kcal/mol, suggesting that the adsorption was attributed to the weak bond by Van der Waals forces. Thus, only physisorption happened on the surface [74]. Although some researchers reported a relatively high arsenic retention on the carbon materals, the mechanism was mainly based on the electrostatic attraction [75–77]. Most researches focused on the modification, loading the metal oxides on the materials to achieve excellent performance. The developed mesoporous Fe/carbon aerogels with the surface areas of 487 m^2/g showed the maximum arsenic adsorption capacity of 216.9 mg/g. Furthermore, the adsorbent had an excellent response to applied external magnetic field, which could be easily separated from water [78]. The iron content loading on the activated carbon was investigated in arsenic adsorption. The maximum uptake of As(V) was obtained with 10% Fe loading. The higher loading resulted in the decrease of adsorption due to the decrease in surface area and pore size [79]. The composite of aluminum-loaded Shirasu-zeolite P_1 (Al–SZO_1) showed

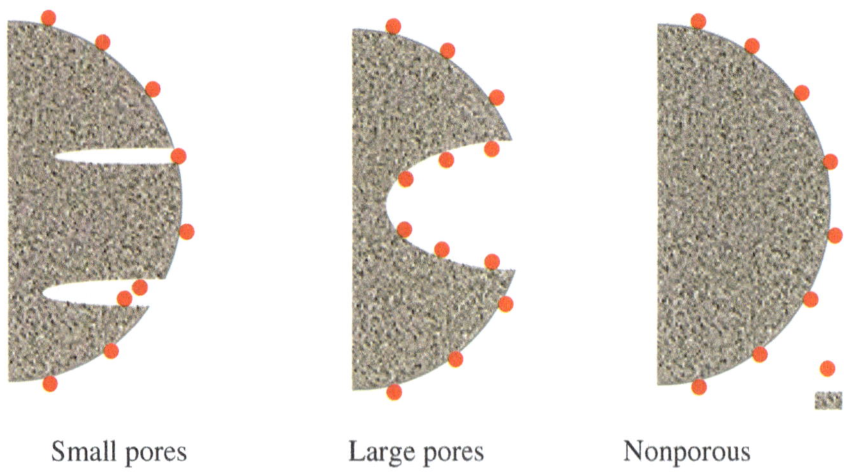

Fig. 2 Schematic representation of As(V) adsorbed on porous hematite

a high As(V) selectivity and uptake, especially advantageous to remove arsenic in low concentration [80]. The montmorillonite modified by Fe polycations and cetyltrimethylammonium bromide (CTMAB) reached the equilibrium within 20 min in adsorption of As(III) and As(V) [81]. Except for the increasing contact surface, the adsorbents loaded on the matrix suitable for the fix bed water reduce the aggregation of bare adsorbents, avoiding further separation. For the separation purpose, the magnetite was used to corporate with other adsorbents for the magnetic performance, such as hematite-coated Fe_3O_4 particle, magnetite-reduced graphene oxide composites [82, 83] (Table 2).

2.2.2 Interference

(i) pH

Arsenic adsorption is highly pH dependent. According to the reports, the As(V) adsorption reached the highest in acidic condition, while the optimum pH for As(III) was slight alkaline [71, 86]. Altundoğan et al. observed the optimum pH range was 1.8–3.5 for As(V) and 5.8–7.5 for As(III) [45]. In the case of the adsorption of arsenic on Zn-Fe binary oxide, the maximum adsorption of arsenite (As(III)) and arsenate (As (V)) was at pH values 8–9 and 3–4, respectively [87]. This was caused by the electrostatic attraction between arsenic ions and adsorbent surface. The pH_{PZC} of the Zn–Fe binary oxide was around 5. At lower pH, the surface of the adsorbent was positive, the arsenate oxyanions electrostatically attracted to the surface. The arsenic species vs pH was shown in Fig. 3. When the pH was higher, the surface charge was negative, and the adsorption capacity was decreased by coulombic repulsions. With the increasing of pH, the negative charge of oxyanions and adsorbent surface increase, while the repulsive force increases, decreasing the adsorption capacity. Thus, except for the surface complexation, electrostatic attraction promoted arsenic removal in suitable pH condition. Arcibar-Orozco et al. studied the influence of iron content, surface area and charge distribution in the arsenic removal on iron modified activated carbon, and found that the most important parameter of AC is the pH_{PZC}, contributing to 52.5% [77]. Most of the researches investigated pH effects ranging from 3 to 12. However, Wei et al. increased the pH value to 13.5 and found that the minimum adsorptions of As(III) and As(V) were both at pH 12. But then, their adsorptions increase again at higher pH values such as 13.0 and 13.5 [88]. However, the mechanism has not yet been clearly understood. In addition, the adsorption on sulfide minerals showed a different tendency.

(ii) Co-existing ions

The above results are commonly based on the single ion system. However, the arsenic is the trace element in natural water. Relatively, there are abounds of cations (Na^+, K^+, Mg^{2+}, Ca^{2+}) and anions (Cl^-, NO_3^-, CO_3^{2-}, SO_4^{2-}, SiO_3^{2-}, PO_4^{3-}) exist simultaneously. The solid-liquid equilibrium between arsenic and adsorbents would change in the present of these ions. Thus, the arsenic adsorption was also conducted

Table 2 Arsenic adsorption among different minerals

Species	pH	C_0 (mg/L)	Adsorption capacity (mg/g)	Adsorption capacity (mg/m^2)	References
Ferryhydrite	6.5	0–100	As(III) 125.5	As(III) 0.40	[23]
			As(V) 67.7	As(V) 0.21	
	3	0–200	As(V) 146.86		[24]
Goethite α−FeOOH	7.5	0.075–74.92		As(III) 0.187	[25]
				As(V) 0.225	
	6.5	0–100	As(III) 7.4	As(III) 0.17	[23]
			As(V) 5.6	As(V) 0.13	
Akaganeite β−FeOOH	7	0–100	As(III) 32.6		[84]
			As(V) 29.1		
Hematite α−Fe_2O_3	7.3	0.075–74.92		As(III) 0.697	[25]
				As(V) 2.173	
	6.5	0–100	As(III) 5.3	As(III) 0.12	[23]
			As(V) 2.9	As(V) 0.06	
	6.5	0.074–14.98		As(V) 0.247	
Magnetite Fe_3O_4	6.5	0.075–74.92		As(III) 0.232	[25]
				As(V) 0.285	
	7	2–75	As(V) 92.85		[34]
Siderite $FeCO_3$	7.0	1–20	As(V) 2.19		[28]
Activated alumina	6.9	0.79–4.9	As(III) 3.48		[37]
	5.2	2.85–11.5	As(V) 15.9		
Birnessite		1–800	As(V) ~ 110		[51]
Gibbsite	5.5	0.15	As(III) 3.3		[39]
			As(V) 4.6		
Hydrated yttrium oxide	5	1–100	As(V) 480.2		[59]
Hydrous cerium oxide	7	1–100	As(III) 170		[61]
			As(V) 107		
Zeolite H-MFI-24	6.5	10	35.80		[67]
AC–Fe	6.0	0–22	As(V) 51.3		[85]
			As(III) 38.8		

in the multicomponent system to investigate the effects of these ions. Commonly, the cations would promote arsenic adsorption while the anions affected the arsenic adversely [89].

The cooperative effects of Mg^{2+}, Ca^{2+}, Cu^{2+}, Ni^{2+}, Co^{2+}, Fe^{2+}, $Fe^{3+}$and Zn^{2+} were found in arsenic adsorption [90–92]. The enhanced adsorption might result

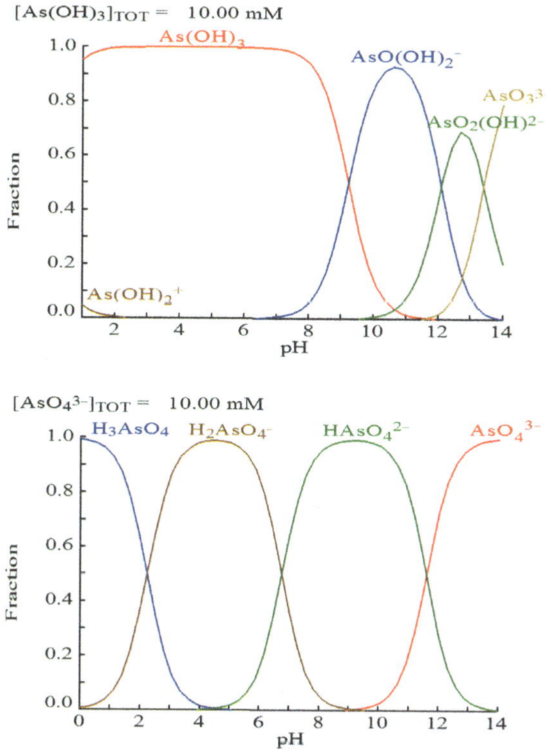

Fig. 3 Arsenic species as a function of pH

from the ternary surface complex formed on the surface [93, 94], as that shown in Fig. 4. The favorable electrostatic effects arose from adsorption of the cations, promoting the later adsorption of arsenic. Same mechanism was found in the adsorption of Ca^{2+}, Cu^{2+} and fulvic acids on the goethite [94].

According to the reports, the enhancement of Fe^{2+}, Zn^{2+} was stronger than the alkaline earth metals. This was caused by the re-adsorption of surface complex at the adsorption sites. The long-time reaction resulted in surface precipitation, which will be discussed later.

However, there are exceptions. The existence of Mg^{2+} and Ca^{2+} was reported to reduce the adsorption of As(III) and As(V) on the CeO_2–ZrO_2 nanospheres [95].

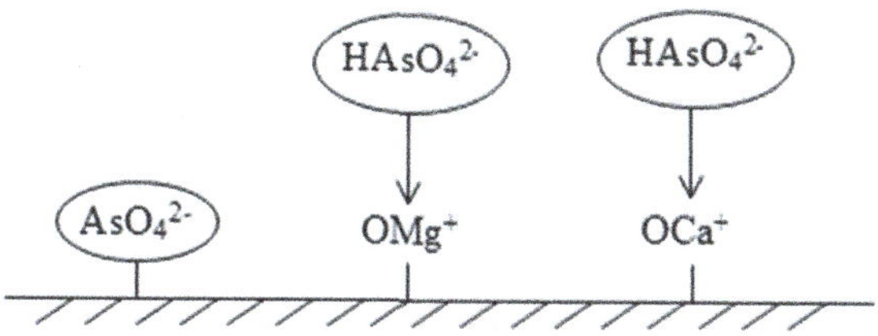

Fig. 4 Schematic representation of formation of ternary surface complex

The two cations had little effects in As(III) adsorption on the goethite-based adsorbent [94, 96]. Fe^{2+} was found to suppress the adsorption of As(III) [97].

In general, the anions followed the order of $NO_3^- < Cl^- < CO_3^{2-} < SO_4^{2-} < SiO_3^{2-} < PO_4^{3-}$ [76, 87, 98–100]. They might directly consume the surface sites that available for arsenic species, or indirectly influence arsenic adsorption by altering the electrostatic charge at adsorbent surfaces [101]. Phosphorus and arsenic belong to one family, they are some alike in nature. Hence, the competition of phosphate in arsenic adsorption was the most significant for all the adsorbents, because it would compete for the same adsorption sites [87, 100]. The adsorption capacity of ~300 mg/g was decreased below 30 mg/g when the concentration of phosphate was increased to 10 mM [59]. At the same phosphate concentration, the arsenic removal was decreased from near 100 to 10% on hydrous zirconium oxide [99]. Therefore, the use of polyoxometalate (POM) to oxidize As (III) was not recommendable, because the hydrolyzed product of POM was PO_4^{3-}, which would decrease arsenic adsorption [31].

Other anions, such as F^-, Br^-, were studied as the competitive anions. All of them decrease the arsenic adsorption capacity moderately [59, 89]. The effects of the organics are complicated due to the functional groups. The one with the functional groups that react with iron would affect arsenic adsorption substantially. Grafe et al. investigated the effects of peat humic acid (HA_p), River fulvic acid (FA), and citric acid (CA) on arsenic removal by goethite and Fh. As(III) adsorption decreased in the order of CA > FA > HA_p, whereas for As(V), the order was HA_p > FA > CA on the goethite. The interference of FA and HA_p was negligible on As(V) adsorption. And HA_p had no effect on As(III) adsorption [102, 103]. Hence, the COOH (CA) and phenol groups (HA_p) had different affinity on the goethite and Fh. Besides, HA was observed to form soluble complex with metallic cations, followed by the formation of aqueous ternary complex with arsenic by metal bridging mechanisms [104, 105].

The suppression of anions also depends on the adsorbents. According to the reports, the hydrated yttrium oxide was insensitive to the SO_4^{2-} in As(V) uptake, so it was with ferrihydrite [99, 106]. The presence of CO_3^{2-} and SO_4^{2-} had no effects on As(III) adsorption on the Fe–Cu binary oxide [107].

In addition, the influence of co-existing ions was found to be relevant to solution pH. The effects of both cations and anions increased with increasing pH [97]. As (III) adsorption was insensitive to NO_3^- at pH 3–4.5, while adsorption was decreased with increasing NO_3^- in the pH range 4.5–9.0. The Zn^{2+} facilitation did not happened when arsenic adsorption was occurred on the magnetite at pH 4.5–6 [108]. The reduction of As(V) by silicate was decreased with decreasing pH [109].

(iii) Temperature

The increase in the temperature would normally decreases arsenic adsorption. The increased temperature suppressed the adsorption affinity and damaged the stability of surface complexes. The effects of temperature on arsenic adsorption indicated that the adsorption was an exothermic process.

But the increment in temperature also increases the driving force and decreases the energy barrier in the adsorption, facilitating the adsorption. Therefore, the adsorption is efficient at higher temperature for some adsorbents, indicating that the adsorption was endothermic [110, 111]. The adsorption capacity on La magnetized carbon increased from 146 to 266 mg/g when temperature was increased from 289 to 309 K [110].

2.3 *Adsorption Mechanism*

2.3.1 Ligand Exchange

The basic mechanism in arsenic adsorption is the surface complexation. Specifically, the hydroxyl groups on the surface undergo a ligand exchange with arsenic ions to form mononuclear bidentate (^{2}E) and binuclear bidentate surface complexes (^{2}C), as well as the unstable monodentate one (^{1}V) [112, 113], shown in Fig. 5. The binuclear bidentate complexes results from AsO_4 tetrahedra bonding to adjacent apexes of edge-sharing pairs of FeO_6 octahedra. If the FeO_6 octahedra share the free edge with AsO_4, they would form the mononuclear bidentate complexes. And the monodentate complex connects AsO_4 and FeO_6 by free corner (goethite was used as the example), Fig. 6 [112, 114].

According to the reports, ^{2}C complexes were the major surface complexes on most of the adsorbents. ^{2}E and ^{2}C complexes were formed with a distance of 0.290 ± 0.005 and 0.335 ± 0.005 nm when adsorbing arsenite on Fh [112]. However, the ^{2}C complex was predominant when reacted with arsenate [115]. The ^{2}C complex occupied the dominant surfaces with a minor ^{1}V complex on the goethite when adsorbing arsenite, same as that on the lepidocrocite. It was demonstrated that the ^{2}C complex was less energetic as compared to ^{1}V complex [112, 115]. Thus, it was favored in the formation. With the increase of surface coverage, ^{2}E complex would replace ^{1}V complex. The complex on the hematite was the same as that on the Fh, including ^{2}C and ^{2}E complexes. The difference might cause by the different crystal faces that reacted with arsenite [112]. For arsenate, the ^{2}C complex was the main production on these ferric oxides [115, 116]. Mn(III)

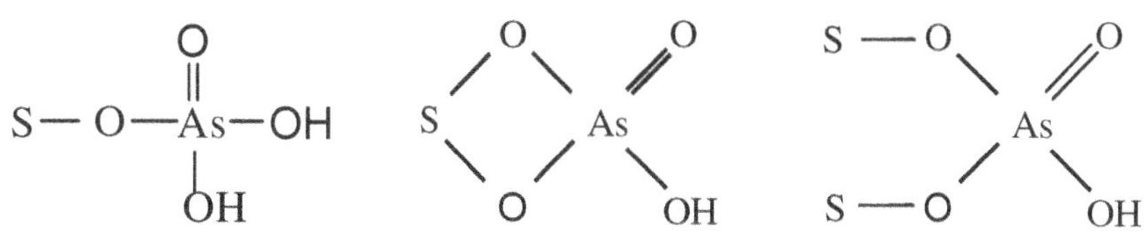

(a) Monodentate (b) Mononuclear bidentate (c) Binuclear bidentate

Fig. 5 Arsenate surface complexes

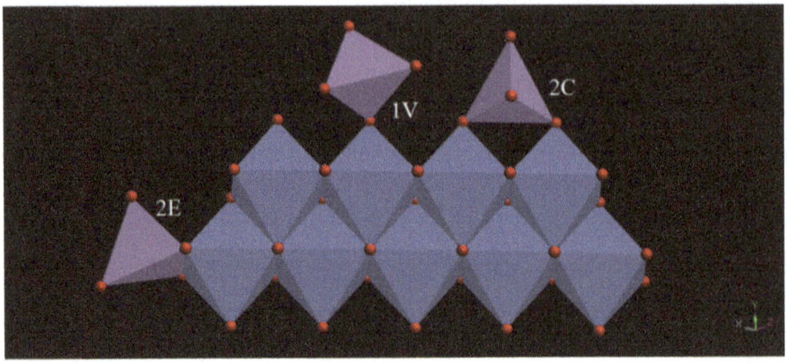

Fig. 6 Possible arsenate surface complexes on goethite

octahedra and arsenate tetrahedral formed both bidentate and monodentate mononuclear complexes on the manganese oxides [117].

Expect for three inner-sphere surface complexes, outer-sphere complex is found when arsenite reacts with some adsorbents. For example, in the case of the adsorption of arsenite on the siderite, only a weak out-surface complex was formed [118]. The inner-sphere complexes form through ligand exchange, while the outer-sphere complexes is through electrostatic interaction. The difference is shown in Fig. 7.

2.3.2 Surface Precipitation

The existence of Fe^{3+} is important for surface precipitation. The mechanism follows three steps: (1) fast adsorption of arsenic via surface complexation; (2) re-adsorption of Fe^{3+} on the surface complex; (3) arsenic adsorption on the cations again and finally slow buildup of a surface precipitate, shown in Fig. 8 [24].

The surface precipitation could be observed by XRD at the Fe/As ratio 2, pH 3, aging for two weeks. Jiang et al. added ferric ions after As(V) adsorption on the Fh. The added Fe^{3+} concentration had a good linear relation with the concentration of re-adsorbed arsenic, indicating that Fe^{3+} was adsorbed on the As(V)-Fh surface as a ternary complex, acting as new adsorption sites for arsenic. The long time reaction resulted in the surface precipitation of ferric arsenate. The high arsenic retention on Fh was considered to be contributed to the surface complexation and precipitation

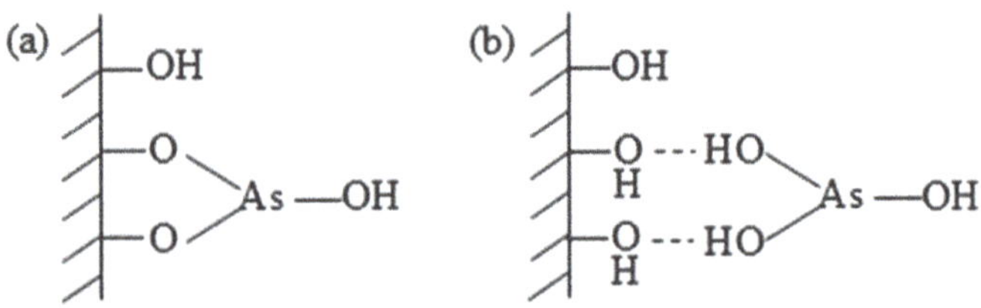

Fig. 7 **a** Inner-sphere complex; **b** outer-sphere complex

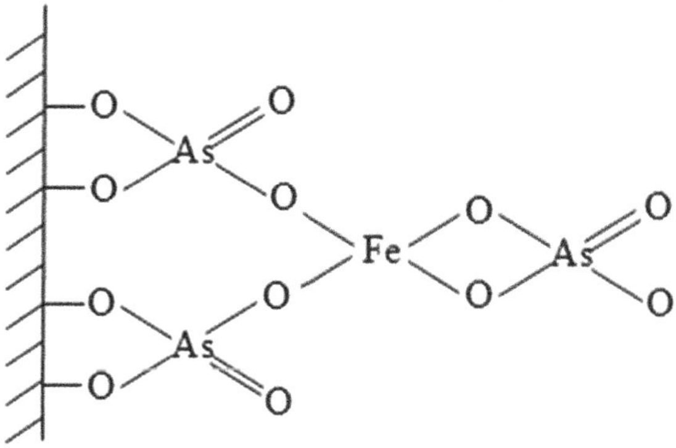

Fig. 8 Schematic diagram of surface complexation and surface precipitation

[24, 119]. The mechanism was explained as the surface complexation followed by the partial dissolution of Fh, followed by the formation of surface precipitate with adsorbed arsenate [119]. Precipitation was observed on the goethite during the adsorption without addition of ferric ions. The partial dissolution of adsorbents was contributed to the precipitation [92, 119, 120].

Similar phenomenon was observed when arsenic was adsorbed on birnessite. High concentration of arsenic contributed to the dissertation of birnessite, resulted in high arsenic adsorption. In addition, the precipitation of manganese arsenate was observed on Mn_3O_4 magnetic [121, 122]. In acid condition, lanthanum arsenate was formed on lanthanum compounds [123]. Zn was reported to form complex with Fe or arsenic on the surface, continuning in the adsorption of As. The As(V) uptake was increased by 500% in the presence of Zn at pH 7 [124].

2.3.3 Ion Exchange

Ion exchange is a process where ions on the adsorbents are replaced by other ions in the solution. It is the main mechanism for ion exchange resin. In this study, the resin is not included in the adsorbents. According to the reports, ion exchange occurred when clay and zeolite were used in the adsorption process. In addition, the exchange of CO_3 with arsenate was reported when lanthanum carbonate was used as an adsorbent [123]. However, when ion exchange was predominant in the adsorption, the interference of the anions was amplified. The adsorption of arsenate was suppressed on the modified bentonite by the addition of anions. In contrast, the adsorption of arsenite was hardly affected, because arsenite was uncharged at neural condition [125].

3 Phosphorus Adsorption

3.1 Contamination, Origin, Harm, Distribution

Phosphorus, with symbol P, is the most abundant pnictogen in the Earth's crust, which mainly exists as inorganic phosphate. The phosphorus in water exists as polyphosphate, organic phosphate and phosphate ions ($H_2PO_4^-$, HPO_4^{2-}, and PO_4^{3-}) with the respective deprotonation dissociation constants (pKa) of 1.68, 6.79 and 11.67, shown in Fig. 9. The polyphosphate and organic phosphate can be hydrolyzed into phosphate ions [124].

Different from arsenic, P is an essential nutrient for organisms, playing an important role in cellular metabolism. However, the excessive phosphorus in aquatic ecosystem would cause the eutrophication and algal blooms [126]. The overgrowth of aquatic plants and algae depletes the oxygen in water, which in turn, endangers the existence of fish and other aquatic life. It is considered that 0.02 mg/g dissolved phosphate could promote the profuse growth of algal [127]. United States Environmental Protection Agency (US EPA) suggests that total P should not exceed 0.05 mg P/L in a stream where it enters a lake or reservoir [128].

P is released into water by means of natural and anthropogenic activities, such as rock weathering, human fertilizer, laundry and mining activities. But recently, the inputs of P by anthropogenic activities far exceed the natural inputs and impair the water quality. As reported, over 40% water bodies in the world have eutrophication problems. Researches have shown that about 50% of all the lakes and reservoirs except Africa and Oceania in the world are eutrophic [129]. Situation is worse in China. According to the reports, 85.4% of the 138 investigated lakes are considered to be in a eutrophic status. Among them, 40.1% are heavily eutrophic [130].

The high occurrence of eutrophication has increased the pressure in water treatment. Because the ratio of P:N in the formation of the organic matter is 1:7, it's more effective to remove P from water to solve eutrophication. The water quality monitoring confirmed that phosphorus was the limiting nutrient factor for water eutrophication in Danjiangkou Reservoir located in China [131].

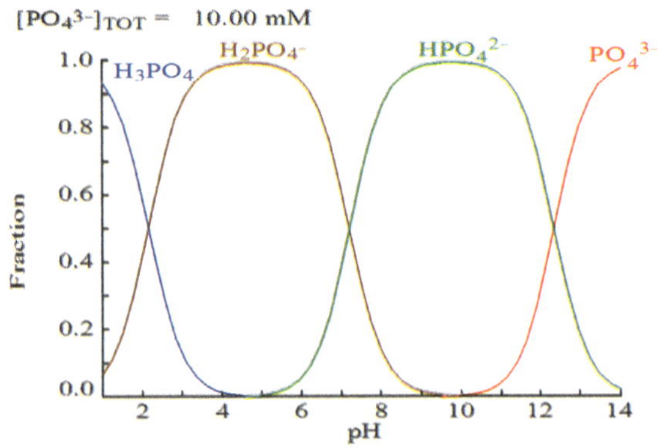

Fig. 9 Phosphorus species as a function of pH

3.2 Adsorption

Various technologies, including chemical precipitation (with aluminium, iron and calcium salts), biological treatment (bacteria, algae, plants), and adsorption, reverse osmosis and ion exchange have been developed to removal P from aquatic ecosystem. Among them, adsorption is a supplement to the biological treatment due to its simple operation, relatively fast removal and stable removal efficiency. In addition, it could be easily applied in small-scale treatment facilities.

Due to the agriculture studies in P retention for fertilizer, the good affinity between P and Al, Fe and Ca are well accepted. Hence, the minerals containing these elements have attracted attention among the abundant adsorbents. The adsorption on minerals, such as iron and aluminum oxides, is a major way in the mobility and transport of P from aqueous phases to solid phases. In addition, the process provides a higher removal rate of phosphate than biologically-based phosphate treatment.

3.2.1 Adsorbents

(i) Iron-based minerals

HFO is prepared by adding alkali into ferric ions. The fresh floc showed a high P retention due to adsorption/complexation reactions on the surface of HFO [132]. The adsorption capacity was decreased with the flock aging, as a result of transformation of HFO from amorphous to crystal [133]. The Freundlich isotherm model fitted well with the adsorption behavior of P on AFO [134].

Ferrihydrite (Fh, $Fe_5HO_8 \cdot 4H_2O$), hematite (α–Fe_2O_3) and goethite (α–FeOOH) were compared in P removal [135, 136]. The adsorption isotherms of P on three minerals agreed better with the Freundlich model than the Langmuir model, which might be contributed to the heterogeneity of surface sites. The Fh exhibited a much higher P adsorption as compared with other two adsorbents. The discrepancy was attributed to the high surface area and amorphous layer [135]. But goethite showed the highest affinity than other two minerals [136]. As reported, the morphology and crystalline properties of the goethite and hematite affected P adsorption. The adsorption capacity of P on the goethite with different morphologies was similar, being approximately 2.51 $\mu mol/m^2$. The phosphate bonded on goethite as the binuclear complex on the (110) face [137]. However, the effects were different on hematite. The adsorption capacity varied from 0.19 to 3.3 $\mu mol/m^2$, decreasing with the increased diameter/thickness ratio of the particle [138].

Iron oxyhydroxides, especially goethite, are abundant in the natural system and have been used in various adsorption studies. The goethite and akaganeite are compared in the adsorption of P [139]. Both the minerals had the maximum adsorption capacity of 10 mg/g at an equilibrium phosphate concentration of 0.3 mg P/L. In addition, the adsorption follows Freundlich isotherm B.

The adsorption capacity reached the maximum value of around 2.6 mg/g at pH 3.8 on hematite, which was decreased with increased pH. The hematite could move into bentonite's interlayer space when mixed with bentonite at pH 6.0. The adsorption capacity of the hematite-bentonite system was lower than that of hematite [140]. Magnetite is attractive for the magnetic properties, contributing to the easy separation from aqueous solution. Daou et al. found that the adsorption capacity of P on magnetite was 5.2 mg/g. The surface complex was a protonated binuclear species. Thetop layer would be in the (111) plane [141]. Yoon et al. reported a similar adsorption capacity of 5.03 mg/g. The uptake of P was decreased with increased pH [142].

(ii) Aluminium-based minerals

Bauxite, of which the major component is $A1(OH)_3$, is an important industrial material for alumina production. Meanwhile, the material shows P retention and is an economical adsorbent for P removal. It was reported that the adsorption mechanism was based on the ligand exchange. The bauxite had the highest P adsorption in slightly acidic conditions (pH 3.2–5.5) [143]. Then the material was treated by acid and heat. Acid treatment decreased the adsorption capacity. When treated by calcination, the adsorption capacity was increased with the increasing surface area, but the adsorption was not proportional to the increase of specific surface area [144].

Bauxite, pseudoboehmite (AlOOH) of low crystallinity on P removal were studied by Rajan et al. and the quantitative relationship between divalent phosphate adsorbed on and hydroxyl ions released from, the minerals at different P concentrations were reported. It was proposed the surface complex included monodentate, binuclear and bidentate [145, 146]. The surface complexes formed on the boehmite (γ–AlOOH) was considered as bidentate, analyzed by NMR spectroscopy [147].

Gibbsite was used in P removal and compared with hematite. The result showed both of the adsorbents followed the Freundlich model. The adsorption capacity of gibbisite was higher than that of hematite. EDTA was the most effective when used in the desorption of phosphate as compared with oxalate, hydroxyl, fluoride [148].

Alunite ($KAl_3(SO_4)_2(OH)_6$) is one of the minerals of the jarosite group. The containment of aluminium oxide makes it a candidate material for P removal. Studies have shown the removal of P increased when alunite is calcined [149, 150]. The decomposition starts at 873 K [149]:

$$KAl_3(SO_4)_2(OH)_6 \rightarrow 2\,KAl(SO_4)_2 + Al_2O_3 + 6H_2O \tag{7}$$

At 973 K, another reaction happens:

$$2KAl(SO_4)_2 \rightarrow K_2SO_4 + Al_2O_3 + 3SO_3 \tag{8}$$

The P uptake reached the maximum of 118 mg/g on alunite calcined at 1073 K. Therefore, the existence of Al_2O_3 was contributed to the adsorption. The adsorption followed first-order rate kinetics and Langmuir model.

Palygorskite is an aluminum–magnesium silicate with high surface area, good porosity and thermal resistance. The good physichemical performance makes it as a promising adsorbent and catalyst supports. This mineral was modified and studied in P adsorption. According to the reports, the palygorskite treated by acid and calcined at 320 °C had a higher surface area than the original ones, but with higher adsorption rate and capacity. The adsorption isotherm data fitted well with the Freundlich model [151].

(iii) Calcium-based minerals

Calcium adsorbents are of increasing interest in P removal. Natural apatite ($Ca_5(PO_4)_3OH$) is used as the filter media in filter-based systems and as bed media in constructed wetlands. The adsorption of P on apatite was reported [152, 153]. The adsorption behavior varied with hydraulic conditions and reaction time. In batch experiment with short reaction time, the complex adsorption was predominant. The Ca addition increased the adsorption rate but did not affect the adsorption capacity [153]. While in column experiment, the removal mechanism consists of adsorption and crystallization. The crystallization mechanism includes the nucleation of hydroxyapatite (HAP: $Ca_{10}(PO_4)_6OH_2$) and crystal growth. The presence of apatite lowers the activation energy barrier and the adsorption enhances the crystallization [154].

Calcite was another choice as a natural adsorbent in P removal. The removal was favorable in both acidic and strong basic conditions. The adsorption started with the dissolution of calcite, followed by the precipitation. Under acidic condition, the precipitation was variable brushite-type while it was it was hydroxyapatite (HAp) under alkaline condition [155].

Calcium peroxide (CaO_2) was synthesized on clinoptilolite in sewage treatment [157]. The dissolution of CaO_2 produced Ca^{2+} and H_2O_2, resulting in the precipitation of P. Meanwhile, H_2O_2 oxidized the organic pollutants in sewage and reduced the concentration of COD. The ammonium could be adsorbed on the clinoptilolite. Hence, the adsorbent could remove P, COD and N simultaneously. The pseudo-second-order kinetic model well-fitted the adsorption process of P and the adsorption capacity was 50.25 mg/g on the adsorbent.

The dolomite was calcinated at 650–750 °C and used as an adsorbent for P. The final products were the mixture of MgO and calcite. The adsorption capacity was 10 mg/g with a formation of calcium phosphate on the surface.

(iv) Manganese, titanium, zirconium and lanthanum based minerals

Similar to arsenic, phosphorus can also be adsorbed by other element-based minerals, such as manganese, titanium, zirconium, lanthanum. But the studies are less focused as compared with Fe-, Al-, Ca-based minerals. Yao et al. suggested that the phosphate adsorption on manganese dioxide through the outer-sphere complex [158]. The zirconium oxide (ZrO_2) had a high P retention than the traditional adsorbents, with an adsorption capacity of 99.01 mg/g and 91.05 mg/g on the amorphous and mesoporous ZrO_2, respectively [159, 160]. Xie et al. compared the removal of P by activated aluminum oxide with lanthanum oxide. The result showed that the lanthanum oxide was more advantageous. The influence of pH on the adsorption on lanthanum oxide was not significant until pH > 10.5. The adsorption capacity was 46.95 mg/g for lanthanum oxide while 20.88 mg/g for aluminum oxide. The desorption on lanthanum oxide could be performed by either acid or alkali [161]. The La–EDTA–Fe_3O_4 showed a faster adsorption rate than EDTA–Fe_3O_4, as well as a higher adsorption capacity. In addition, the adsorbent could be easily removed by permanent magnet [162].

(v) Clay, Zeolite and Sand

Clays are used in many applications for water treatment due to their high surface area and relative low cost. Many researches focused on P removal by clays. The Ca- and K-kaolinite and -montmorillonite were applied in P removal [163, 164]. The adsorption behavior could be described by the modified Langmuir model. The adsorption capacity of Ca-kaolinite was higher than that of K-kaolinite. The addition of citrate, bicarbonate at pH > 8 decreased P retention on clays as a result of the competition of active Al sites. However, the presence of acetate and amino acids increased P adsorption. The H-bonding of the amino groups decreased the interlamellar spacing and produced quasicrystals, which reduced the spillover of negative charge from the surface to the edges of the clays. Hence, the phosphate ion could approach to the Al site.

The bentonite consists mostly of calcium montmorillonite. The layer structure enables cations to intercalate inside. The raw bentonite had an adsorption capacity of 4.12 mg/g. After modification, the adsorption capacity on the adsorbents exceeded 10 mg/g. The adsorption rate of P on the adsorbents fitted pseudo-second-order kinetic models. Both the Langmuir and Freundlich models described the adsorption isotherm data well. The thermodynamic studies indicated the adsorption process was endothermic and spontaneous in nature.

Sepiolite is a hydrated magnesium silicate clay mineral with a fibrous chain structure. The Ca-rich sepiolite had the maximum phosphorus adsorption capacity of 32 mg/g which was estimated by Freundlich equation. The removal mechanism was contributed the calcium-bound phosphorus precipitation [130].

Zeolites, as the hydrated aluminosilicate minerals, is used in water treatment for the well-developed three-dimensional framework structure. The isomorphous replacement of Si^{4+} by Al^{3+} produces a negative charge in the lattice which is balanced by the cations (sodium, potassium, or calcium) [167]. Therefore, the material can be used in P removal.

The sand is a major media as subsurface flow reed beds in constructed wetlands, removing phosphorus by adsorption and precipitation [169, 170]. The adsorption capacity was correlated to the contents of Ca and Mg, grain size, porosity of the sands. The content of Ca was more important than that of Al and Fe, because the precipitation of calcium phosphates was favored under the pH of domestic sewage. The precipitation of iron and aluminium phosphates (strengite, variscite) were predominant at a relatively lower pH [171]. According to the reports, the quartz sand showed the lowest P retention. The Langmuir model fitted well the data when precipitation was absent.

(vi) Modified minerals

The physiochemical properties of adsorbents, such as surface area, surface charge, porosity, influence the adsorption performance. Activation by acid and heat treatment is carried out to achieve a higher P retention on the minerals. After calcination, the adsorbent exhibited a higher surface area and more flourish porosity than the original one, contributing to a higher P adsorption. For example, the surface area of the bauxite calcined at 600 °C was eight times higher after dehydration [144]. The surface area of alunite was increased from 29.2 to 123 m^2/g when temperature was increased from 373 to 1173 K [149]. Porous hematite could be prepared from goethite by calcination at 300 °C. The further increase in temperature led to the decrease in the hydroxyl groups and surface area of hematite, decreasing phosphorus adsorption [172]. The red mud and palygorskite had a higher phosphorus adsorption after acid treatment [151, 173]. As reported, the red mud removed about 99% phosphate with an initial concentration of 155 mg/L, presenting a stablely high sorption capacity over broader pH [173]. But the bauxite showed an opposition effect [144]. To increase the surface area, the metal oxides were coated on the activated carbon, clays. As reported, the nano-magnetite was immobilized onto granular activated carbon. Then phosphate bonded on the surface as the bidentate complexes. The adsorption capacity was about 400 mg P/g Fe. The adsorption first happened on the active sites on the magnetite surface, followed by the diffusion into the interior of the magnetite layer. then the outer binding sites was available again [174]. Considering the magnetic properties, magnetite was used as matrix and hydrous lanthanum oxide was loaded. The synthesized Fe–Si–La adsorbent had the uptake of 27.8 mg/g, and nearly 99% of phosphate could be removed within 10 min [175]. To increase the surface charge, the hydrated manganese oxide was encapsulated inside the basic anion exchanger. Then the pHpzc value increased from 6.2 to 10.5. the adsorption capacity of P increased about 10 times correspondingly [176] (Table 3).

3.2.2 Interference

(i) pH

Commonly, the removal of synthetic phosphate solution decreases when pH increases duo to the electronic repulsion [143, 151, 165], e.g., The higher pH of solution enables more negative charges on mineral surface, resulting in a stronger repulsion between the surface and charged ions. However, the P removal from the secondary effluent was maintained at approximately 80% in the pH range of 5–10 [135]. This was in close association with the interaction of phosphorus with Ca and Mg ions in the solution. With the increasing of pH, the apatite could be formed around pH 6.5–8. Then the chemical precipitation of dolomite ($CaMg(CO_3)_2$),

Table 3 Phosphorus adsorption among different minerals

Species	pH	C_0 (mg/L)	Adsorption capacity (mg/g)	References
Goethite	7.8	2	10	[139]
Akaganeite				
Ferryhydrite	7	0–7	19.4	[135]
Goethite				
Hematite			2.85	
			0.977	
Magnetite		2–20	5.03	[142]
Apatite	7.0	0–500	4.76	[153]
Quartz sand	~8.5	0–320	0.272	[169]
Darup sand			3.941	
Natural iron oxide coated sand	5	4.0	1.8	[177]
Calcined alunite	5	0–200	118	[149]
CaO_2@clinoptilolite	7	5–150	50.25	[157]
Bentonite	7	0.05–5	4.12	[165]
Zenith/Fe			11.15	
Al–Bentonite	3	25–60	12.7	[166]
Fe–Bentonite			11.2	
Fe–Al–Bentonite			10.5	
Sepiolite		5–1000	32	[130]
Zeolite		3–40	0.30	[168]
Natural palygorskite	7.0	5–150	4	[151]
Modified palygorskite			9	
Aluminum oxide		5–200	20.88	[161]
Lanthanum oxide		5–500	46.95	
Am–ZrO_2	6.2		99.01	[159]

calcite ($CaCO_3$), and brucite ($Mg(OH)_2$) lead to further P adsorption. The presence of Fe, Al would play positive effects when pH < 6 [171].

Based on the unique precipitation mechanism, the adsorption on the calcite in terms of pH was quite different from others. The removal efficiency increased when pH increased from 7.6 to 12 [156, 167].

(ii) Co-ions

The cations could promote the adsorption of P on the adsorbents by changing point of zero charge (PZC) on the surface. In addition, Ca^{2+} and Mg^{2+} could form ion pairs with phosphate at pH 4.5 and 4.0 respectively [158]. Then the complexes were adsorbed on the surface of adsorbents, enhancing the retention of phosphorus.

The existence of chloride, nitrate, sulfate showed insiginficantly negative effects on P adsorption. But hydrogen carbonate and fluorinion decreased the adsorption

[130]. The selectivity of anion adsorption onto palygorskite was in the order of chloride, nitrate, hydrogen carbonate, sulfate and phosphate [151]. Similar tendency was reported on goethite and akaganeite [139]. However, the influence of sulfate on calcite was different. It was found that the presence of sulfate enhanced the P retention by promoting the dissolution of calcite in both acidic and basic conditions. Only when the S:P ratio was greater than 6, the adsorption would be inhibited as a result of formation of calcium sulfate in acidic solution [155].

Arsenic ions are the most competitive anion for phosphate as they belong to the same group and have similar electronic structures. When arsenate and phosphate were added simultaneously, the adsorption was about equal on the goethite [178].

Similarly, the existence of organic matters (OM), such as humic acid, fulvic acid, and citric acid, significantly changes the surface properties and reactivity of the minerals. The changes further inhibit the P adsorption by site competition, electrostatic effects, and steric hindrance [179–181]. Yan et al. compared the phosphorus adsorption on ferrihydrite/goethite-humic acid with raw ferric oxides [180] and found that the complex with humic acid decreased the phosphorus retention. Thus, the immobilization of P on soils in humic-rich areas should consider the limited effects. However, the limitation was negative on goethite as reported by Borggaard et al. [179]. And the bentonite intercalated with cetyltrimethyl ammonium bromide (CTMAB) adsorbed organic compounds and phosphate from water simultaneously [165], with a removal of more than 98% at an initial P concentration of 20 mg/L, higher than that on Al-bentonite. The coexisting organic compounds had little impact on the adsorption.

(iii) Temperature

The adsorption of P increases with the increasing temperature, indicating that the adsorption reaction is of endothermic nature [177, 182]. The adsorption standard free energy (ΔG), the average standard enthalpy change (ΔH) and entropy change (ΔS) derived from the Van't Hoff thermodynamic equations provide information on the adsorption mechanism. The negative value of ΔG indicates a spontaneous adsorption processwhile a positive ΔH confirms an endothermic adsorption process. A positive ΔS suggests the increased randomness at the solid-solution interface during phosphate adsorption [126, 166].

3.3 *Adsorption Mechanism*

3.3.1 Ligand Exchange

This mechanism is significant when P is adsorbed on metal oxides/oxyhydroxides [147]. The adsorption mechanism in molecular level can be deduced from spectroscopic techniques and quantum mechanical calculation. The amount of OH^- released and phosphate adsorbed on the ferric oxides was determined as a linear correlation by Wang et al. [136]. It is well accepted that the ligand exchange is predominant in the adsorption process. And the inner sphere complex is formed at the sorbent surface. The reaction for phosphate ions on the hydroxylated mineral surface can be expressed as follows [183].

$$aSOH_{(s)} + H_bPO_{4(aq)}^{b-3} + CH_{(aq)}^{+} = S_aH_cPO_{4(s)} + bH_2O_{(l)} + (a-b)OH_{(aq)}^{-} \quad (9)$$

where, S is a metal ion in the hydroxylated mineral, OH^- is the surface hydroxyl, and b is the degree of protonation of P ions, which is ≤ 3. The formed surface complexes include three different species [145, 146].

The bidentate binuclear surface complex was identified on the surfaces of goethite and Fh at low pH, while the monodentate complex was formed at high pH and low phosphate concentration [184, 185]. Based on the adsorption mechanism, the desorption could be performed by the ligand exchange reaction between hydroxyl groups and adsorbed phosphates.

3.3.2 Surface Precipitation

Surface precipitation is based on surface complexation. The dissolved cations further react with phosphate and finally form the surface precipitate. It is the basic mechanism for Ca-based adsorbents. Calcite is a more complex mineral with a higher dissolution rate, leading to the hydrolysis and complex formation reactions [156]. Consequently, the adsorption of P was based on the hydrolysis effect. And the process could be described by the following reactions [186, 187]:

$$Ca^{2+} + H_2PO_4^- = CaH_2PO_4^+,\ pK = -1.08 \quad (10)$$

$$Ca^{2+} + HPO_4^{2-} = CaHPO_4,\ pK = -7.0 \quad (11)$$

$$Ca^{2+} + PO_4^{3-} = CaPO_4^-,\ pK = -6.5 \quad (12)$$

$$3Ca^{2+} + 2PO_4^{3-} = Ca_3(PO_4)_2,\ pK = -26 \quad (13)$$

$$2Ca^{2+} + HPO_4^{2-} + HCO_3^- = Ca_2HPO_4CO_3 + H^+,\ pK = 1.33 \quad (14)$$

$$Ca_2HPO_4CO_3 = Ca_2HPO_4CO_3^- + H^+,\ pK = 8.3 \quad (15)$$

$$10CaCO_3 + 6HPO_4^{2-} + 2H_2O = Ca_{10}(PO_4)_6(OH)_2 + 10HCO_3^-,\ pK = -32 \quad (16)$$

The hydrolysis of ferric oxides was also observed on P adsorption [188, 189]. After adsorption on Fh and goethite, the remaining P was 10.6 and 11.6%, which might be bonded with the ferric ions [188]. The combination of surface complexation and precipitation was observed in the adsorption on gibbsite and corundum conducted at low pH with high P concentration, as reported by Van Emmerik and Del Nero [190, 191].

4 Fluoride Adsorption

4.1 Fluoride Contamination

Management of fluoride (F^-) in drinking water is of great interest for public health. Fluoride can be beneficial or detrimental for both humans and animals depending on the concentration in drinking water and the total amount ingested. When fluorine in drinking water is within a permissible concentration range of 0.5–1.0 mg/L, it is commonly considered to have a beneficial effect on the maintenance of healthy bones and teeth, particularly for children [190, 191]. However, when the concentration of fluoride excesses the allowable value, it would interfere with DNA synthesis, carbohydrates, lipids, proteins, vitamins and mineral metabolism [185], and then causes various diseases such as fluorosis, osteoporosis, arthritis, brittle bones, cancer, infertility, brain damage, Alzheimer syndrome and thyroid disorder [186].

Given the effects of fluoride on human and animal health, the World Health Organization (WHO) stipulates a fluoride limit of 1.5 mg/L for drinking water. However, it is estimated that more than 200 million people worldwide are drinking groundwater with fluoride concentration higher than 1.5 mg/L [187]. Figure 10 shows the world distribution of major occurrence of high fluoride contents in groundwater above the WHO standard. It indicates that high-fluoride groundwater contamination is widely distributed worldwide. It is reported that the majority of serious cases occur in the developing countries. Some regions even have more than 30 mg/L of fluoride in groundwater [188].

Fluoride pollution of water may result from both natural geological reasons and human activities. In many minerals, such as fluorite, biotites, topaz, granite, basalt, syenite, the fluoride can be released by rainwater or volcanic eruption, etc., and thereby contaminating ground and surface water. On the other hand, fluoride contamination may occur due to the fluoride chemical industry, as well as the semiconductor, aluminium smelter, metal processing, electroplating, fertilizer and glass manufacturing industries [190, 189].

Due to the harm of fluoride to mankind, it is essential to reduce the fluoride concentration in water systems to a suitable level. Fluoride removal from water can be achieved by precipitation, adsorption, ion exchange, electrodialysis and membrane processes. In the precipitation of fluoride, calcium and aluminium salts are

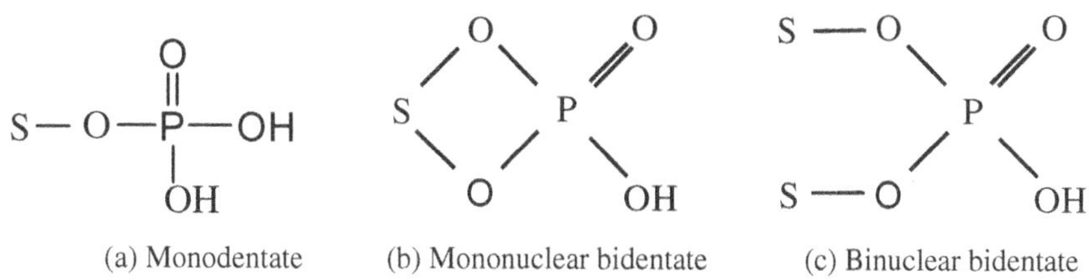

Fig. 10 Phosphorus surface complexes

commonly used to remove fluoride from industrial wastewater [191]. The Ca^{2+} ions released from calcium salts could interact with fluoride to reduce the concentration of fluoride down to 10–20 mg/L through the formation of CaF_2 precipitate. After that, aluminium salts are further used to reduce the concentration of fluoride to about 2 mg/L through the interaction between Al^{3+} and F^- through the formation of AlF_n^{3-n} and $Al(OH)_{3-m}F_m$, etc. The final concentration of fluoride in the water after precipitation largely depends on the solubility of fluorite precipitates. For example, the theoretical solubility of CaF_2 is 8 mg/L, therefore, the concentration of fluoride in water is still much higher than 2 mg/L even at a high dosage of calcium salts [192]. In addition, the associated problems such as generation of alkali water, residual calcium and aluminium ions, soluble fluoride complexes, generation of sludges are of major concerns [193]. Ion exchange, electrodialysis and membrane processes are effective and can remove the fluoride to a lower level, however they suffer from the high price and the frequent regeneration of resin beads or membrane and cleaning of the scaling and fouling [194].

Removal of fluoride via adsorption is regarded as one promising technology compared with other defluoridation methods in terms of its low cost, simplicity of design and operation. In addition, it can be applied over a wide pH range and reduce the concentration of fluoride to a lower value than precipitation. The efficiency of this technique mainly depends on the adsorbents used. Various adsorbents, such as activated carbon, activated alumina, charcoal, clay, zeolite, bleaching earth, have been investigated for the removal of fluoride from water [191, 195]. However, some of them would lose the fluoride adsorption capacity when the concentration of fluoride is below 2 mg/L and some of them can only work at an extreme pH value, for example, activated carbon is only effective at pH lower than 3.0. In recent years, considerable attention has been devoted to the mineral adsorbents due to their high efficiency and low cost.

4.2 *Adsorbents*

The common adsorbents for the removal of fluoride are activated alumina, activated carbon, ion exchange resins and fibres, minerals (clay, zeolite), etc. Among them, minerals feature at low cost, high stability, good adsorption capacity, etc. The effectiveness of fluoride adsorption from water using minerals as adsorbents will be presented.

4.2.1 Calcium-Based Minerals

Calcium-based minerals are important defluoridation adsorbents due to the good affinity of calcium for fluoride anion. Turner et al. used crushed limestone (99% pure calcite) as adsorbent to remove fluoride from solutions with a concentration of fluoride ranging from 3 to 2100 mg/L [196]. A high fluoride removal adsorption on limestone was attributed to the combination of surface adsorption and precipitation of CaF_2. 80.6% of the fluoride was removed using activated quick lime at an initial concentration of 50 mg/L fluoride [197]. Scanning electron microscopy (SEM) and X-ray diffraction (XRD) revealed that chemisorption and precipitation resulted in a high Langmuir maximum sorption capacity (16.67 mg/g) of activated quick lime. The maximum adsorption capacity of limestone and limestone impregnated by aluminium hydroxide was found to be 43.10 and 84.03 mg/g, respectively [198].

4.2.2 Iron-Based Minerals

Iron-based minerals have also been extensively studied as fluoride adsorbents due to the good affinity of iron towards fluoride. The Langmuir maximum adsorption capacity of schwertmannite for fluoride was found to be 50.2–55.3 mg/g at different temperatures [199]. The adsorption of fluoride on granular ferric hydroxide (GFH) through batch and mini-column scale experiments revealed an adsorption capacity of 34.2 mg/g for fluoride [199]. The adsorption mechanism was described as an exchange of F^- against OH^- in singly coordinated FeOH surface groups [200]. Goethite has also been widely studied as a fluoride adsorbent. A high adsorption capacity of goethite (59 mg/g) was found by Mohapatra [201].

4.2.3 Zeolites

The Na-zeolite had no PZC because it has negative surface charges at all pH values (3–13), and therefore it has high adsorption capacity for cations, but low capacity for anions, for example fluoride, due to the charge repulsion. Nonetheless, the adsorption capacity of zeolites for anions could be improved by modifying the zeolite surface with cationic surfactants or multivalent metallic cations [202]. The surface modification with metallic cations was commonly used for the adsorption of fluoride on zeolites.

The surface of zeolite was modified by exchanging the internal Na^+ with Al^{3+} and La^{3+} to create active sites for fluoride adsorption [203]. It was found that the introduction of Al and La largely increased the adsorption capacity of zeolite. The adsorption of fluoride on Al-zeolite was dominated by a chemical adsorption process (ligand exchange), while the adsorption on La-zeolite was mostly by a physical adsorption. Samatya et al. studied the removal of fluoride on La-, Al-, and Zr-incorporated zeolite, respectively. It was found that the highest and lowest adsorption capacities of fluoride at the equilibrium fluoride concentration range of 0–12 mg/L was on Zr-zeolite and La-zeolite. Even though, all the three metal-zeolites removed 95% of fluoride from the aqueous solution containing 2.5 mg/L of fluoride at an adsorbent dosage of 6 g/L.

4.2.4 Clays

Various clays such as montmorillonite, kaolinite, illite, and bauxite were also used as adsorbents to remove fluoride from water. The structure of gibbsite or aluminium oxides enables clays as potential fluoride adsorbents. The hydroxyl group on the surface of clay was also considered as one contributor in the adsorption of fluoride from water.

Moges et al. reported that the adsorption capacity of fluoride with fired clay chips as adsorbent was 0.2 mg/g [204]. Tor studied the defluoridation of water with montmorillonite and found that the saturation capacity for fluoride was 0.263 mg/g at room temperature [205]. The Langmuir adsorption capacity of fluoride on laterite was 0.8461 mg/g at 30 °C [206]. Clays are cost less materials, however they exhibit low fluoride adsorption capacity, therefore, modification might be necessary for the improvement of fluoride removal.

Chemical treatment with Na_2CO_3 and HCl for example and thermal treatment with less than 600 °C could improve the adsorption capacity of some clays according to the literature. Modification with positive atoms is a more effective method than the above ones. Thakre et al. found that bentonite modified with magnesium chloride showed higher fluoride removal capacity than the unmodified bentonite over a wide pH range of 3–10 [207]. The magnesium incorporated bentonite reached a maximum adsorption capacity of 2.26 mg/g at an initial fluoride concentration of 5 mg/L. The easy hydroxylation of bentonite in the presence of Mg in the interlayer and surface might be the reason for higher fluoride removal capacity because fluoride could replace those hydroxyl groups. Kamble et al. found that 10% La loaded bentonite showed a better fluoride removal property for drinking water than those of Mg or Mn loaded bentonite [208]. The Langmuir maximum adsorption capacity of 10% La-bentonite for fluoride was around 4.24 mg/g. Magnesium and aluminium modified attapulgite with a mass ratio of attapulgite: $MgCl_2 \cdot 6H_2O$:$AlCl_3 \cdot 2H_2O$ = 2:1:2 exhibited an extremely high fluoride adsorption capacity of 41.5 mg/g [209].

4.2.5 Layered Double Hydroxides (LDHs)

Layered double hydroxides (LDHs), which are also known as anionic hydrotalcite-like clays, are typical layered minerals consisting of positively charged brucite-like host layers and hydrated exchangeable anions located in the interlayer gallery for charge balance. The chemical composition of LDHs can be expressed by the general formula of $\left[M_{1-x}^{2+}M_x^{3+}(OH)_2\right]\left[A_{x/n}^{n-}\cdot mH_2O\right]$, where M^{2+} and M^{3+} represent divalent metal cations (Mg^{2+}, Ni^{2+}, Cu^{2+}, Zn^{2+}, etc.) and trivalent metal cations (Fe^{3+}, Al^{3+}, Cr^{3+}, La^{3+}, etc.), respectively; x is the molar ratio of M^{3+}/(M^{2+}+M^{3+}), A^{n-} is the intercalated anion (CO_3^{2-}, NO_3^-, Cl^-, etc.). The charge of LDHs results from the isomorphous substitution of a part of the divalent metal cations with trivalent ones. Charges can also be arisen from the ionisation of the

OH^- groups at the surface of LDH. The PZC of LDHs occurs at pH of 9–12, therefore LDHs carry positive charges at neutral and acid solutions, which would attract anions to its surface, fluoride as an example. In addition, the anions in the interlayers of LDHs can also be exchanged by fluoride ions, further improving the adsorption capacity of LDHs.

The adsorption capacity of fluoride largely depends on the metallic constitution in the LDHs structure. Lv found that Mg/Al LDH has the highest fluoride adsorption capacity compared with Ni/Al and Zn/Al LDH because of the lowest atomic weight of Mg [210]. In addition, the fluoride adsorption capacity was highest when the molar ratio of Mg/Al in Mg/Al LDH was 2.

The literature revealed that LDHs can have higher fluoride adsorption capacity (17–213 mg/g) if being calcined. The optimum temperature of calcination was between 450 and 500 °C [211, 212]. It was reported that the adsorption capacity of fluoride on a Mg/Al LDH increased from 65 to 70 mg/g and then to 80 mg/g when the calcination temperature was increased from 200 to 400 °C and then to 500 °C, respectively. However, the removal capacity decreased to 62 and 50 mg/g if the temperature was further increased to 600 and 800 °C, respectively [210]. The increased fluoride adsorption on calcinated LDHs might be resulted from the higher specific surface area and porosity, as well as the hihger adsorption reactivity of LDHs produced by calcination [211]. The decreased fluoride adsorption capacity at calcining temperatures higher than 500 °C was considered due to the fact that LDHs was transformed into a spinel structure, which could not effectively adsorb fluoride [210].

The adsorption property of some representative mineral adsorbents is given in Table 4.

Table 4 Characteristics of adsorptive removal of fluoride from water by mineral adsorbents

Species	pH	C_0 (mg/L)	Adsorption capacity (mg/g)	References
Calcite	6.0	2.5×10^{-5} to 6.34×10^{-2}	0.39	[213]
Hydroxyapatite	6.0	2.5×10^{-5} to 6.3×10^{-2}	4.54	[213]
Zeolite F-9		10–80	28–41	[203]
Calcium chloride modified natural zeolite	6.0	25–100	1.766	[214]
Acid activated kaolinite clay	Low pH	3.0	0.045	[215]
Chemically modified bentonite clay	7.0		4.24	[208]
Calcined Mg–Al–CO_3 LDHs	6.0	50	213.2	[210]
Calcined Zn/Al LDHs	6	2–60	17	[216]

4.3 Adsorption Behaviour

Fluoride adsorption kinetics studies have shown that the removal rate is high in the initial 5–120 min during which more than 80% of fluoride could be adsorbed. After that, the adsorption rate significantly decreases and eventually approaches to zero when the adsorption equilibrium is reached. The fast adsorption is attributed to the fact that the adsorption sites are vacant and the concentration gradient difference of fluoride in solution and on the surface of adsorbent. The subsequent decrease of the adsorption rate mainly results from the decrease of vacant sites on the adsorbents and the decreased concentration of fluoride in the solution.

The adsorption rate of fluoride on adsorbents largely depends on the amount of adsorbents used in the adsorption process, the initial concentration of fluoride, the structural property of the adsorbents and the interaction between F and the adsorption sites on adsorbents. The adsorption proceeds at a faster rate when increasing the amount of adsorbents and decreasing the initial concentration of fluoride [217]. The relationship between the adsorption rate and the F-adsorbent interaction will be discussed in detail in the adsorption mechanism.

The adsorption of fluoride on mineral adsorbents normally includes three essential steps: (i) diffusion or transport of fluoride ions to the external surface of the adsorbent from bulk solution; (ii) adsorption of fluoride ions on the surfaces of the adsorbent; (iii) the adsorbed fluoride ions might be exchanged with the structural elements inside the adsorbent particles, or being transferred into the internal surfaces of porous adsorbents [213]. The transport of the adsorbate to the external surface of adsorbent is by film diffusion, which is fast, while the internal transport of the adsorbate to the internal surfaces of adsorbent through a system of pores is by slow Knudsen and surface diffusions [218]. When the active sites primarily locate in the internal pores of the adsorbents, the adsorption time is mainly determined by the internal diffusion. It has been found that the internal diffusion rate is directly related to the square root of adsorption time. Such a relationship was commonly obtained in previous research for the adsorption of fluoride on granular ferric hydroxide [200], activated alumina [219] and manganese oxide-coated alumina [220]. In the case of that the adsorption sites expose outside, fluoride could be attracted on the surface of adsorbents quickly with the help of film diffusion.

4.4 Affecting Factors

The removal of fluoride from water with minerals as adsorbents can be influenced by several factors. Among these affecting factors, pH, co-existing ions, temperature, adsorption time are considered to be the main ones, which will be introduced below.

4.4.1 pH

Although the adsorption capacity of fluoride is determined by the adsorbent characteristics, pH is an important influential factor. The effect of pH on the adsorption of fluoride could be explained form the views of the surface charge of adsorbent and the existing form of fluoride.

When the pH of solution is below PZC, the surface of the adsorbent is positively charged, therefore fluoride existing as F^- could be well adsorbed. While, when the pH is above PZC, the surface of the adsorbent is negatively charged, which would repel the fluoride. In this case, fluoride is not easy to be attracted [221]. In addition to the increased number of positive surface charges at pH lower than PZC, surface protonation would occur at low pH, which provides an increased number of H atoms on the surface of the adsorbent. The H-bonding between the H atoms on the adsorbent surface and F in solution increases F adsorption. Therefore, pH below PZC is beneficial F^- adsorption. However, F^- forms weakly ionised neutral HF at pH lower than 4, which reduced the removal of fluoride from water [222]. In some situations, F may present in the form of positively charged AlF complexes at low pH, leading to the reduction of fluoride adsorption [223]. Therefore, the highest adsorption capacity of fluoride is commonly found at a neutral pH. F removal is generally lowest at extremely low and high pH values. For example, the adsorption of fluoride on Zn/Al oxide increased with the increase of pH, reaching the maximum at pH 6.0, and then decreased with further increase of pH [216]. The maximum removal capacity of Fe and Al oxides for fluoride was obtained at pH of 6–8 [219]. Both montmorillonite and Mg–Al–CO_3–LDH reached the maximum adsorption of fluoride at pH of 6 [210]. Hence, prior pH adjustment is not normally required for the effective removal of F in treatment plants.

4.4.2 Co-existing Anions

Several anions including PO_4^{3-}, Cl^-, SO_4^{2-}, BrO_3^-, NO_3^-, and CO_3^{2-} are simultaneously present with F^- in natural water. The competitive extent of these anions depends on their relative concentration and affinity to the adsorbent. When fluoride is non-specifically adsorbed on the adsorbent, other anions could be adsorbed on the adsorption sites, reducing the removal of fluoride. When fluoride is specifically adsorbed on the adsorbent, non-specifically adsorbing anions do not compete with F^- for adsorption, only anions adsorbed by specific adsorption that will compete with F^- for adsorption.

Granular ferric hydroxide is known for having specific adsorption on F^-. Kumar et al. investigated its adsorption to F^- in the presence of competing anions Cl^-, SO_4^{2-}, BrO_3^-, NO_3^-, CO_3^{2-} and PO_4^{3-} [200]. The initial concentration of F^- is 20 mg/L, while that of the competing anions varies from 20 to 100 mg/L. It was found that no significant influence of competing anions occurred on F^- removal when the adsorbent dose was 10 g/L, which was attributed to the sufficient sorption sites. However, the adsorption capacity of F^- decreased to 65%, 75%, and 80% in the presence of 100 mg/L PO_4^{3-}, CO_3^{2-}, and SO_4^{2-}, respectively when the adsorbent concentration was reduced to 5 g/L. Das et al. also found that SO_4^{2-} and PO_4^{3-} had profound effects on the removal of fluoride due to their affinity towards Zn/Al oxide [216]. The other anions (Cl^-, BrO_3^-, NO_3^-) did not significantly reduce the F^- adsorption capacity. Raichur and Basu also found that F^- adsorption by a mixture of natural rare earth oxides was not significantly affected by the presence of NO_3^- in water at a concentration equal to that of F^- (100 mg/L) [222]. F^- adsorption was reduced from 85% to 40% and 62%, respectively, at 50 mg/L SO_4^{2-} and NO_3^- from a solution containing 20 mg/L F^-. In contrast, F^- adsorption was reduced to 20% at only 20 mg/L PO_4^{3-} and selenite [224]. It could be obtained from the literature review that phosphate, selenate, and arsenate strongly competed with fluoride for adsorption on adsorbents that specifically adsorbed fluoride, while nitrate and chloride usually did not compete with fluoride.

4.4.3 Temperature

The effect of temperature on the adsorption of fluoride depends on the nature of the adsorption. If the adsorption is endothermic, the removal capacity of fluoride increases with increasing temperature. In contrast, if it is exothermic, the capacity decreases with the increase of temperature.

The adsorption of fluoride on many adsorbents, for example granular ferric hydroxide [200], calcined Mg/Al/CO_3 LDH [210], increased with increasing temperature. However, the significant increase usually occurs in the temperature range of 15–45 °C. It was reported that the adsorption was low at extremely low temperatures (5, 10 °C) because of the slow movement of fluoride to the adsorption sites, while the adsorption of fluoride increased significantly when the temperature increased from 5 to 25 °C, but increased slightly when the temperature rose to 50 °C [221].

The adsorption of fluoride on several adsorbents (calcined Zn/Al LDH [216], geomaterials [225], Zn/Al oxide [216]) decreased with increasing temperature because the tendency for fluoride to escape from these adsorbents increased when the temperature rose. It has also been reported that temperature has no significant effect on the adsorption of fluoride on some other adsorbents, for example zeolite [203].

4.5 Adsorption Mechanism

The adsorption behaviour (capacity, kinetics, etc.) of fluoride on minerals are dominated by the adsorption mechanism. It can optimize the adsorption process and give useful information on the modification of the adsorbents by a full understanding on the adsorption mechanism of fluoride on minerals with adsorbents. The mechanism can be classified into three types: physical adsorption, chemical adsorption, and combined physical and chemical adsorption.

Physical adsorption, mainly including van der Waals forces and electrostatic force, is a non-specific adsorption. Therefore, fluoride cannot be easily adsorbed on the minerals in the presence of other anions in water. Van der Waals forces are weak short range forces acting between the mineral adsorbent and fluoride. The larger adsorbate size, the greater attraction force. Therefore, the Van der Waals forces of mineral adsorbents to fluoride are weak, leading to a low adsorption capacity of fluoride. Fluoride was considered to be adsorbed on manganese oxide coated alumina by van der Waals forces [220]. When the adsorbents are positively charged, the anions can be attracted through electrostatic force. This type of physical adsorption was considered stronger than that of van der Waals forces. Although the adsorbents have low adsorption capacity when the adsorption is governed by physical attraction, fluoride can desorb easily from the adsorbents and the adsorbents can be easily regenerated and recycled.

Hydrogen bonding (H-bonding) and coordination are principal chemical adsorptions, which are specific to fluoride. Due to the specific attraction, fluoride can be selectively removed by the adsorbents from water even in the presence of other anions except for those competitive anions which can also be specifically adsorbed on the adsorbents. H-bonding is a strong dipole–dipole attractive force between the electropositive H atom on the surface of adsorbent and an electronegative fluorine atom [226]. The adsorption energy of H-bonding is stronger than that of van der Waals forces and electrostatic force but weaker than that of coordination. H-bonding usually occurs in the adsorption of fluoride on coal-based adsorbents [227]. In a coordination mechanism, the anion F^- forms a strong covalent chemical bond with the metallic cation on the surface of adsorbent. Adsorbents have high adsorption capacity and high selectivity for F^- with this mechanism. Therefore, these adsorbents can efficiently remove large proportions of F^- with a higher selectivity at a very low initial concentration even in the presence of higher concentrations of competing anions with lower selectivity [228]. It can be concluded that coordination is the main mechanism of fluoride adsorption on natural adsorbents with high F^- adsorption capacity, while H-bonding is also a great contributor when the surface of natural adsorbents is modified by organic molecules.

5 Nitrate Adsorption

5.1 Contamination, Origin, Harm and Treatment Technology of Nitrate

The pollution of ground and surface water due to nitrates has become a common concern of industrial and developing countries, since these inorganic anions are toxic and harmful to humans and animals at very low concentrations (ppb). Keeping with the view that serious health problems are associated with excess nitrate concentrations in drinking water, World Health Organization (WHO) has stipulated the drinking water standard for nitrate at 50 mg/L. However, many sources of water, especially in the areas of intensive agriculture, contain intolerable levels of nitrate ion. Some industrial sectors, e.g. the nuclear industry, also produce large amounts of nitrate wastes [229]. In addition, nitrate is a by-product of many industrial processes, such as paper and munitions manufacturing which will generate greater environmental risks [230]. Therefore, in order to protect the water ecological environment and maintain the healthy development of humans, removal of the nitrate in water has theoretical and practical significances [231–233]. Based on this, the nitrate (NO_3^-) removal in waste water has attracted a large amount of concerns on a global scale.

Nitrate removal can be carried out by membrane separation [234], ion exchange [235], adsorption process [236] and chemical treatment [237], etc. Removal of NO_3^- by ion exchange method is feasible from an economic standpoint, but it requires longer period. Membrane separation techniques are cost-prohibitive and can yield concentrated brines. Biological denitrification can selectively remove NO_3^-, but this process is complex and requires monitoring of a carbon source. Recent researches have also demonstrated that an electrochemical reduction can offer an attractive method to treat the nitrate waste water [238, 239]. However, this process is limited in practice due to the formation of by-products of nitrite and ammonia [240]. At present, most NO_3^- removal processes have various limitations which particularly are not suitable for small communities. In contrast, adsorption is a widely accepted technology for removing both organic and inorganic contaminants [241], especially for nitrate removal. Compared to other methods, adsorption is a simple, effective and economical method for NO_3^- removal.

5.2 *Mineral Adsorbents*

A large amount of adsorbents for NO_3^- removal can be found in the market. Among of these adsorbents, mineral-based adsorbents are versatile which are widely used in the removal of various pollutants including organic matters and inorganic ions in aqueous solutions [246\2, 243]. In other words, the mineral-based adsorbents can remove not only NO_3^- ions but also heavy metal ions. Recently, various forms of mineral-based adsorbent have been investigated to improve the adsorption efficacy [244]. The removal capacity and mechanisms of NO_3^- are shown in Table 4.

5.2.1 Zeolites

Natural zeolites are porous aluminosilicate minerals with excellent physicochemical properties. Besides their natural forms, zeolite-based materials have showed enormous potential as mineral adsorbents in the separation and purification processes of nitrate ions. Zeolite protonated with HCl, which was coated on the surface of chitosan, had a comparable capacity to other weak anion exchangers with an exchange capacity of 45.87 mg/g for NO_3^- [245]. Zeolite modified with hexadecyltrimethyl ammonium bromide (HDTMABr) surfactant (SMZ) was also used to deal with nitrate ions [246]. It was quick (less than 1 h) and efficient to remove the NO_3^- in batches using the as-prepared SMZ in aqueous solutions, and the removal rate was more than 80%. In this case, the liquid/solid weight ratios were less than 10 and NO_3^- initial concentrations were ranged from 5 to 150 mg/L. In addition, the presence of competing anions (chloride, sulphate and bicarbonate) at equivalent concentrations showed negligible effects on the nitrate removal rate. Therefore, the zeolites coated with different surfactants showed a considerate adsorption capacity of nitrate and were potential to deal with nitrate contaminations in the surface or ground water.

5.2.2 Clays

Clays are known as the hydrous aluminosilicates, including bentonite, sepiolite, halloysite, kaolinite, etc. Clays and clay-based materials were often used as adsorbents for nitrate ions removal.

It was found that untreated bentonite and kaolinite could not remove NO_3^- while halloysite showed a NO_3^- removal capacity of 0.54 mg/g [247]. Considering that the untreated clays showed a low adsorption capacity for NO_3^-. These clays were modified with non-functional surfactants, and then different kinds of organoclays were obtained. Among all of these organoclays, bentonite modified using hexadecyltrimethylammonium bromide (HDTMA) showed the highest adsorption capacity. The CEC of modified bentonite showed an exchange capacity of 66.67 meq/100 g), much higher than that of kaolinite and halloysite (9.78 and 10 meq/100 g, respectively). In addition, sepiolite also showed good adsorption properties for NO_3^- due to its channel structure, high surface area and high sorption capacities. Sepiolite and surfactant-modified sepiolite were also used to remove NO_3^- in aqueous solution [248]. The adsorption capacities of natural sepiolite and sepiolite modified with dodecylethyldimethylammonium calculated by the pseudo-second-order model were 25.32 and 28.11 mg/g, respectively.

Not only surfactants but also acids could modify clays used for NO_3^- removal. Experimental data showed that the adsorption rate of sepiolite activated by hydrochloric acid (HCl) was very high especially at early stage of treatment [249]. The calcium bentonite has also been modified by acid thermoactivation with HCl and sulphuric acid (H_2SO_4) to deal with NO_3^- in the waste water [250]. It was found that calcium bentonite activated by HCl showed a higher NO_3^- removal capacity than that of H_2SO_4. The removal capacity could reach up to 22.28% due to the ionic exchange between chlorine and NO_3^- ions (hydrochloric acid treatment). These results demonstrated that clays and clay-based materials could be as effective adsorbents for NO_3^- removal.

5.2.3 Layered Double Hydroxides (LDHs)

Hydrotalcite-like compounds, layered double hydroxides (LDHs), showed desirable properties to remove nitrate from water. A laboratory study was conducted to investigate the ability of Zn–Al–Cl layered double hydroxide for NO_3^- removal in synthetic nitrate solution [251]. The removal efficiency of NO_3^- was found to be 85.5% under neutral conditions when the LDH and NO_3^- solution concentrations were 0.03 g/L and 10 mg/L, respectively. The Langmuir adsorption capacity was calculated to be 40.26 mg/g. LDHs with several interlayer metal ions (Mg–Al, Co–Fe, Ni–Fe, and Mg–Fe) were prepared and used in the removal of NO_3^-, and their anion exchange properties were studied by measurements of distribution coefficient (K_d) and ion exchange capacity [252]. It was found that The K_d of NO_3^- ions was a function of the basal spacing of the LDHs samples, while the basal spacing of LDHs varied depending on the interlayer metal ions. A relatively high K_d values for NO_3^- ions was observed on Ni–Fe type LDHs. The uptake capacities of NO_3^- in solution were 0.024, 0.062, 7.44, 10.42, and 5.02 mg/g for Mg–Al, Co–Fe, Ni–Fe, Ni–Fe (Hydrothermally-treated) and Mg–Fe, respectively. In other words, the higher K_d, the higher adsorption capacity of NO_3^-.

In addition, the calcined temperature of hydrotalcite-type compounds showed an influence on the removal of NO_3^- ions. The hydrotalcite-type compounds were calcined at 550 °C (HT550), 650 °C (HT650), and 850 °C (HT850), respectively, and then the sorption of NO_3^- on these materials was conducted [253]. The removal efficiency of these calcined materials was 70.5% for HT550, 85.3% for HT650 and 99.5% for HT850, respectively (at 25 °C). Thus, the increasing of calcined temperature of the hydrotalcite (from 550 to 850 °C) improved its adsorption capacity of NO_3^- ions. It should be noticed that the loss of CO_2 and formation of amorphous $Mg_{1-x}Al_xO_{(1+x/2)}$ mixed oxide would be taken place during the calcination, which would bring the change of hydrotalcite structure. When these materials were put into aqueous solution, this mixed oxide tended to recover its original structure and re-hydrate, leading to the incorporation of NO_3^- in the interlayer. As the calcination temperature increased to 850°C, a greater loss of CO_2 was generated, thus increasing the adsorption capacity of NO_3^-. The adsorption of NO_3^- using M^{II}–Al–CO_3 (Mg–Al and Zn–Al) was also investigated [254]. The nature and content of divalent cations in LDHs had strong influence on the adsorption process. Calcined Mg–Al LDH with an Mg/Al molar ratio of 3.0 showed higher adsorption capacity compared to other calcined LDHs. The removal of NO_3^- was found to be taken place via an adsorption process followed by a reconstruction of the calcined materials.

These mineral or mineral-based adsorbents showed considerate removal amount for nitrate, which were given in Table 5.

Table 5 The adsorption capacity and mechanisms of nitrate using mineral adsorbents

Adsorbents	Modification	Maximum adsorption capacity (mg/g)	Mechanisms	References
Zeolite	Cetylpyridinium bromide (CPB)	9.36	Anion exchange and electrostatic interaction	[255]
Natural zeolite	Chitosan	45.87	Anion exchange	[245]
Mg/Al chloride hydrotalcite-like compound	Thermally activated	40.26	Anion exchange	[250]
Sepiolite	HCl	Original sample: 3.46; Modified sample: 9.8	Anion exchange and electrostatic interaction	[256]
Calcium bentonite	HCl; H_2SO_4	–	Ionic exchange	[250]
MII–Al–CO_3 Layered Double Hydroxide	–	Mg3P10-500: 35; Zn3P10-500: 20	Anion exchange	[254]
Red mud	HCl	Original sample: 115; Modified sample: 363	Ligand exchange	[257]
Layered double hydroxides (LDHs) with different types of metal ions	–	Mg–Al type: 0.025; Co–Fe type: <0.062; Ni–Fe type: 7.44; Mg–Fe type: 5.02	Ion exchange	[252]

5.3 *Interference*

Some parameters, such as pH, competitive anions and temperature were regarded as the most important factors in the adsorption process of nitrate.

5.3.1 pH

The pH of the aqueous solutions is an important variable, which controls the adsorption between the adsorbent and aqueous interface. The pH for NO_3^- removal was usually in acid solution ranging from 2.0 to 7.0 [237, 251, 258]. For instance, sepiolite was modified with nonfunctional surfactant (dodecylethyldimethylammonium (DEDMA) bromide) in order to increase the positive surface charge. After modification, the maximum NO_3^- adsorption capacity occurred at pH 2.0 [248]. In another case, the removal of nitrate fluctuated very little in the pH range of 2–7 and decreased at a pH above 7 due to stronger competition with hydroxide ions on the surface of red mud [257]. Consequently, lower pH could result in the higher NO_3^- adsorption capacity.

5.3.2 Competitive Anions

The existence of competitive anions (carbonate, phosphate, chloride and sulfate) was usually considered to inhibit the adsorption of NO_3^-. Since NO_3^- is one of the most important ions for plant growth, its concentration is variable in the soil solution. The function of SO_4^{2-} as a regulator showed a very significant effect on the adsorption of NO_3^-. The interaction between NO_3^- and SO_4^{2-} for the adsorption could be interpreted on the basis of the structure of the Four-plane model shown in the [259]. The decrease in NO_3^- adsorption caused by the increase in SO_4^{2-} concentration was expected because of the competitive relation between NO_3^- and SO_4^{2-} [260]. When the Zn–Al–Cl LDH was used as an adsorbent for NO_3^-, the adsorption was inhibited by competitive anions and the inhibiting effect of these competitive anions followed the order of carbonate > phosphate > chloride > sulphate [251]. Therefor, the presence of competitive ions considerably lowered the removal capacity of NO_3^-.

5.3.3 Temperature

In most studies, the adsorption of nitrate on the surface of minerals possessed positive enthalpy change (ΔH), suggesting that it was an exothermic processes. This was because that the adsorption of nitrate occurred on the surface of minerals and modified minerals were usually dominated by the ions exchange process, which was believed as a physical process. For instance, with regard to the removal of NO_3^- using zeolite modified with cetylpyridinium bromide (CPB) as the adsorbent at 15, 25 and 35 °C, the enthalpy change of adsorption was calculated to be −13.7 kJ/mol, indicating that the adsorption occurred at 15 °C was more favourable than that in other two circumstances [255]. Thereby, temperature and adsorption capacity was usually negatively correlated.

5.4 Mechanism

The mechanisms of NO_3^- adsorption on the minerals were mainly reported to be ion exchange between the nitrate and other anions in solution, which could be concluded in Table 5. For instance, the anionic exchange between nitrate and counterion bromide on the surface of zeolite could be expressed as follows:

$$\text{Zeolite}-\text{CP}-\text{CP}-\text{Br}+NO_3^- = \text{Zeolite}-\text{CP}-\text{CP}-NO_3+\text{Br}^- \quad (17)$$

where, CP is the abbreviation of the cetylpyridinium cations.

In another case, the interaction between nitrate ion and metal oxide was modelled by ligand exchange reactions (Eqs. 18 and 19) [257]. At the beginning, hydroxylated surfaces of these oxides were protonated in a humid environment, and then the exchange between negatively charged NO_3^- and H_2O occurred on the surface of the adsorbents.

$$> MOH + H^+ \longleftrightarrow > MOH_2^+ \quad (18)$$

$$> MOH_2^+ + NO_3^- \longleftrightarrow > MNO_3 + H_2O \quad (19)$$

where, M presents metal ions (Al, Fe or Si).

References

1. Sarquis, M.: Arsenic and old myths. RI Med. **77**, 233–234 (1994)
2. Mandal, B.K., Suzuki, K.T.: Arsenic round the world: a review. Talanta **58**, 201–235 (2002)
3. Smedley, P.L., Kinniburgh, D.G.: A review of the source, behaviour and distribution of arsenic in natural waters. Appl. Geochem. **17**, 517–568 (2002)
4. Mohan, D., Pittman, C.U.: Arsenic removal from water/wastewater using adsorbents—a critical review. J. Hazard. Mater. **142**, 1–53 (2007)
5. Lafferty, B.J., Kramer, T. Methyl arsenic adsorption and desorption behavior on iron oxides methyl arsenic adsorption and desorption behavior on iron oxides (2004)
6. Rahman, M.A., Hassler, C.: Is arsenic biotransformation a detoxification mechanism for microorganisms? Aquat. Toxicol. **146**, 212–219 (2014)
7. Tangahu, B.V., Sheikh Abdullah, S.R., Basri, H., Idris, M., Anuar, N., Mukhlisin, M.: A review on heavy metals (As, Pb, and Hg) uptake by plants through phytoremediation. Int. J. Chem. Eng. (2011)
8. Huang, J.H., Matzner, E.: Dynamics of organic and inorganic arsenic in the solution phase of an acidic fen in Germany. Geochim. Cosmochim. Acta **70**, 2023–2033 (2006)
9. Cullen, W.R., Reimer, K.J.: Arsenic speciation in the environment. Chem. Rev. **89**, 713–764 (1989)
10. Wen, W., Wen, J., Lu, L., Liu, H., Yang, J., Cheng, H., Che, W., Li, L., Zhang, G.: Metabolites of arsenic and increased DNA damage of p53 gene in arsenic plant workers. Toxicol. Appl. Pharm. **254**, 41–47 (2011)

11. Singh, R., Singh, S., Parihar, P., Singh, V.P., Prasad, S.M.: Arsenic contamination, consequences and remediation techniques: a review. Ecotox. Environ. Safe **112**, 247–270 (2015)
12. WHO: Guidelines for Drinking Water Quality, 4th edn. World Health Organisation, Geneva, Switzerland (2011)
13. Chakraborti, D., Das, B., Rahman, M.M., Chowdhury, U.K., Biswas, B., Goswami, A.B., Nayak, B., Pal, A., Sengupta, M.K., Ahamed, S., Hossain, A., Basu, G., Roychowdhury, T., Das, D.: Status of groundwater arsenic contamination in the state of West Bengal, India: a 20-year study report. Mol. Nutr. Food Res. Int. **53**, 542–551 (2009)
14. Ungureanu, G., Santos, S., Boaventura, R., Botelho, C.: Arsenic and antimony in water and wastewater: overview of removal techniques with special reference to latest advances in adsorption. J. Environ. Manag. **151**, 326–342 (2015)
15. Smedley, P.L., Zhang, M., Zhang, G., Luo, Z.: Mobilisation of arsenic and other trace elements in fluviolacustrine aquifers of the Huhhot Basin, Inner Mongolia. Appl. Geochem. **18**, 1453–1477 (2003)
16. Sa, D., Go, L., Gira, I., Velasco, A., Morales, E.: Arsenic speciation in river and estuarine waters from southwest Spain. Sci. Total Environ. **345**, 207–217 (2005)
17. Mandal, B., Chowdhury, T., Samanta, G., Basu, G., Chowdhary, P., Chanda, C., Lodh, D., Karan, N., Dhar, R., Tamili, D., Das, D., Saha, K., Chakraborti, D.: Arsenic in groundwater in seven districts of West Bengal, India-The biggest aresenic calamity in the world. Curr. Sci. **70**, 976–986 (1996)
18. Peter, R., Hugh, B., Keith, R.: Arsenic Pollution: A global problem Leaflet. Wiley (2009)
19. Mohapatra, M., Anand, S.: Synthesis and applications of nano-structured iron oxides/ hydroxides—a review. Int. J. Eng. Sci. Technol. **2**, 127–146 (2010)
20. Cornell, R.M., Schwertmann, U.: The Iron Oxides Structure, Properties, Reactions, Occurrences and Uses. Wiley (2003)
21. Ohnalkova, A.L.C.D., Ussell, C.O.K.R.: Transformation of 2-line ferrihydrite to 6-line ferrihydrite under oxic and anoxic conditions. Am. Mineral. **88**, 1903–1914 (2003)
22. Raven, K.P., Jain, A., Loeppert, R.H.: Arsenite and arsenate adsorption on ferrihydrite: kinetics, equilibrium, and adsorption envelopes. Environ. Sci. Technol. **32**, 344–349 (1998)
23. Martin, M., Violante, A., Ajmone-Marsan, F., Barberis, E.: Surface interactions of arsenite and arsenate on soil colloids. Soil Sci. Soc. Am. J. **78**, 157 (2014)
24. Jiang, X., Peng, C., Fu, D., Chen, Z., Shen, L., Li, Q., Ouyang, T., Wang, Y.: Removal of arsenate by ferrihydrite via surface complexation and surface precipitation. Appl. Surf. Sci. **353**, 1087–1094 (2015)
25. Giménez, J., Martínez, M., de Pablo, J., Rovira, M., Duro, L.: Arsenic sorption onto natural hematite, magnetite, and goethite. J. Hazard. Mater. **141**, 575–580 (2007)
26. Silva, J., Mello, J.W.V., Gasparon, M., Abrahão, W.A.P., Ciminelli, V.S.T., Jong, T.: The role of Al-Goethites on arsenate mobility. Water Res. **44**, 5684–5692 (2010)
27. Mamindy-Pajany, Y., Hurel, C., Marmier, N., Roméo, M.: Arsenic (V) adsorption from aqueous solution onto goethite, hematite, magnetite and zero-valent iron: Effects of pH, concentration and reversibility. Desalination **281**, 93–99 (2011)
28. Zhao, K., Guo, H.: Behavior and mechanism of arsenate adsorption on activated natural siderite: Evidences from FTIR and XANES analysis. Environ. Sci. Pollut. Res. **21**, 1944–1953 (2014)
29. Zheng, Y.M., Lim, S.F., Chen, J.P.: Preparation and characterization of zirconium-based magnetic sorbent for arsenate removal. J. Colloid Interf. Sci. **338**, 22–29 (2009)
30. Pang, S.Y., Jiang, J., Ma, J.: Oxidation of sulfoxides and arsenic(III) in corrosion of nanoscale zero valent iron by oxygen: evidence against ferryl ions (Fe(IV)) as active intermediates in fenton reaction. Environ. Sci. Technol. **45**, 307–312 (2011)
31. Wu, C.C., Hus, L.C., Chiang, P.N., Liu, J.C., Kuan, W.H., Chen, C.C., Tzou, Y.M., Wang, M.K., Hwang, C.E.: Oxidative removal of arsenite by Fe(II)- and polyoxometalate (POM)-amended zero-valent aluminum (ZVAl) under oxic conditions. Water Res. **47**, 2583–2591 (2013)

32. Morin, G., Wang, Y., Foster, A.L., Juillot, F., Calas, G.: XANES evidence for rapid arsenic (III) oxidation at magnetite and ferrihydrite surfaces by dissolved O_2 via Fe^{2+}-mediated reactions. Environ. Sci. Technol. **44**, 5416–5422 (2010)
33. Mayo, J.T., Yavuz, C., Yean, S., Cong, L., Shipley, H., Yu, W., Falkner, J., Kan, A., Tomson, M., Colvin, V.L.: The effect of nanocrystalline magnetite size on arsenic removal. Sci. Technol. Adv. Mater. **8**, 71–75 (2007)
34. Lin, Y.F., Chen, J.L., Xu, C.Y., Chung, T.W.: One-pot synthesis of paramagnetic iron(III) hydroxide nanoplates and ferrimagnetic magnetite nanoparticles for the removal of arsenic ions. Chem. Eng. J. **250**, 409–415 (2014)
35. Chandra, V., Park, J., Chun, Y., Lee, J.W., Hwang, I.C., Kim, K.S.: Water-dispersible magnetite-reduced graphene oxide composites for arsenic removal. ACS Nano **4**, 3979–3986 (2010)
36. Kim, Y., Kim, C., Choi, I., Rengaraj, S., Yi, J.: Arsenic removal using mesoporous alumina prepared via a templating method. Environ. Sci. Technol. **38**, 924–931 (2004)
37. Lin, T.F., Wu, J.K.: Adsorption of arsenite and arsenate within activated alumina grains: equilibrium and kinetics. Water Res. **35**, 049–2057 (2001)
38. Guan, X.H., Su, T., Wang, J.: Quantifying effects of pH and surface loading on arsenic adsorption on NanoActive alumina using a speciation-based model. J. Hazard. Mater. **166**, 39–45 (2009)
39. Ladeira, A.C.Q.A.-C.V.S.T.: Adsorption and desorption of arsenic on an oxisol and its constituents. Water Res. **38**, 2087–2094 (2004)
40. Ladeira, A.C.Q., Ciminelli, V.S.T., Duarte, H.A., Alves, M.C.M., Ramos, A.Y.: Mechanism of anion retention from EXAFS and density functional calculations: arsenic (V) adsorbed on gibbsite. Geochim. Cosmochim. Acta **65**, 1211–1217 (2001)
41. Arai, Y., Elzinga, E.J., Sparks, D.L.: X-ray Absorption spectroscopic investigation of arsenite and arsenate adsorption at the aluminum oxide-water interface. J. Colloid Interf. Sci. **235**, 80–88 (2001)
42. Arai, Y., Sparks, D.L., Davis, J.A.: Arsenate adsorption mechanisms at the allophane—water interface. Environ. Sci. Technol. **39**, 2537–2544 (2005)
43. Masue, Y., Loeppert, R.H.: Arsenate and arsenite adsorption and desorption behavior on coprecipitated aluminum: iron hydroxides. Environ. Sci. Technol. **41**, 837–842 (2007)
44. Poulin, É., Blais, J.F., Mercier, G.: Transformation of red mud from aluminium industry into a coagulant for wastewater treatment. Hydrometallurgy **92**, 16–25 (2008)
45. Altundogan, H.S., Altundogan, S., Tümen, F., Bildik, M.: Arsenic adsorption from aqueous solutions by activated red mud. Waste Manag. **22**, 357–363 (2002)
46. Burke, I.T., Peacock, C.L., Lockwood, C.L., Stewart, D.I., Mortimer, R.J.G., Ward, M.B., Renforth, P., Gruiz, K., Mayes, W.M.: Behavior of aluminum, arsenic, and vanadium during the neutralization of red mud leachate by HCl, gypsum, or seawater. Environ. Sci. Technol. **47**, 6527–6535 (2013)
47. Amirbahman, A., Kent, D.B., Curtis, G.P., Davis, J.A.: Kinetics of sorption and abiotic oxidation of arsenic(III) by aquifer materials. Geochim. Cosmochim. Acta **70**, 533–547 (2006)
48. Nesbitt, H.W., Canning, G.W., Bancroft, G.M.: XPS study of reductive dissolution of 7Å-birnessite by H_3AsO_3, with constraints on reaction mechanism. Geochim. Cosmochim. Acta **62**, 2097–2110 (1998)
49. Manning, B.A., Fendorf, S.E., Bostick, B.: Arsenic (III) oxidation and arsenic (V) adsorption reactions on synthetic birnessite arsenic (III) oxidation and arsenic (V) (2002)
50. Zhang, G., Liu, F., Liu, H., Qu, J., Liu, R.: Respective role of Fe and Mn oxide contents for arsenic sorption in iron and manganese binary oxide: an X-ray absorption spectroscopy investigation. Environ. Sci. Technol. **48**, 10316–10322 (2014)
51. Dias, A., Sá, R.G., S pitale, M.C., Athayde, M., Ciminelli, V.S.T.: Microwave-hydrothermal synthesis of nanostructured Na-birnessites and phase transformation by arsenic(III) oxidation. Mater. Res. Bull. **43**, 1528–1538 (2008)

52. Saleh, T.A., Agarwal, S., Gupta, V.K.: Synthesis of MWCNT/MnO_2 and their application for simultaneous oxidation of arsenite and sorption of arsenate. Appl. Catal. B-Environ. **106**, 46–53 (2011)
53. Pena, M.E., Korfiatis, G.P., Patel, M., Lippincott, L., Meng, X.: Adsorption of As(V) and As (III) by nanocrystalline titanium dioxide. Water Res. **39**, 2327–2337 (2005)
54. Bang, S., Patel, M., Lippincott, L., Meng, X.: Removal of arsenic from groundwater by granular titanium dioxide adsorben. Chemosphere **60**, 389–397 (2005)
55. Pena, M., Meng, X., Korfiatis, G.P., Jing, C.: Adsorption mechanism of arsenic on nanocrystalline titanium dioxide. Environ. Sci. Technol. **40**, 1257–1262 (2006)
56. Yu, Y., Yu, L., Chen, J.P.: Introduction of an yttrium-manganese binary composite that has extremely high adsorption capacity for arsenate uptake in different water conditions. Ind. Eng. Chem. Res. **54**, 3000–3008 (2015)
57. Goswami, A., Raul, P.K., Purkait, M.K.: Arsenic adsorption using copper (II) oxide nanoparticles. Chem. Eng. Res. Des. **90**, 1387–1396 (2012)
58. Martinson, C.A., Reddy, K.J.: Adsorption of arsenic(III) and arsenic(V) by cupric oxide nanoparticles. J. Colloid Interf. Sci. **336**, 406–411 (2009)
59. Yu, Y., Yu, L., Sun, M., Paul Chen, J.: Facile synthesis of highly active hydrated yttrium oxide towards arsenate adsorption. J. Colloid Interface Sci. **474**, 216–222 (2016)
60. Wasay, S.A., Haron, M.J., Uchiumi, A., Tokunaga, S.: Removal of arsenite and arsenate ions from aqueous solution by basic yttrium carbonate. Water Res. **30**, 1143–1148 (1996)
61. Li, R., Li, Q., Gao, S., Shang, J.K.: Exceptional arsenic adsorption performance of hydrous cerium oxide nanoparticles: part A. Adsorption capacity and mechanism. Chem. Eng. J. **185–186**, 127–135 (2012)
62. Cui, H., Li, Q., Gao, S., Shang, J.K.: Strong adsorption of arsenic species by amorphous zirconium oxide nanoparticles. J. Ind. Eng. Chem. **18**, 1418–1427 (2012)
63. Uddin, M.K.: A review on the adsorption of heavy metals by clay minerals, with special focus on the past decade. Chem. Eng. J. **308**, 438–462 (2017)
64. Mohapatra, D., Mishra, D., Chaudhury, G.R., Das, R.P.: Arsenic(V) adsorption mechanism using kaolinite, montmorillonite and illite from aqueous medium. J. Environ. Sci. Heal A **42**, 463–469 (2007)
65. Eloussaief, M., Bouaziz, S., Kallel, N., Benzina, M.: Valorization of El Haria clay in the removal of arsenic from aqueous solution. Desalin. Water Treat **52**, 2220–2224 (2014)
66. Bentahar, Y., Hurel, C., Draoui, K., Khairoun, S., Marmier, N.: Adsorptive properties of Moroccan clays for the removal of arsenic(V) from aqueous solution. Appl. Clay Sci. **119**, 385–392 (2016)
67. Chutia, P., Kato, S., Kojima, T., Satokawa, S.: Arsenic adsorption from aqueous solution on synthetic zeolites. J. Hazard. Mater. **162**, 440–447 (2009)
68. Shevade, S., Ford, R.G.: Use of synthetic zeolites for arsenate removal from pollutant water. Water Res. **38**, 3197–3204 (2004)
69. Bostick, B.C., Fendorf, S.: Arsenite sorption on troilite (FeS) and pyrite (FeS_2). Geochim. Cosmochim. Acta **67**, 909–921 (2003)
70. Bostick, B.C., Fendorf, S., Manning, B.A.: Arsenite adsorption on galena (PbS) and sphalerite (ZnS). Geochim. Cosmochim. Acta **67**, 895–907 (2003)
71. Ramirez-Muiz, K., Jia, F., Song, S.: Adsorption of AsV in aqueous solutions on porous hematite prepared by thermal modification of a siderite-goethite concentrate. Environ. Chem. **9**, 512–520 (2012)
72. Yang, X., Xia, L., Li, J., Dai, M., Yang, G., Song, S.: Adsorption of As(III) on porous hematite synthesized from goethite concentrate. Chemosphere **169** (2017)
73. Dai, M., Xia, L., Song, S., Peng, C., Lopez-Valdivieso, A.: Adsorption of As(V) inside the pores of porous hematite in water. J. Hazard. Mater. **307**, 312–317 (2016)
74. Eguez, H.E., Cho, E.H.: Adsorption of arsenic on activated charcoal. J. Miner. Met. Mater. **39**, 38–41 (1987)
75. Minovic, T.Z., Gulicovski, J.J., Stoiljkovic, M.M., Jokic, B.M., Zivkovic, L.S., Matovic, B. Z., Babic, B.M.: Surface characterization of mesoporous carbon cryogel and its application

in arsenic (III) adsorption from aqueous solutions. Micropor. Mesopor. Mat. **201**, 271–276 (2015)
76. Li, W.G., Gong, X.J., Wang, K., Zhang, X.R., Fan, W.B.: Adsorption characteristics of arsenic from micro-polluted water by an innovative coal-based mesoporous activated carbon. Bioresour. Technol. **165**, 166–173 (2014)
77. Ja, Arcibar-Orozco, Josue, D.B., Rios-Hurtado, J.C., Rangel-Mendez, J.R.: Influence of iron content, surface area and charge distribution in the arsenic removal by activated carbons. Chem. Eng. J. **249**, 201–209 (2014)
78. Wu, Z., Li, W., Webley, P.A., Zhao, D.: General and controllable synthesis of novel mesoporous magnetic iron oxide@carbon encapsulates for efficient arsenic removal. Adv. Mater. **24**, 485–491 (2012)
79. Sigrist, M.E., Brusa, L., Beldomenico, H.R., Dosso, L., Tsendra, O.M., González, M.B., Pieck, C.L., Vera, C.R.: Influence of the iron content on the arsenic adsorption capacity of Fe/GAC adsorbents. J. Environ. Chem. Eng. **2**, 927–934 (2014)
80. Xu, Y.H., Nakajima, T., Ohki, A.: Adsorption and removal of arsenic (V) from drinking water by aluminum-loaded Shirasu-zeolite. J. Hazard. Mater. **92**, 275–287 (2002)
81. Ren, X., Zhang, Z., Luo, H., Hu, B., Dang, Z., Yang, C., Li, L.: Adsorption of arsenic on modified montmorillonite. Appl. Clay Sci. **97–98**, 17–23 (2014)
82. Simeonidis, K., Gkinis, T., Tresintsi, S., Martinez-Boubeta, C., Vourlias, G., Tsiaoussis, I., Stavropoulos, G., Mitrakas, M., Angelakeris, M.: Magnetic separation of hematite-coated Fe3O4 particles used as arsenic adsorbents. Chem. Eng. J. **168**, 1008–1015 (2011)
83. Chandra, V., Park, J., Chun, Y., Lee, J.W., Hwang, I., Kim, K.S.: Removal **4**, 3979–3986 (2010)
84. Kolbe, F., Weiss, H., Morgenstern, P., Wennrich, R., Lorenz, W., Schurk, K., Stanjek, H., Daus, B.: Sorption of aqueous antimony and arsenic species onto akaganeite. J. Colloid Interf. Sci. **357**, 460–465 (2011)
85. Chen, W., Parette, R., Zou, J., Cannon, F.S., Dempsey, B.A.: Arsenic removal by iron-modified activated carbon. Water Res. **41**, 1851–1858 (2007)
86. Manning, B.A., Fendorf, S.E., Goldberg, S.: Surface structures and stability of arsenic(III) on goethite: spectroscopic evidence for inner-sphere complexes. Environ. Sci. Technol. **32**, 2383–2388 (1998)
87. Ren, Z., Zhang, G., Paul Chen, J.: Adsorptive removal of arsenic from water by an iron-zirconium binary oxide adsorbent. J. Colloid Interf. Sci. **358**, 230–237 (2011)
88. Wei, Z., Liang, K., Wu, Y., Zou, Y., Zuo, J., Arriagada, D.C., Pan, Z., Hu, G.: The effect of pH on the adsorption of arsenic(III) and arsenic(V) at the TiO_2 anatase [101] surface. J. Colloid Interf. Sci. **462**, 252–259 (2016)
89. Yu, L., Ma, Y., Ong, C.N., Xie, J., Liu, Y.: Rapid adsorption removal of arsenate by hydrous cerium oxide-graphene composite. Rsc. Adv. **5**, 64983–64990 (2015)
90. Wilkie, J.A., Hering, J.G.: Adsorption of arsenic onto hydrous ferric oxide: effects of adsorbate/adsorbent ratios and co-occurring solutes. Colloid Surf. A **107**, 97–110 (1996)
91. Chakravarty, S., Dureja, V., Bhattacharyya, G., Maity, S., Bhattacharjee, S.: Removal of arsenic from ground water using low cost ferruginous manganese ore. Water Res. **36**, 625–632 (2002)
92. Mohapatra, M., Sahoo, S.K., Anand, S., Das, R.P.: Removal of As(V) by Cu(II)-, Ni(II)-, or Co(II)-doped goethite samples. J. Colloid Interf. Sci. **298**, 6–12 (2006)
93. Gao, Y., Mucci, A.: Individual and competitive adsorption of phosphate and arsenate on goethite in artificial seawater. Chem. Geol. **199**, 91–109 (2003)
94. Weng, L., Van Riemsdijk, W.H., Hiemstra, T.: Cu^{2+} and Ca^{2+} adsorption to goethite in the presence of fulvic acids. Geochim. Cosmochim. Acta **72**, 5857–5870 (2008)
95. Xu, W., Wang, J., Wang, L., Sheng, G., Liu, J., Yu, H., Huang, X.J.: Enhanced arsenic removal from water by hierarchically porous CeO_2-ZrO_2 nanospheres: role of surface- and structure-dependent properties. J. Hazard. Mater. **260**, 498–507 (2013)
96. Kanematsu, M., Young, T.M., Fukushi, K., Green, P.G., Darby, J.L.: Arsenic(III, V) adsorption on a goethite-based adsorbent in the presence of major co-existing ions: modeling

competitive adsorption consistent with spectroscopic and molecular evidence. Geochim. Cosmochim. Acta **106**, 404–428 (2013)

97. Zhu, H., Jia, Y., Wu, X., Wang, H.: Removal of arsenic from water by supported nano zero-valent iron on activated carbon. J. Hazard. Mater. **172**, 1591–1596 (2009)
98. Bagherifam, S., Komarneni, S., Lakzian, A., Fotovat, A., Khorasani, R., Huang, W., Ma, J., Wang, Y.: Evaluation of Zn-Al-SO_4 layered double hydroxide for the removal of arsenite and arsenate from a simulated soil solution: isotherms and kinetics. Appl. Clay Sci. **95**, 119–125 (2014)
99. Pan, B., Li, Z., Zhang, Y., Xu, J., Chen, L., Dong, H., Zhang, W.: Acid and organic resistant nano-hydrated zirconium oxide (HZO)/polystyrene hybrid adsorbent for arsenic removal from water. Chem. Eng. J. **248**, 290–296 (2014)
100. Bhowmick, S., Chakraborty, S., Mondal, P., Van Renterghem, W., Van den Berghe, S., Roman-Ross, G., Chatterjee, D., Iglesias, M.: Montmorillonite-supported nanoscale zero-valent iron for removal of arsenic from aqueous solution: kinetics and mechanism. Chem. Eng. J. **243**, 14–23 (2014)
101. Burns, P.E., Hyun, S., Lee, L.S., Murarka, I.: Characterizing As(III, V) adsorption by soils surrounding ash disposal facilities. Chemosphere **63**, 1879–1891 (2006)
102. Grafe, M., Eick, M.J., Grossl, P.R., Saunders, A.M.: Adsorption of arsenate and arsenite on ferrihydrite in the presence and absence of dissolved organic carbon. J. Environ. Qual. **31**, 1115–1123 (2002)
103. Grafe, M., Eick, M.J., Grossl, P.R.: Adsorption of Arsenate (V) and Arsenite (III) on goethite in the presence and absence of dissolved organic carbon. Soil Sci. Soc. Am. J. **65**, 1680–1687 (2001)
104. Buschmann, J., Kappeler, A., Lindauer, U., Kistler, D., Berg, M., Sigg, L.: Arsenite and arsenate binding to dissolved humic acids: Influence of pH, type of humic acid and aluminum. Environ. Sci. Technol. **40**, 6015–6020 (2006)
105. Liu, T., Tsang, D.C.W., Lo, I.M.C.: Chromium(VI) reduction kinetics by zero-valent iron in moderately hard water with humic acid: iron dissolution and humic acid adsorption. Environ. Sci. Technol. **42**, 2092–2098 (2008)
106. Jain, A., Loeppert, R.H.: Effect of competing anions on the adsorption of arsenate and arsenite by ferrihydrite. J. Environ. Qual. **29**, 1422–1430 (2000)
107. Zhang, G., Ren, Z., Zhang, X., Chen, J.: Nanostructured iron(III)-copper(II) binary oxide: a novel adsorbent for enhanced arsenic removal from aqueous solutions. Water Res. **47**, 4022–4031 (2013)
108. Yang, W., Kan, A.T., Chen, W., Tomson, M.B.: PH-dependent effect of zinc on arsenic adsorption to magnetite nanoparticles. Water Res. **44**, 5693–5701 (2010)
109. Bang, S., Korfiatis, G.P., Meng, X.: Removal of arsenic from water by zero-valent iron. J. Hazard. Mater. **121**, 61–67 (2005)
110. Jais, F.M., Ibrahim, S., Yoon, Y., Jang, M.: Enhanced arsenate removal by lanthanum and nano-magnetite composite incorporated palm shell waste-based activated carbon. Sep. Purif. Technol. **169**, 93–102 (2016)
111. M-BRKHG-KMHMS, S.J.: Evaluation of some natural zeolites and their relevant synthetic types as sorbents for removal of arsenic from drinking water. TT. Iran. Public Health **33**, 36–44 (2004)
112. Ona-Nguema, G., Morin, G., Juillot, F., Calas, G., Brown, G.E.: EXAFS analysis of arsenite adsorption onto two-line ferrihydrite, hematite, goethite, and lepidocrocite. Environ. Sci. Technol. **39**, 9147–9155 (2005)
113. Fendorf, S., Eick, M.J., Grossl, P., Sparks, D.L.: Arsenate and chromate retention mechanisms on goethite. 1 Surface structure. Environ. Sci. Technol. **31**, 315–320 (1997)
114. Waychunas, G.A., Rea, B.A., Fuller, C.C., Davis, J.A.: Surface chemistry of ferrihydrite: Part 1. EXAFS studies of the geometry of coprecipitated and adsorbed arsenate. Geochim. Cosmochim. Acta **57**, 2251–2269 (1993)

115. Sherman, D.M., Randall, S.R.: Surface complexation of arsenic(V) to iron(III) (hydr)oxides: Structural mechanism from ab initio molecular geometries and EXAFS spectroscopy. Geochim Cosmochim Ac **67**, 4223–4230 (2003)
116. Müller, K., Ciminelli, V.S.T., Dantas, M.S.S., Willscher, S.: A comparative study of As(III) and As(V) in aqueous solutions and adsorbed on iron oxy-hydroxides by Raman spectroscopy. Water Res. **44**, 5660–5672 (2010)
117. Manuscript, A., Cytokines, P.P.: Transplantation, arsenite oxidation by a poorly-cystalline manganese oxide. Environ. Sci. Technol. **44**, 1–14 (2010)
118. Jönsson, J., Sherman, D.M.: Sorption of As(III) and As(V) to siderite, green rust (fougerite) and magnetite: implications for arsenic release in anoxic groundwaters. Chem. Geol. **255**, 173–181 (2008)
119. JiaY, XuL, Fang, Z., Demopoulos, G.P.: Observation of surface precipitation of arsenate on ferrihydrite. Environ. Sci. Technol. **40**, 3248–3253 (2006)
120. Wang, X., Demopoulos, G.P., Jia, Y.: Infrared spectroscopic and X-ray diffraction characterization of the nature of adsorbed arsenate on ferrihydrite Infrared spectroscopic and X-ray diffraction characterization of the nature of adsorbed arsenate on ferrihydrite. Geochim. Cosmochim. Acta **71**, 1643–1654 (2007)
121. Silva, G.C., Almeida, F.S., Dantas, M.S.S., Ferreira, A.M., Ciminelli, V.S.T.: Raman and IR spectroscopic investigation of As adsorbed on Mn_3O_4 magnetic composites. Spectrochim. Acta A Mol. Biomol. Spectro. **100**, 161–165 (2013)
122. Tournassat, C., Charlet, L., Bosbach, D., Manceau, A.: Arsenic(III) oxidation by birnessite and precipitation of manganese(II) arsenate. Environ. Sci. Technol. **36**, 493–500 (2002)
123. Tokunaga, S., Wasay, S.A., Park, S.W.: Removal of arsenic(V) ion from aqueous solutions by lanthanum compounds. Water Sci. Technol. **35**, 71–78 (1997)
124. Gräfe, M., Nachtegaal, M., Sparks, D.L.: Formation of metal-arsenate precipitates at the goethite-water interface. Environ. Sci. Technol. **38**, 6561–6570 (2004)
125. Su, J., Huang, H.G., Jin, X.Y., Lu, X.Q., Chen, Z.L.: Synthesis characterization and kinetic of a surfactant-modified bentonite used to remove As(III) and As(V) from aqueous solution. J. Hazard. Mater. **185**, 63–70 (2011)
126. Tu, Y.J., You, C.F.: Phosphorus adsorption onto green synthesized nano-bimetal ferrites: Equilibrium, kinetic and thermodynamic investigation. Chem. Eng. J. **251**, 285–292 (2014)
127. USEPA Ecological restoration: a tool to manage stream quality, Rep EPA 841–F–95–007 pp. 1–2 (1995)
128. Mueller, D.K., Helsel, D.R., Kidd, M.A.: Nutrients in the nation's waters: too much of a good thing? US Government (1996)
129. Fertilizer, S., Sengupta, S., Pandit, A.: Selective removal of phosphorus from wastewater combined with its recovery as a solid-phase fertilizer. Water Res. **45**, 3318–3330 (2011)
130. Yin, H., Yun, Y., Zhang, Y., Fan, C.: Phosphate removal from wastewaters by a naturally occurring, calcium-rich sepiolite. J. Hazard. Mater. **198**, 362–369 (2011)
131. Li, S., Liu, W., Gu, S., Cheng, X., Xu, Z., Zhang, Q.: Spatio-temporal dynamics of nutrients in the upper Han River basin, China. J. Hazard. Mater. **162**, 1340–1346 (2009)
132. Smith, G.T.D., Szabo, A.S., Takács, I., Murthy, S.: Phosphate complexation model and its implications for chemical phosphorus removal. Water Environ. Res. **80**, 633–646 (2008)
133. Mao, Y., Ninh Pham, A., Xin, Y., David Waite, T.: Effects of pH, floc age and organic compounds on the removal of phosphate by pre-polymerized hydrous ferric oxides. Sep. Purif. Technol. **91**, 38–45 (2012)
134. Conidi, D., Parker, W.J.: The effect of solids residence time on phosphorus adsorption to hydrous ferric oxide floc. Water Res. **84**, 323–332 (2015)
135. Kang, S., Choo, K., Lim, K.: Use of iron oxide particles as adsorbents to enhance phosphorus removal from secondary wastewater effluent. Sep. Sci. Technol. **38**, 3853–3874 (2003)
136. Wang, X., Liu, F., Tan, W., Li, W., Feng, X., Sparks, D.L.: Characteristics of phosphate adsorption-desorption onto ferrihydrite. Soil Sci. **178**, 1–11 (2013)

137. Torrent, J., Barron, V., Schwertmann, U.: Phosphate adsorption by goethites differing in crystal morphology. Soil Sci. Soc. Am. J. **54**, 1007–1012 (1990)
138. Barron, V., Herruzo, M., Torrent, J.: Phosphate adsorption by aluminous hematites of different shapes. Soil Sci. Soc. Am. J. **52**, 647 (1988)
139. Chitrakar, R., Tezuka, S., Sonoda, A., Sakane, K., Ooi, K., Hirotsu, T.: Phosphate adsorption on synthetic goethite and akaganeite. J. Colloid Interf. Sci. **298**, 602–608 (2006)
140. Dimirkou, A., Ioannou, A., Doula, M.: Preparation characterization and sorption properties for phosphates of hematite, bentonite and bentonite-hematite systems. Adv. Colloid Interf. Sci. **97**, 37–61 (2002)
141. Daou, T.J., Begin-Colin, S., Grenèche, J.M., Thomas, F., Derory, A., Bernhardt, P., Legaré, P., Pourroy, G.: Phosphate adsorption properties of magnetite-based nanoparticles. Chem. Mater. **19**, 4494–4505 (2007)
142. Yoon, S.Y., Lee, C.G., Park, J.A., Kim, J.H., Kim, S.B., Lee, S.H., Choi, J.W.: Kinetic, equilibrium and thermodynamic studies for phosphate adsorption to magnetic iron oxide nanoparticles. Chem. Eng. J. **236**, 341–347 (2014)
143. Altundoan, H.S., Tmen, F.: Removal of phosphates from aqueous solutions by using bauxite. I: Effect of pH on the adsorption of various phosphates. J. Chem. Technol. Biotechnol. **77**, 77–85 (2002)
144. Altundoğan, H.S., Tümen, F.: Removal of phosphates from aqueous solutions by using bauxite II: the activation study. J. Chem. Technol. Biot. **78**, 824–833 (2003)
145. Rajan, S.S.S.: Changes in net surface charge of hydrous alumina with phosphate adsorption. Nature **262**, 45–46 (1976)
146. Rajan, S.S.S.: Adsorption of divalent phosphate on hydrous aluminum-oxide. Nature **253**, 434 (1975)
147. Li, W., Feng, J., Kwon, K.D., Kubicki, J.D., Phillips, B.L.: Surface speciation of phosphate on Boehmite (γ-AlOOH) determined from NMR spectroscopy. Langmuir **26**, 4753–4761 (2010)
148. Kuo, S., Lotse, G.: Kinetics of phosphate adsorption and desorption by hematite and gibbsite. Soil Sci. **116**, 400–406 (1974)
149. Özacar, M.: Adsorption of phosphate from aqueous solution onto alunite. Chemosphere **51**, 321–327 (2003)
150. Özacar, M.: Equilibrium and kinetic modelling of adsorption of phosphorus on calcined alunite. Adsorpt. Sci. Technol. **9**, 125–132 (2003)
151. Ye, H., Chen, F., Sheng, Y., Sheng, G., Fu, J.: Adsorption of phosphate from aqueous solution onto modified palygorskites. Sep. Purif. Technol. **50**, 283–290 (2006)
152. Bellier, N., Chazarenc, F., Comeau, Y.: Phosphorus removal from wastewater by mineral apatite. Water Res. **40**, 2965–2971 (2006)
153. Molle, P., Liénard, A., Grasmick, A., Iwema, A., Kabbabi, A.: Apatite as an interesting seed to remove phosphorus from wastewater in constructed wetlands. Water Sci. Technol. **51**, 193–203 (2005)
154. Sposito, G.: The Surface Chemistry of Solids. Oxford University Press (1984)
155. Liu, Y., Sheng, X., Dong, Y., Ma, Y.: Removal of high-concentration phosphate by calcite: effect of sulfate and pH. Desalination **289**, 66–71 (2012)
156. Karageorgiou, K., Paschalis, M., Anastassakis, G.N.: Removal of phosphate species from solution by adsorption onto calcite used as natural adsorbent. J. Hazard. Mater. **139**(2007), 447–452 (2007)
157. Zhou, K., Wu, B., Su, L., Gao, X., Chai, X., Dai, X.: Development of nano-CaO_2-coated clinoptilolite for enhanced phosphorus adsorption and simultaneous removal of COD and nitrogen from sewage. Chem. Eng. J. **328**, 35–43 (2017)
158. Yao, W., Millero, F.J.: Adsorption of phophate on manganese dioxide in seawater. Environ. Sci. Technol. **30**, 536–541 (1996)
159. Su, Y., Cui, H., Li, Q., Gao, S., Ku, J.: Strong adsorption of phosphate by amorphous zirconium oxide nanoparticles. Water Res. **7**, 1–9 (2013)

160. Liu, H.L., Sun, X.F., Yin, C.G., Hu, C.: Removal of phosphate by mesoporous ZrO_2. J. Hazard. Mater. **151**, 616–622 (2008)
161. Xie, J., Lin, Y., Li, C., Wu, D., Kong, H.: Removal and recovery of phosphate from water by activated aluminum oxide and lanthanum oxide. Powder Technol. **269**, 351–357 (2014)
162. Yang, J., Zeng, Q., Peng, L., Lei, M., Song, H., Tie, B., Gu, J.: La-EDTA coated Fe_3O_4 nanomaterial: preparation and application in removal of phosphate from water. J. Environ. Sci. (China) **25**, 413–418 (2013)
163. Bar-Yosef, B., Rosenberg, R., Kafkafi, U., Sposito, G.: Phosphorus adsorption by kaolinite and montmorillonite: I. effect of time, ionic strength, and pH. Soil Sci. Soc. Am. J. **52**, 1580 (1988)
164. Kafkafi, U., Bar-Yosef, B., Rosenberg, R., Sposito, G.: Phosphorus adsorption by kaolinite and montmorillonite: II. Organic anion competition. Soil Sci. Soc. Am. J. **52**, 1585 (1988)
165. Zamparas, M., Gianni, A., Stathi, P., Deligiannakis, Y., Zacharias, I.: Removal of phosphate from natural waters using innovative modified bentonites (2012)
166. Yan, L.G., Xu, Y.Y., Yu, H.Q., Xin, X.D., Wei, Q., Du, B.: Adsorption of phosphate from aqueous solution by hydroxy-aluminum, hydroxy-iron and hydroxy-iron-aluminum pillared bentonites. J. Hazard. Mater. **179**, 244–250 (2010)
167. Karapinar, N.: Application of natural zeolite for phosphorus and ammonium removal from aqueous solutions. J. Hazard. Mater. **170**, 1186–1191 (2009)
168. Jiang, C., Jia, L., He, Y., Zhang, B., Kirumba, G., Xie, J.: Adsorptive removal of phosphorus from aqueous solution using sponge iron and zeolite. J. Colloid Interf. Sci. **402**, 246–252 (2013)
169. Arias, C.A., Del Bubba, M., Brix, H.: Phosphorus removal by sands for use as media in subsurface flow constructed reed beds. Water Res. **35**, 1159–1168 (2001)
170. Del Bubba, M., Arias, C.A., Brix, H.: Phosphorus adsorption maximum of sands for use as media in subsurface flow constructed reed beds as measured by the Langmuir isotherm. Water Res. **37**, 3390–3400 (2003)
171. Gerritse, R.G.: Prediction of travel times of phosphate in soils at a disposal site for wastewater. Water Res. **27**, 263–267 (1993)
172. Liu, H., Chen, T., Chang, J., Zou, X., Frost, R.L.: The effect of hydroxyl groups and surface area of hematite derived from annealing goethite for phosphate removal. J. Colloid Interf. Sci. **398**, 88–94 (2013)
173. Li, Y., Liu, C., Luan, Z., Peng, X., Zhu, C., Chen, Z., Zhang, Z., Fan, J., Jia, Z.: Phosphate removal from aqueous solutions using raw and activated red mud and fly ash. J. Hazard. Mater. **137**, 374–383 (2006)
174. Zach-Maor, A., Semiat, R., Shemer, H.: Adsorption-desorption mechanism of phosphate by immobilized nano-sized magnetite layer: interface and bulk interactions. J. Colloid Interf. Sci. **363**, 608–614 (2011)
175. Lai, L., Xie, Q., Chi, L., Gu, W., Wu, D.: Adsorption of phosphate from water by easily separable $Fe_3O_4@SiO_2$ core/shell magnetic nanoparticles functionalized with hydrous lanthanum oxide. J. Colloid Interf. Sci. **465**, 76–82 (2016)
176. Pan, B., Han, F., Nie, G., Wu, B., He, K., Lu, L.: New strategy to enhance phosphate removal from water by hydrous manganese oxide. Environ. Sci. Technol. **48**, 5101–5107 (2014)
177. Boujelben, N., Bouzid, J., Elouear, Z., Feki, M., Jamoussi, F., Montiel, A.: Phosphorus removal from aqueous solution using iron coated natural and engineered sorbents. J. Hazard. Mater. **151**, 103–110 (2008)
178. Stanforth, R.: Competitive adsorption of phosphate and arsenate on goethite. Environ. Sci. Technol. **35**, 4753–4757 (2001)
179. Borggaard, O.K., Raben-Lange, B., Gimsing, A.L., Strobel, B.W.: Influence of humic substances on phosphate adsorption by aluminium and iron oxides. Geoderma **127**, 270–279 (2005)
180. Yan, J., Jiang, T., Yao, Y., Lu, S., Wang, Q., Wei, S.: Preliminary investigation of phosphorus adsorption onto two types of iron oxide-organic matter complexes. J. Environ. Sci(China) **42**, 152–162 (2016)

181. Almeelbi, T., Bezbaruah, A.: Aqueous phosphate removal using nanoscale zero-valent iron. J. Nanopart. Res., 14 (2012)
182. Chen, N., Feng, C., Zhang, Z., Liu, R., Gao, Y., Li, M., Sugiura, N.: Preparation and characterization of lanthanum(III) loaded granular ceramic for phosphorus adsorption from aqueous solution. J. Taiwan Inst. Chem. E **43**, 83–789 (2012)
183. Goldberg, S., Sposito, G.: On the mechanism of specific phosphate adsorption by hydroxylated mineral surfaces: a review. Commun. Soil Sci. Plant Anal. **16**, 801–821 (1985)
184. Arai, Y., Sparks, D.L.: ATR-FTIR spectroscopic investigation on phosphate adsorption mechanisms at the ferrihydrite-water interface. J. Colloid Interf. Sci. **241**, 317–326 (2001)
185. Tejedor-Tejedor, M.I., Anderson, M.A.: Protonation of phosphate on the surface of goethite as studied by cir-ftir and electrophoretic mobility. Langmuir **6**, 602–611 (1990)
186. Wu, L., Forsling, W., Holmgren, A.: Surface complexation of calcium minerals in aqueous solution. J. Colloid Interf. Sci. **224**, 211–218 (2000)
187. Somasundaran, P.: Zeta potential of apatite in aqueous solutions and its change during equilibration. J. Colloid Interf. Sci. **27**, 659–666 (1968)
188. Yan, J., Jiang, T., Yao, Y., Wang, J., Cai, Y., Green, N.W., Wei, S.: Underestimation of phosphorus fraction change in the supernatant after phosphorus adsorption onto iron oxides and iron oxide–natural organic matter complexes. J. Environ. Sci. (China) **55**, 197–205 (2017)
189. Pratt, C., Shilton, A., Pratt, S., Haverkamp, R.G., Bolan, N.S.: Phosphorus removal mechanisms in active slag filters treating waste stabilization pond effluent. Environ. Sci. Technol. **41**, 3296–3301 (2007)
190. Van Emmerik, T.J., Sandström, D.E., Antzutkin, O.N., Angove, M.J., Johnson, B.B.: 31P solid-state nuclear magnetic resonance study of the sorption of phosphate onto gibbsite and kaolinite. Langmuir **23**, 3205–3213 (2007)
191. Del Nero, M., Galindo, C., Barillon, R., Halter, E., Madé, B.: Surface reactivity of α-Al2O3 and mechanisms of phosphate sorption: in situ ATR-FTIR spectroscopy and ζ potential studies. J. Colloid Interf. Sci. **342**, 437–444 (2010)
192. Lounici, H., Addour, L., Belhocine, D., Grib, H., Nicolas, S., Bariou, B., Mameri, N.: Study of a new technique for fluoride removal from water. Desalination **114**, 241–251 (1997)
193. Mahramanlioglu, M., Kizilcikli, I., Bicer, I.O.: Adsorption of fluoride from aqueous solution by acid treated spent bleaching earth. J. Fluor. Chem. **115**, 41–47 (2002)
194. Islam, M., Patel, R.: Thermal activation of basic oxygen furnace slag and evaluation of its fluoride removal efficiency. Chem. Eng. J. **169**, 68–77 (2011)
195. Harrison, P.T.C.: Fluoride in water: a UK perspective. J. Fluor. Chem. **126**, 1448–1456 (2005)
196. Gorchev, H.G., Ozolins, G.: Guidelines for drinking-water quality, B World Health Organ (2004)
197. Chernet, T., Travi, Y., Valles, V.: Mechanism of degradation of the quality of natural water in the lakes region of the Ethiopian Rift Valley. Water Res. **35**, 2819–2832 (2001)
198. Amor, Z., Malki, S., Taky, M., Bariou, B., Mameri, N., Elmidaoui, A.: Optimization of fluoride removal from brackish water by electrodialysis. Desalination **120**, 263–271 (1998)
199. Hasany, S.M., Chaudhary, M.H.: Sorption potential of Haro river sand for the removal of antimony from acidic aqueous solution. Appl. Radiat. Isot. **47**, 467–471 (1996)
200. Hichour, M., Persin, F., Sandeaux, J., Gavach, C.: Fluoride removal from water by Donnan dialysis. Sep. Purif. Technol. **18**, 1–11 (2000)
201. Wang, Y., Reardon, E.J.: Activation and regeneration of a soil sorbent for defluoridation of drinking water. Appl. Geochem. **16**, 531–539 (2001)
202. Maliyekkal, S.M., Sharma, A.K., Philip, L.: Manganese-oxide-coated alumina: a promising sorbent for defluoridation of water. Water Res. **40**, 3497–3506 (2006)
203. Pervov, A.G., Dudkin, E.V., Oa, Sidorenko, Antipov, V.V., Sa, Khakhanov, Makarov, R.I.: RO and NF membrane systems for drinking water production and their maintenance techniques. Desalination **132**, 315–321 (2000)

204. Rubel, F., Woosley, R.D.: The removal of excess fluoride from drinking water by activated alumina. J. Am. Water Works Assoc. **71**, 45–49 (1979)
205. Turner, B.D., Binning, P., Stipp, S.L.S.: Fluoride removal by calcite: Evidence for fluorite precipitation and surface adsorption. Environ. Sci. Technol. **39**, 9561–9568 (2005)
206. Islam, M., Patel, R.K.: Evaluation of removal efficiency of fluoride from aqueous solution using quick lime. J. Hazard. Mater. **143**, 303–310 (2007)
207. Jain, S., Jayaram, R.V.: Removal of fluoride from contaminated drinking water using unmodified and aluminium hydroxide impregnated blue lime stone waste. Sep. Sci. Technol. **44**, 1436–1451 (2009)
208. Streat, M., Hellgardt, K., Newton, N.L.R.: Hydrous ferric oxide as an adsorbent in water treatment. Part 3: Batch and mini-column adsorption of arsenic, phosphorus, fluorine and cadmium ions. Process Saf. Environ. **86**, 21–30 (2008)
209. Kumar, E., Bhatnagar, A., Ji, M., Jung, W., Lee, S.-H., Kim, S.-J., Lee, G., Song, H., Choi, J.-Y., Yang, J.-S.: Defluoridation from aqueous solutions by granular ferric hydroxide (GFH). Water Res. **43**, 490–498 (2009)
210. Mohapatra, M., Rout, K., Gupta, S.K., Singh, P., Anand, S., Mishra, B.K.: Facile synthesis of additive-assisted nano goethite powder and its application for fluoride remediation. J. Nanopart. Res. **12**, 681–686 (2010)
211. Diaz-Nava, C., Olguin, M.T., Solache-Ríos, M.: Water defluoridation by Mexican heulandite–clinoptilolite. Sep. Sci. Technol. **37**, 3109–3128 (2002)
212. Onyango, M.S., Kojima, Y., Aoyi, O., Bernardo, E.C., Matsuda, H.: Adsorption equilibrium modeling and solution chemistry dependence of fluoride removal from water by trivalent-cation-exchanged zeolite F-9. J. Colloid Interf. Sci. **279**, 341–350 (2004)
213. Moges, G., Zewge, F., Socher, M.: Preliminary investigations on the defluorida-tion of water using fired clay chips. J. Afr. Earth Sci. **22**, 479–482 (1996)
214. Tor, A.: Removal of fluoride from an aqueous solution by using montmorillonite. Desalination **201**, 267–276 (2006)
215. Sarkar, M., Banerjee, A., Pramanick, P.P., Sarkar, A.R.: Use of laterite for the removal of fluoride from contaminated drinking water. J. Colloid Interf. Sci. **302**, 432–441 (2006)
216. Thakre, D., Rayalu, S., Kawade, R., Meshram, S., Subrt, J., Labhsetwar, N.: Magnesium incorporated bentonite clay for defluoridation of drinking water. J. Hazard. Mater. **180**, 122–130 (2010)
217. Kamble, S.P., Dixit, P., Rayalu, S.S., Labhsetwar, N.K.: Defluoridation of drinking water using chemically modified bentonite clay. Desalination **249**, 687–693 (2009)
218. Zhang, J., Xie, S., Ho, Y.-S.: Removal of fluoride ions from aqueous solution using modified attapulgite as adsorbent. J. Hazard. Mater. **165**, 218–222 (2009)
219. Fan, J., Xu, Z., Zheng, S.: Comment on "Factors influencing the removal of fluoride from aqueous solution by calcined Mg-Al-CO3 layered double hydroxides". J. Hazard. Mater. **139**, 175–177 (2007)
220. Wang, H., Chen, J., Cai, Y., Ji, J., Liu, L., Teng, H.H.: Defluoridation of drinking water by Mg/Al hydrotalcite-like compounds and their calcined products. Appl. Clay Sci. **35**, 59–66 (2007)
221. Diaz-Nava, C., Solache-Rios, M., Olguin, M.T.: Sorption of fluoride ions from aqueous solutions and well drinking water by thermally treated hydrotalcite. Sep. Sci. Technol. **38**, 131–147 (2003)
222. Fan, X., Parker, D.J., Smith, M.D.: Adsorption kinetics of fluoride on low cost materials. Water Res. **37**, 4929–4937 (2003)
223. Zhang, Z., Tan, Y., Zhong, M.: Defluorination of wastewater by calcium chloride modified natural zeolite. Desalination **276**, 246–252 (2011)
224. Gogoi, P.K., Baruah, R.: Fluoride removal from water by adsorption on acid activated kaolinite clay (2008)
225. Das, D.P., Das, J., Parida, K.: Physicochemical characterization and adsorption behavior of calcined Zn/Al hydrotalcite-like compound (HTlc) towards removal of fluoride from aqueous solution. J. Colloid Interf. Sci. **261**, 213–220 (2003)

226. Lv, L., He, J., Wei, M., Duan, X.: Kinetic studies on fluoride removal by calcined layered double hydroxides. Ind. Eng. Chem. Res. **45**, 8623–8628 (2006)
227. Ibragimova, P.I., Grebennikov, S.F., Gur'yanov, V.V., Fedyukevich, V.A., Vorob'ev-Desyatovskii, N.V.: Effect of the porous structure of activated carbon on the adsorption kinetics of gold(I) cyanide complex. Russ. J. Phys. Chem. A **88**, 1037–1041 (2014)
228. Ghorai, S., Pant, K.K.: Equilibrium, kinetics and breakthrough studies for adsorption of fluoride on activated alumina. Sep. Purif. Technol. **42**, 265–271 (2005)
229. Teng, S.X., Wang, S.G., Gong, W.X., Liu, X.W., Gao, B.Y.: Removal of fluoride by hydrous manganese oxide-coated alumina: performance and mechanism. J. Hazard. Mater. **168**, 1004–1011 (2009)
230. Lai, Y.D., Liu, J.C.: Fluoride removal from water with spent catalyst. Sep. Sci. Technol. **31**, 2791–2803 (1996)
231. Raichur, A.M., Jyoti Basu, M.: Adsorption of fluoride onto mixed rare earth oxides. Sep. Purif. Technol. **24**, 121–127 (2001)
232. Ku, Y., Chiou, H.: The adsorption of fluoride ion from aqueous solution. Water Air Soil Pollut. **133**, 349–360 (2002)
233. Çengeloğlu, Y., Kır, E., Ersöz, M.: Removal of fluoride from aqueous solution by using red mud. Sep. Purif. Technol. **28**, 81–86 (2002)
234. Sujana, M.G., Pradhan, H.K., Anand, S.: Studies on sorption of some geomaterials for fluoride removal from aqueous solutions. J. Hazard. Mater. **161**, 120–125 (2009)
235. Weiner, E.R.: Applications of Environmental Aquatic Chemistry: A Practical Guide. CRC press (2012)
236. Sivasamy, A., Singh, K.P., Mohan, D., Maruthamuthu, M.: Studies on defluoridation of water by coal-based sorbents. J. Chem. Technol. Biotechnol. **76**, 717–722 (2001)
237. Tokunaga, S., Haron, M.J., Wasay, S.A., Wong, K.F., Laosangthum, K., Uchiumi, A.: Removal of fluoride ions from aqueous solutions by multivalent metal compounds. Int. J. Environ. Stud. **48**, 17–28 (1995)
238. Bockris, J., Kim, J.: Electrochemical treatment of low-level nuclear wastes. J. Appl. Electrochem. **27**, 623–634 (1997)
239. Elmidaoui, A., Elhannouni, F., Menkouchi Sahli, M.A., Chay, L., Elabbassi, H., Hafsi, M., Largeteau, D.: Pollution of nitrate in Moroccan ground water: removal by electrodialysis. Desalination **136**, 325–332 (2001)
240. Carpenter, S.R., Caraco, N.F., Correll, D.L., WHowarth, R., Sharpley, A.N., Smith, V.H.: Nonpoint pollution of surface waters with phosphorus and nitrogen. Appl. Soil Ecol. **8**, 559–568 (1998)
241. Fu, B., Liu, Y., Lü, Y., He, C., Zeng, Y., Wu, B.: Assessing the soil erosion control service of ecosystems change in the Loess Plateau of China. Ecol. Complex **8**, 284–293 (2011)
242. Vitousek, P.M., Aber, J.D., Howarth, R.H., Likens, G.E., Matson, P.A., Schindler, D.W., Schlesinger, W.H., Tilman, D.G.: Human alteration of the global nitrogen cycle: source and consequences. Ecol. Appl. **7**, 737–750 (1997)
243. Kimura, K., Nakamura, M., Watanabe, Y.: Nitrate removal by a combination of elemental sulfur-based denitrification and membrane filtration. Water Res. **36**, 1758–1766 (2002)
244. Samatya, S., Kabay, N., Yüksel, Ü., Arda, M., Yüksel, M.: Removal of nitrate from aqueous solution by nitrate selective ion exchange resins. React. Funct. Polym. **66**, 1206–1214 (2006)
245. Chatterjee, S., Woo, S.H.: The removal of nitrate from aqueous solutions by chitosan hydrogel beads. J. Hazard. Mater. **164**, 1012–1018 (2009)
246. Ayyasamy, P.M., Shanthi, K., Lakshmanaperumalsamy, P., Lee, S.-J., Choi, N.-C., Kim, D.-J.: Two-stage removal of nitrate from groundwater using biological and chemical treatments. J. Biosci. Bioeng. **104**, 129–134 (2007)
247. Katsounaros, I., Ipsakis, D., Polatides, C., Kyriacou, G.: Efficient electrochemical reduction of nitrate to nitrogen on tin cathode at very high cathodic potentials. Electrochim. Acta **52**, 1329–1338 (2006)

248. Li, M., Feng, C., Zhang, Z., Sugiura, N.: Efficient electrochemical reduction of nitrate to nitrogen using Ti/IrO_2–Pt anode and different cathodes. Electrochim. Acta **54**, 4600–4606 (2009)
249. Polatides, C., Dortsiou, M., Kyriacou, G.: Electrochemical removal of nitrate ion from aqueous solution by pulsing potential electrolysis. Electrochim. Acta **50**, 5237–5241 (2005)
250. Gupta, V.K., Ali, I., Saini, V.K.: Adsorption studies on the removal of Vertigo Blue 49 and Orange DNA[13] from aqueous solutions using carbon slurry developed from a waste material. J. Colloid Interf. Sci. **315**, 87–93 (2007)
251. Goyal, M., Bhagat, M., Dhawan, R.: Removal of mercury from water by fixed bed activated carbon columns. J. Hazard. Mater. **171**, 1009–1015 (2009)
252. Demirbas, A.: Agricultural based activated carbons for the removal of dyes from aqueous solutions: a review. J. Hazard. Mater. **167**, 1–9 (2009)
253. Ruparelia, J.P., Duttagupta, S.P., Chatterjee, A.K., Mukherji, S.: Potential of carbon nanomaterials for removal of heavy metals from water. Desalination **232**, 145–156 (2008)
254. Arora, M., Eddy, N.K., Mumford, K.A., Baba, Y., Perera, J.M., Stevens, G.W.: Surface modification of natural zeolite by chitosan and its use for nitrate removal in cold regions. Cold Reg. Sci. Technol. **62**, 92–97 (2010)
255. Schick, J., Caullet, P., Paillaud, J.L., Patarin, J., Mangold-Callarec, C.: Batch-wise nitrate removal from water on a surfactant-modified zeolite. Micropor. Mesopor. Mat. **132**(3), 395–400 (2010)
256. Xi, Y., Mallavarapu, M., Naidu, R.: Preparation, characterization of surfactants modified clay minerals and nitrate adsorption. Appl. Clay Sci. **48**, 92–96 (2010)
257. Ozcan, A., Sahin, M., Ozcan, A.S.: Adsorption of nitrate ions onto sepiolite and surfactant-modified sepiolite. Adsorpt. Sci. Technol. **23**, 323–333 (2005)
258. Öztürk, N., Bektaş, T.E.: Nitrate removal from aqueous solution by adsorption onto various materials. J. Hazard. Mater. **112**, 155–162 (2004)
259. Mena-Duran, C.J., Sun Kou, M.R., Lopez, T., Azamar-Barrios, J.A., Aguilar, D.H., Domínguez, M.I., Odriozola, J.A., Quintana, P.: Nitrate removal using natural clays modified by acid thermoactivation. Appl. Surf. Sci. **253**, 5762–5766 (2007)
260. Islam, M., Patel, R.: Synthesis and physicochemical characterization of Zn/Al chloride layered double hydroxide and evaluation of its nitrate removal efficiency. Desalination **256**, 120–128 (2010)

Adsorption of Cations on Minerals

Feng Rao, Zhili Li, and Ramiro Escudero Garcia

Abstract The adsorption of cations on minerals plays an important role theoretically in surface and colloidal chemistry, and applicably in various engineering processes such as water treatment, agriculture and flotation of minerals. This chapter introduces the fundamentals of the adsorption of cations on minerals, the measurements, and classifies the applications. Then, structure and surface chemistry of clay minerals, which have been proved to be the most workable mineral in cations adsorption, are introduced. Other oxide minerals, such as goethite, are present for their structure and adsorption behavior as well. After that, the chapter summarizes and differentiates the adsorption on the basis of different types of cations, namely lead, mercury, copper, chromium and other cations. It tries to present the research results through a novel classification, rather than the types of adsorbents. The gaps between fundamental research and engineering applications are discussed, as well as the further research opportunities are present.

Keywords Ion exchange · Heavy metals · Clay minerals · PILC

1 Introduction

In aqueous solution, the adsorption of cations on minerals is a prevailing interaction in natural processes, which is utilized in various engineering processes such as contaminated water treatment, mineral flotation and materials engineering. This

F. Rao (✉)
School of Zijin Mining, Fuzhou University, Fuzhou 350108, Fujian, China
e-mail: fengrao@umich.mx

F. Rao · R. E. Garcia
CONACYT Instituto de Investigación en Metalurgia y Materiales,
Universidad Michoacana de San Nicolás de Hidalgo, 58030 Morelia, Michoacán, Mexico

Z. Li
School of Xingfa Mining Engineering, Wuhan Institute of Technology,
Wuhan 430073, Hubei, China

S. Song and B. Li (eds.), *Adsorption at Natural Minerals/Water Interfaces*,
Engineering Materials,
https://doi.org/10.1007/978-3-030-54451-5_5

chapter introduces the fundamentals of cations adsorption on minerals, the measurements, and the classification of applications. Then, exemplary studies of cations adsorption on minerals are presented based on the engineering process. The gaps between fundamental research and engineering applications, as well as the further research opportunities, will be discussed.

2 Fundamentals of Cation Adsorption on Minerals

2.1 Types of Mineral Adsorbents and Adsorption Mechanisms

The interactions of cations with ligands on mineral surfaces are determined by the properties of mineral surfaces and solution conditions. Based on the surface functional groups, mineral adsorbents can be classified into four groups: (1) metal oxide mineral adsorbents containing functional groups of Me–OH (Me represents metals), e.g., goethite, hematite and aluminum oxide; (2) aluminosilicate mineral adsorbents without permanent charge possessing functional group of Me–OH, e.g., feldspar and garnet, and with permanent charge resulted from isomorphous substitution in their crystallines for ion exchange, e.g., zeolite, kaolinite and montmorillonite; (3) mineral salt adsorbents containing functional group of Me–OH; (4) sulfide mineral adsorbents possessing functional groups of Me–SH and S–H, which are reactive to cations, e.g., pyrite.

The proposed adsorption mechanisms of minerals to cations are ion exchange, chemisorption, and complexation. Figure 1 shows the examplery adsorption mechanisms of minerals to cations. Cation exchange is determined by isomorphous substitution of the mineral and chemical composition of solution. The cation exchange capacity vary with the type of mineral, nature and concentration of replacing cation, pH and associated cations in the solution and cations in the exchange positions of the mineral. It is a reversible reaction that takes place between cations held near the mineral surface by unbalanced electrical charge from the mineral framework and cations in the solution in contact with the mineral [1]. For example, cation exchange in montmorillonite interlayers takes place between interlayer cations (e.g., Na^{+}) and cations in the solution. Kaolinite clay obtained from Longyan, China was investigated to remove Pb^{2+}, Cd^{2+}, Ni^{2+} and Cu^{2+} in aqueous solution. It was reported that the increasing pH favors the removal of metal ions till they are precipitated as the insoluble hydroxides, removing 92% of Pb^{2+}, 68% of Cd^{2+}, 71% of Ni^{2+} and 53% of Cu^{2+} [2]. The different ion exchange capacities for these cations can be attributed to the diameter difference ($Cu^{2+} > Ni^{2+} > Cd^{2+} > Pb^{2+}$) since cation with lower diameter is of higher preference in the adsorption process [3]. Chemisorption involves reactions and new chemical bonds between cations and mineral surface. Depending on the property of cations and the surface structure of minerals, chemisorption can greatly differ in process and capacity. For example, cations can be adsorbed at calcite surface chemically through substituting calcium sites [4]. Complexion is the process that cations adsorb at mineral surface by forming complex at the surface. The complexing between hydrolyzed goethite surface and hydrolyzed lead ion belongs to this mechanism [5].

Table 1 lists the differences between physisorption and chemisorption. Physical adsorption is due to weak forces of attraction between molecules or ions (van Der Waals forces and electrostatic attraction). The adsorbed material is not fixed to a specific site and the adsorption is reversible. Whereas chemical adsorption, takes place as a result of a chemical bond being formed between the molecule of the solute and the adsorbent. The adsorbed molecules are localized at specific sites and therefore are not free to migrate on the surface. Chemical adsorption is generally irreversible and exothermic. Physical adsorption caused by van Der Waals forces is

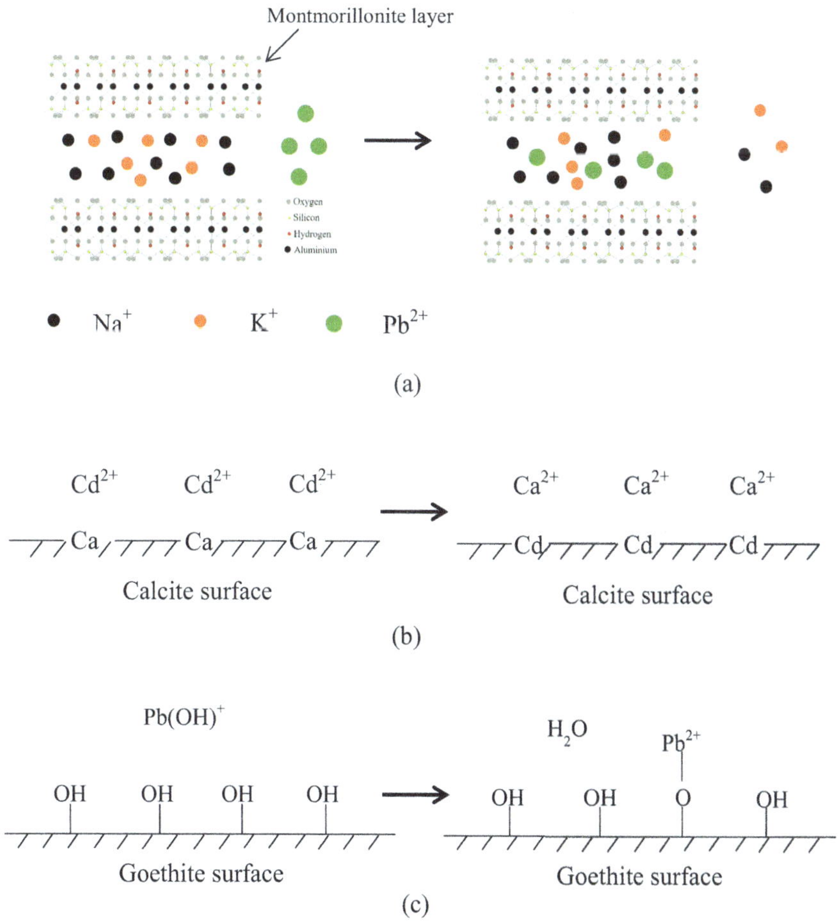

Fig. 1 Examplery presentation of adsorption mechanism of minerals to cations: **a** ion exchange, **b** chemisorption and **c** complexation

Table 1 Comparison of physisorption and chemisorption

	Physisorption	Chemisorption
Adsorbate/ adsorbent	No electron transfer, van der Waals force long range force	Electron transfer, chemical bond, short range force
Adsorption heat	<20 kJ/mol	>40 kJ/mol
	Heat of liquefaction	Formation of chemical bonds
Activation energy	Without activation	High activation energy
Adsorption magnitude	Multilayers	Monolayer
Adsorption nature	Reversible	Irreversible
Adsorption products	The same as the adsorbate	Novel derivatives
Specifics	It is not specific	On specific sites of the surface

universal. Exchange adsorption or ion exchange is a process in which ions of one substances concentrate at a surface as a result of electrostatic attraction to charged sites at the surface. For two potential ionic adsorbates, in the absence of other specific adsorption effects, the charge of the ion is the determining factor for exchange adsorption (an ion with a high valence will be adsorbed faster). For ions with equal charge, molecular size determines order of preference for adsorption, the smaller being able to come closer to the adsorption site and thus being favored.

2.2 *Determining Cation Adsorption Amount and Phenomenon*

2.2.1 Adsorption Amount

Different from other adsorption processes, the adsorption of cations on minerals in aqueous solution is usually determined through a batch depletion method at a given temperature. The amount of cations adsorbed by minerals is calculated from the concentration differences before and after adsorption, because cations concentrations are determined quantitatively in aqueous solutions. The instruments for this application can be atomic absorption spectroscopy (AAS), ultraviolet-visible spectroscopy (UV-vis) or inductively coupled plasma spectrometry (ICP). AAS is a spectroanalytical technique for quantitative determination of chemical elements using the absorption of optical radiation by free atoms in the gaseous state. It can be used to determine over 70 different elements in aqueous solutions [6, 7]. For example, Zhao et al. [8] utilized AAS spectroscopy to determine the adsorption of Pb^{2+} on β-MnO_2. In addition, although no obvious peaks can be obtained for caions in UV-vis spectroscopy measurements, they can be detected after forming complexes. For example, in the adsorption of diatomite to Pb^{2+}, UV-vis spectroscopy was used to determine the Pb^{2+} concentration by applying Pb^{2+}-chlorophosphonazo^{3-} complex at wavelength of 616 nm [9].

ICP can be divided into inductively coupled plasma atomic emission spectroscopy (ICP-AES) and inductively coupled plasma mass spectrometry (ICP-MS). ICP-AES is one method of optical emission spectrometry. When plasma energy is given to an analysis sample from outside, the component elements (atoms) is excited. When the excited atoms return to low energy position, emission rays (spectrum rays) are released and the emission rays that correspond to the photon wavelength are measured. The element type is determined based on the position of the photon rays, and the content of each element is determined based on the rays' intensity. Liang et al. [10] studied the adsorption behavior of heavy metal ions on nanometer-size titanium dioxide with ICP-AES. The apparatus equipped with a 2 kW power, 27 ± 3 MHz ICP generator, a WPG100 plane grating spectrometer with 1200 grooves mm^{-1} (Beijing Broadcast Equipment Factory, Beijing, China) and a conventional plasma silica torch. ICP-MS is a type of mass spectrometry which is capable of detecting metals and several non-metals at concentrations as low as one part in 10^{15} (part per quadrillion, ppq) on non-interfered low-background isotopes. This is achieved by ionizing the sample with inductively coupled plasma and then using a mass spectrometer to separate and quantify those ions. In a study concerning removal of Pb^{2+} from a freshwater lake by Mn oxides in heterogeneous surface coating materials, concentration of Pb^{2+} was determined by analysis of Pb^{208} using a sector field mass spectrometer with an inductively coupled plasma ion source (ICP-MS) (Element 2, Finnigan MAT, Bremen, Germany) [11].

2.2.2 Adsorption Phenomenon

Studies have been carried out to reveal the phenomenon in the adsorption of cations on mineral, such as adsorption process, interaction between cations and adsorbents, ionic status of adsorbed cations and the impact of cation adsorption on mineral surface. Infrared spectroscopy (IR) was employed to study the impact of cations adsorption on sites of mineral surface through comparing the spectra with and without cations adsorption. Figure 2 gives IR spectra of raw bentonite, acid-activated bentonite, Cu^{2+} adsorbed raw bentonite and Cu^{2+} adsorbed acid-activated bentonite. After adsorption of Cu^{2+} on bentonite, the stretching OH band was shifted from 3635 to 3643 cm^{-1} and moreover, a new band, which is assigned to AlMgCuOH vibration, appeared near 3514 cm^{-1} in the spectra of all bentonite samples after Cu^{2+} adsorption. It confirms the presence of Cu^{2+} in the former vacant octahedral sites. The broad band near 1038 cm^{-1}, assigned to complex Si–O stretching vibrations in the tetrahedral sheet, upon adsorption process moved to 1063 cm^{-1} for the raw bentonite sample. The position of the Si–O bending vibration at 527 cm^{-1}, due to Si–O–Al remained basically unchanged after the adsorption of Cu^{2+}, but some broadening and a decrease in intensity of the Si–O–Al band were observed. The positions and shapes of fundamental vibrations of the OH and Si–O groups were influenced by the adsorbed Cu^{2+}, indicating that Cu^{2+} cations replaced the original metal ions in the interlayer or locate into hexagonal cavities of Si–O sheet in the Cu^{2+}-saturated bentonite samples [12].

X-ray photoelectron spectroscopy (XPS) measures the elemental composition, chemical state and electronic state of the elements, thus it can provide the surface characteristics of cation adsorption on minerals. In a study of adsorption of Pb^{2+} from aqueous solution to goethite, Fig. 3 shows the plot of Pb 4f XPS signal (in arbitrary unites) and the amount of Pb^{2+} removed from solution (in mM as measured by AAS) [13]. The XPS signal from goethite surface increased dramatically as increasing pH, which agreed well with adsorbed Pb^{2+} amount measured from AAS. Thus the Pb 4f XPS signal was utilized to identify the adsorbed concentration of Pb^{2+} on goethite, and the calculated surface Pb/Fe atomic ratio was 0.93 in the pH range of saturated adsorption (e.g., pH 8). In addition, Li et al. [14] studied the adsorption of Pb^{2+} at rutile/water interface by XPS, and reported that hydrolyzed

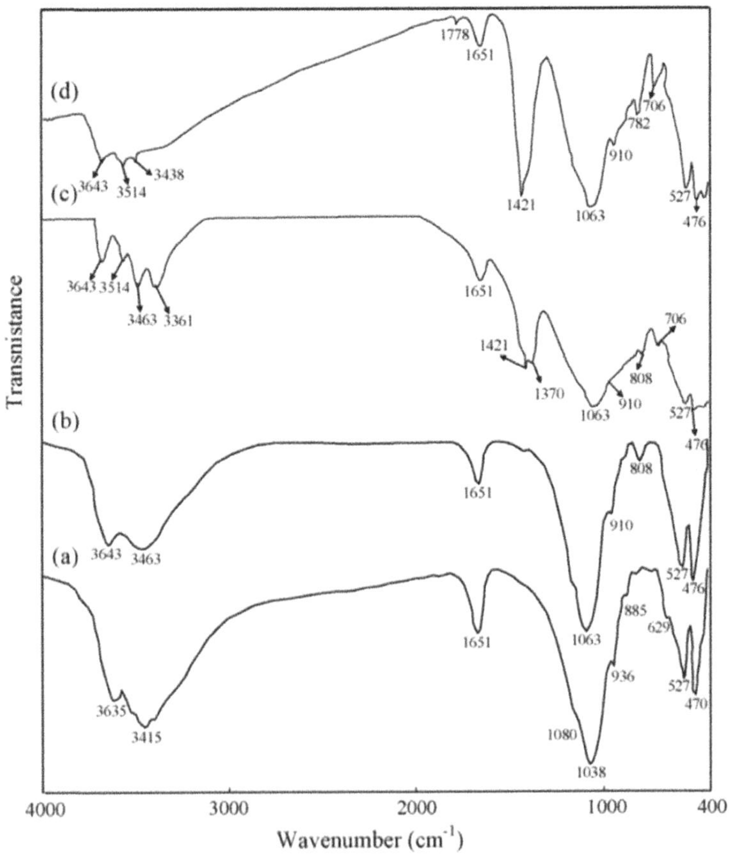

Fig. 2 IR spectra of raw bentonite (**a**), acid-activated bentonite (**b**), Cu^{2+} adsorbed raw bentonite (**c**) and Cu^{2+} adsorbed acid-activated bentonite (**d**) [12]

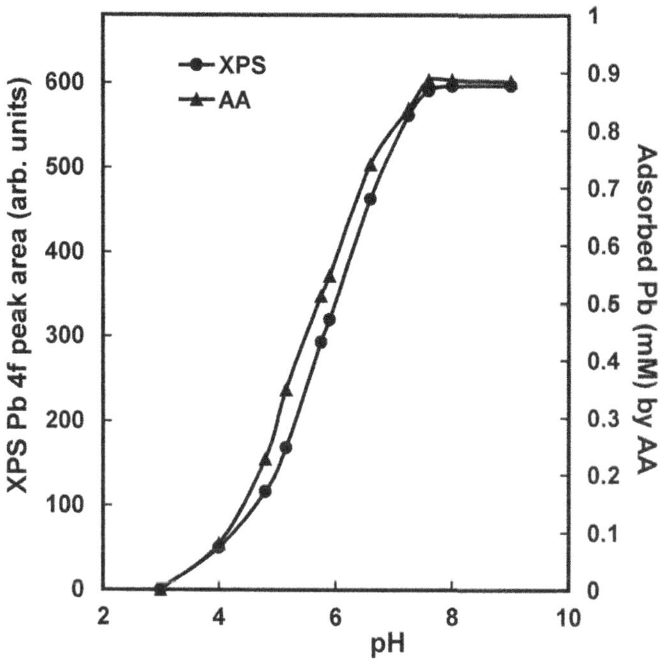

Fig. 3 Plot of Pb 4f XPS signal (in arbitrary unites) and the amount of Pb^{2+} removed from solution (in mmol as measured by AAS) for adsorption of Pb^{2+} from aqueous solution to goethite as a function of pH

Pb^{2+} species (e.g., $PbOH^+$) interacted with the Ti–OH at the rutile surface forming a surface Ti–O–Pb^+ complex.

X-ray absorption spectroscopy (XAS) is a widely used technique for determining the local geometric and/or electronic structure of surface. It can provide information of target atom on bond length, coordination number, coordination atoms, disorder degree factor, distribution symmetry and chemical state, which is important in confirming the adsorption site and structure of complexion products after adsorption of cations on minerals. Strawn et al. utilized XAS to study the mechanisms of Pb^{2+} adsorption at aluminum oxide surface [15].

3 Adsorption of Heavy Metal Cations on Minerals

Heavy metal cations, such as lead, copper, cadmium, mercury and chromium, are prevailing inorganic contaminants for water resources. As summarized in Table 2, these cations are widely produced in modern industries and can cause various diseases once their concentrations exceed the contaminant level. Therefore, the treatment of contaminated water by heavy metals has been extensively studied. The technologies used in investigations include [16]: (1) Chemical precipitation, chemical precipitation is effective and by far the most widely used process in industry because it is relatively simple and inexpensive to operate. In precipitation processes, chemicals react with heavy metal ions to form insoluble precipitates and then separated from the water by sedimentation or filtration. (2) Ion-exchange, ion-exchange processes have been widely used to remove heavy metals from wastewater due to their advantages, such as high treatment capacity, high removal efficiency and fast kinetics. The most widely used material is synthetic resins. (3) Adsorption, adsorption is now recognized as an effective and economic method for heavy metal wastewater treatment. Although activated carbon is the most popular and widely used adsorbent in wastewater treatment applications throughout the world, because of its high price, research interest into the production of alternative adsorbents has intensified in recent years [17]. (4) Membrane filtration, membrane filtration technologies applying ultrafiltration, reverse osmosis, nanofiltration and electrodialysis process with different types of membranes show great promise for heavy metal removal for their high efficiency, easy operation and space saving. (5) Flotation, flotation has been employed to separate heavy metal from a liquid phase using bubble attachment, originated in mineral processing. Dissolved air flotation (DAF), ion flotation and precipitation flotation are the main flotation processes for the removal of metal ions from solution. (6) Electrochemical methods, electrochemical methods involve the plating-out of metal ions on a cathode surface and can recover metals in the elemental metal state. Electrochemical wastewater technologies involve relatively large capital investment and the expensive electricity supply, so they haven't been widely applied.

Although these technologies are effective for treatment of contaminated water by heavy metals, because of its low-cost, availability, profitability, eases of operation and efficiency, adsorption of minerals to heavy metal cations has gained significant attention in contaminated water treatment [18]. And the most common mineral adsorbents are clay minerals, carbonate minerals, metal oxide minerals and sulfide minerals.

3.1 Adsorption of Lead (Pb), Copper (Cu), Cadmium (Cd) and Mercury (Hg) on Mineral Surface

Among these heavy metals, lead (Pb), copper (Cu), cadmium (Cd) and mercury (Hg) possess similar properties. Figures 4, 5, 6 and 7 illustrate the distribution of Pb (II), Cu(II), Cd(II) and Hg (II) species calculated from the hydrolysis constants. At low pH values, Pb^{2+}, Cu^{2+}, Cd^{2+} and Hg^{2+} are the predominant species respectively and their concentration decreases with increasing pH. The concentration of hydrolyzed species increases as increasing pH, especially for $PbOH^{+}$, $CuOH^{+}$, $CdOH^{+}$ and $Hg(OH)^{+}$, which are the main hydrolyzed species. At high pH values, precipitations of $Pb(OH)_2$, $Cu(OH)_2$, CuO, $Cd(OH)_2$, $Hg(OH)_2$ and HgO are

Table 2 Sources and health effects of some heavy metals

Metal	Uses	Health effects	Sources
Lead	Car batteries, pigments, radiation protection, architecture [18]	Anaemia, kidney disease and mental retardation [19]	Battery manufacturing, printing, painting, dying, smelting operations, ceramics, mining, bangle industry [19, 20]
Copper	Copper alloys and coins, transmission wires electrical wiring [18]	Kidney damage, high fever, hemolysis and vomiting [21]	Electrical industry, antifouling paintings, mining [22]
Cadmium	Architecture, electroplating of steel, nickel-cadmium batteries, cellular telephones, laptop computers and camcorders [18]	Renal disturbances, lung insufficiency, bone lesions, cancer and hypertension [23]	Fossil fuels, metal production, electroplating, manufacturing of batteries, pigments and screens [24]
Mercury	Barometers, thermometers, pumps, lamps, Architecture [25]	Paralysis, serious intestinal and urinary complications, disfunction of the central nervous system [25]	Chlor-alkali, paint, oil refining, rubber processing and fertilizer industries [25]
Chromium	Electroplating, stainless steel production, leather tanning, textile manufacturing and wood preservation [18]	Dermatitis, damage to liver, kidney circulation, nerve tissue damage [26]	Textile, leather tanning, electroplating, metal finishing industries [27]

formed respectively. Thus the removal of Pb(II), Cu(II), Cd(II) and Hg (II) are realized by both adsorption and precipitation at high pH values. The adsorption of Pb(II), Cu(II), Cd(II) and Hg (II) on minerals presents a typical cationic adsorption behavior. At low pH, metal ions are removed by adsorption and the removal of Pb (II), Cu(II), Cd(II) and Hg (II) are realized by both adsorption and precipitation at high pH values.

The most commonly used clay minerals in adsorption are kaolinite, montmorillonite, illite, zeolite and bentonite [18, 28, 29]. Clay minerals are built up of tetrahedrally (Si, Al, Fe^{3+}) and octahedrally (Al, Fe^{3+}, Fe^{2+}, Mg) coordinated cations organized to form either sheets or chains. The substitution of a cation of a lower charge for a cation of a higher charge in both the octahedral (e.g., Mg^{2+} replacing Al^{3+}) and tetrahedral (e.g., Al^{3+} replacing Si^{4+}) sheets, gives the basal surface a net negative charge which is satisfied by the interlayer cations [30]. These interlayer ions, such as Ca^{2+}, Mg^{2+}, H^+, K^+, NH^{4+} and Na^+, can be exchanged with other ions relatively easily without affecting the clay mineral structure [31]. Clay

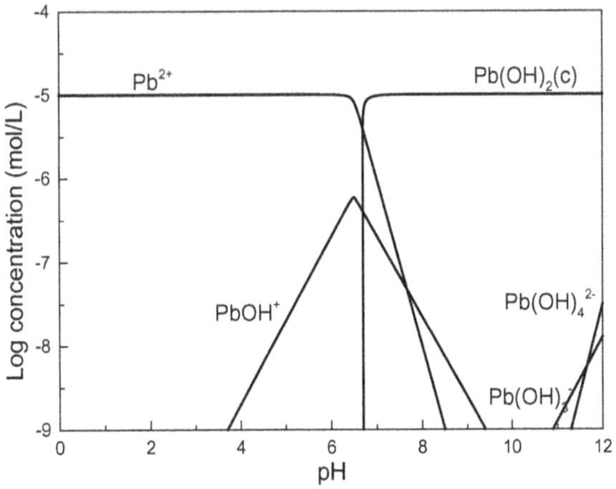

Fig. 4 Speciation diagram for 1 × 10^{-3} mol/L Pb^{2+}

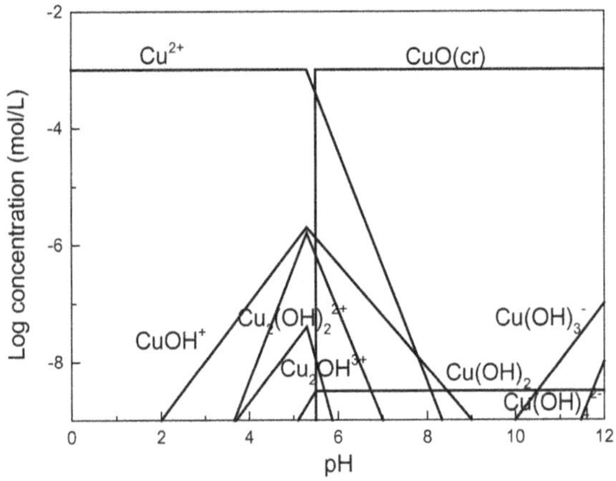

Fig. 5 Speciation diagram for 1 × 10^{-3} mol/L Cu^{2+}

minerals are excellent adsorbent materials because of its large specific surface area, chemical and mechanical stability [31]. Clay minerals can adsorb heavy metals via two different mechanisms:

(i) Cation exchange in the interlayers resulting from the interactions between ions and negative permanent charge. This kind of non-specific ion exchange reactions on the permanently negatively charged basal surface sites was greatly affected by ionic strength. The increase of ionic strength makes the zeta potential of the adsorbent surface less negative and thus would decrease metal ion adsorption [32]. At lower pH, adsorption is mainly occurred at these sites [33].
(ii) Formation of inner-sphere complexes through Si-O^- and Al-O^- groups at the clay particle edges. It can be presented by the following equations. With the increasing pH, the charged edge site became the major adsorption sites [34].

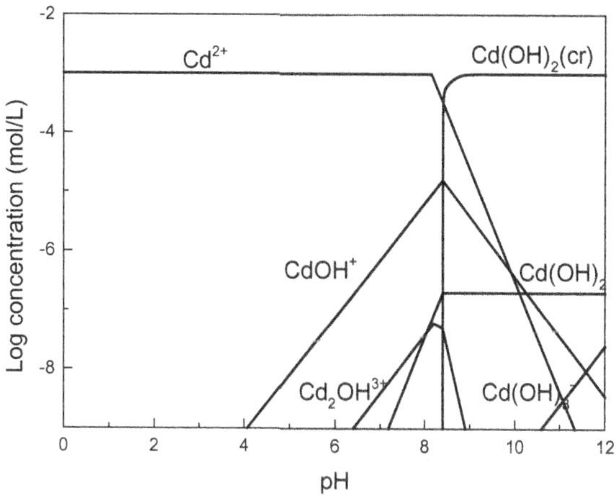

Fig. 6 Speciation diagram for 1×10^{-3} mol/L Cd^{2+}

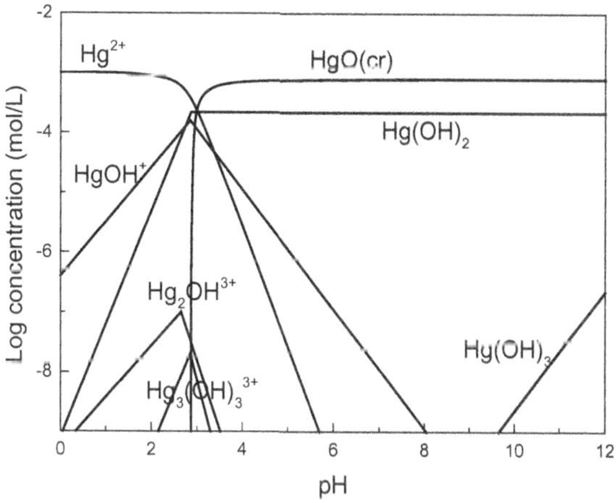

Fig. 7 Speciation diagram for 1×10^{-3} mol/L Hg^{2+}

$$Si{-}O^- \longrightarrow H_2OSi{-}OH \longrightarrow Me(OH)^+ Si{-}O{-}Me^+ + H_2O \quad (1)$$

$$Al{-}O^- \longrightarrow H_2OAl{-}OH \longrightarrow Me(OH)^+ Al{-}O{-}Me^+ + H_2O \quad (2)$$

$$Si{-}O^- \longrightarrow H_2OSi{-}OH \longrightarrow Me^{2+} Si{-}O{-}Me^+ + H^+ \quad (3)$$

$$Al{-}O^- \longrightarrow H_2OAl{-}OH \longrightarrow Me^{2+} Al{-}O{-}Me^+ + H^+ \quad (4)$$

Two-dimensional (2D) molybdenum disulfide shows great adsorption capability of heavy metal removal from wastewater due to its huge sulfur-rich surface area, including a high capture capacity, a fast adsorption rate and an extraordinary affinity [35]. Natural molybdenum disulfide collected from Wuzhou Mine, China, was investigated in order to remove Hg(II) and Pb(II) in aqueous solution. The adsorption rapidly increased at first, then reached a peak of ~183.579 mg/g after ~60 min. And the maximum adsorption capacity of Hg^{2+} at 20 and 35 °C was ~245 and 305 mg/g respectively, which was much higher than that of active carbon which ranging from 50 to 95 mg/g. In addition, a 750 mg/g Hg^{2+} uptake capacity was obtained on MoS_2 after thermal treatment of 500 °C at a Hg^{2+} equilibrium concentration of around 110 mg/L [35]. Furthermore, it also had a dramatic efficiency for Pb(II) removal from water with a 1479 mg/g adsorption capacity, which is much higher than conventional adsorbents less than 300 mg/g [36] (Fig. 8).

The adsorption followed the Freundlich isotherm model and fitted well with pseudo-second-order kinetics model. The excellent cation capture property was mainly attributed to the complexation of cation with intrinsic S and oxidation-induced O atom exposed on 2D molybdenum disulfide surfaces, as well as the electrostatic interaction between negatively charged 2D molybdenum disulfide and cation Hg^{2+}(Pb^{2+}). It was also observed that heavy metal ion adsorbed on molybdenum disulfide surfaces appeared in the form of multilayers [37]. When 2D molybdenum disulfide was dispersed into cation solution as the adsorbent, positively charged Hg^{2+}(Pb^{2+}) would approach the negatively charged surface of adsorbent quickly. The strong complexation between Hg(Pb) and S or O sites then played an important role to bind Hg through the formation of –S–Hg and –S–O, forming the first layer on the adsorbent. Though part of surface was neutralized, it was still negatively charged, which was proved through zeta potential analysis. And the electrostatic interaction caused more Hg adsorbed on the first layer, which could not be washed away by water. Therefore, a multilayer adsorption was observed on 2D molybdenum disulfide. Additionally, thermal treatment enabled molybdenum

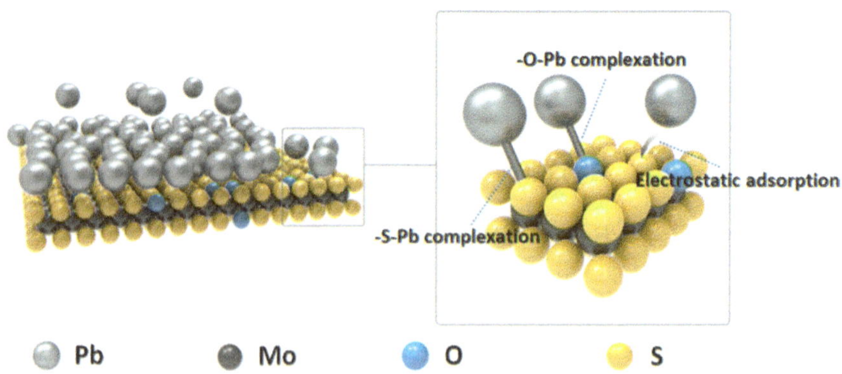

Fig. 8 Digarammatic illustration of the mechanism for Pb(II) adsorption on 2D molybdenum disulfide

disulfide to achieve a tremendous enhancement of adsorption at all pH values, which seemed greatly related to the higher activity of the surface and the formation of molybdenum oxide [38].

The adsorption mechanism of cation on aluminum oxide [39], goethite [5], hematite, magnetite and birnessite [40] are the same. For aluminum oxide, it possess terminal –OH groups which are more likely to accept an additional proton in acidic solution compared to a bridging –OH group. The terminal –OH group will form a positively charged $=Al{-}OH_2^+$ site. The terminal –OH group bonds more strongly to metals than the bridging –OH group [41]. Goethite possesses four types of surface hydroxyls. Its reactivity depends on the coordination environment of the oxygen atom in the =Fe–OH group. Cation surface complexation by hydrous oxides involves formation of bonds with surface oxygen atoms and the release of protons from the surface, which can be illustrated by the following equations,

$$=S{-}OH + Me^{2+} \leftrightarrow =S{-}OMe^{+} + H^{+} \quad (5)$$

$$=S{-}OH + Me^{2+} + H_2O \leftrightarrow =S{-}OMeOH_2^{+} + H^{+} \quad (6)$$

where =S–OH and Me^{2+} is a hydrated oxide surface and a divalent cation respectively [42].

In addition to chemical adsorption, metal oxide minerals can also adsorb cation through electrostatic attraction. Metal oxide minerals are usually positively charged at pH below PZC and negatively charged above PZC, so ions can adsorb at metal oxide minerals surface by electrostatic attraction when the signs of the charges of ions are opposite. At pH below PZC, metal oxide minerals are positively charged, so the adsorption of cations are suppressed because the repulsion of the electrical double layer. At pH above PZC, the adsorption of cations are facilitated because the attraction of the electrical double layer.

Calcite is the dominant sorbent for a variety of metals in carbonate minerals. Investigations have shown that divalent Mn, Co, Ni, Zn and Cd cations are strongly adsorbed by the calcite surface, whereas Sr and Ba are weakly adsorbed by the calcite surface [43]. Cations can substitutes Ca in the calcite lattice and adsorb at calcite surface by chemical adsorption [4, 44]. The adsorption process can be illustrated by the following equation,

$$CaCO_3 + Me^{2+} \rightarrow MeCO_3 + Ca^{2+} \quad (7)$$

Because sulfide minerals are used to adsorb these cations due to it's high affinity to metal ions, such as Pb, Cu, Hg, and Cd [45]. The adsorption process of cations at sulfide mineral surface presents by the following equation. At pH below 3, Me^{2+} precipitates in the presence of H_2S generated in the acidic media [46]. At pH range 3 to precipitation ($Me(OH)_2$) pH, Me^{2+} replaced the Fe^{2+} at pyrite surface and MS_2 is formed. At pH higher than precipitation pH, part of Me^{2+} is removed by forming $Me(OH)_2$ in the solution. The following equations present the adsorption process:

pH < 3

$$4FeS_2 + 6H^+ + 4H_2O \rightarrow 7H_2S + SO_4^{2-} + 4Fe^{2+} \quad (8)$$

$$Me^{2+} + H_2S \rightarrow MeS + 2H^+ \quad (9)$$

3 < pH < Precipitate pH

$$FeS_2 + Me^{2+} \rightarrow MeS_2 + Fe^{2+} \quad (10)$$

Precipitate < pH

$$FeS_2 + Me^{2+} \rightarrow MeS_2 + Fe^{2+} \quad (11)$$

$$2OH^- + Me^{2+} \rightarrow Me(OH)_2 \quad (12)$$

3.2 *Cr Adsorption on Mineral Surface*

Figures 9 and 10 present the distribution of Cr(III) and Cr(VI) species at different pH respectively. It demonstrates that Cr(III) present in the forms of Cr^{3+}, $Cr(OH)^{2+}$, $Cr_2(OH)_2^{4+}$, $Cr_3(OH)_4^{5+}$, $Cr(OH)_2^+$, $Cr(OH)_4^-$, Cr_2O_3 and $Cr(OH)_3$ at various pH values. At pH < 3, Cr^{3+} is the predominant Cr(III) species and then decreases with increasing pH. The concentration of hydrolyzed species increases as increasing pH. At pH higher than 3.8, Cr(III) mainly exists as the forms of precipitated Cr_2O_3 and $Cr(OH)^{2+}$. $HCrO_4^-$ or CrO_4^{2-} are the main forms of Cr (VI) depending on the pH as reported elsewhere [47]. Cr (III) exhibits a typical cationic adsorption behavior. Its adsorption increases with pH and decreases when competing cations are present. While Cr(VI) presents a typical anionic adsorption behavior. Its adsorption decreases with increasing pH and concentration of competing anionic ion [47]. Protonated adsorbent surface can adsorb Cr (VI) by the following equation [48]:

$$=SOH_2^+ + HCrO_4^- \leftrightarrow =SOH_2^+ - HCrO_4^- \quad (13)$$

$$=SOH_2^+ + HCrO_4^- \leftrightarrow =SOH_2^+ - HCrO_4^- \quad (14)$$

Clay minerals and metal oxide minerals possess such protonated surfaces, so they can adsorb Cr (VI) by chemical adsorption. These anionic ions can also adsorbed by attraction of electrical double layer at low pH, since most of oxide minerals and clay minerals are positively charged at pH below PZC.

As for the adsorption of Cr (VI) by sulfide minerals, Cr (VI) reduces to Cr (III) and precipitates in both acid solution and basic solution. Taking pyrrhotite as an example, the following equation can be used to describe the process [49].

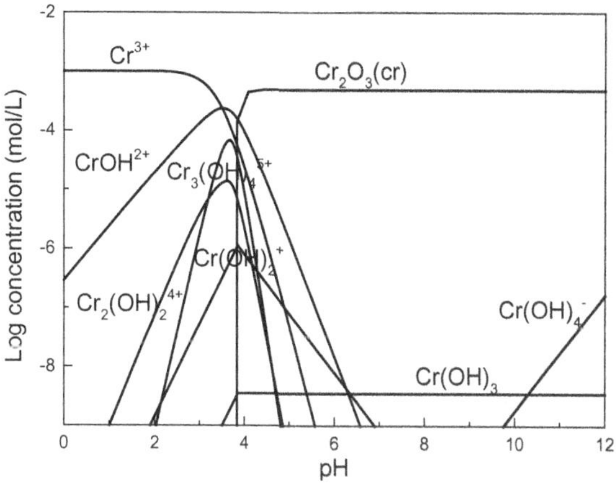

Fig. 9 Speciation diagram for 1×10^{-3} mol/L Cr^{3+}

In acid solution

$$FeS(s) + 2H^+ \rightarrow Fe^{2+} + H_2S(aq) \quad (15)$$

$$4Cr_2O_7^{2-} + 15H_2S(aq) + 2H^+ \rightarrow 4Cr_2S_3(s) + 3SO_4^{2-} + 16H_2O \quad (16)$$

$$Cr_2O_7^{2-} + 4H_2S(aq) \rightarrow Cr_2S_3(s) + SO_3^{2-} + 4H_2O \quad (17)$$

$$Cr_2O_7^{2-} + 6Fe^{2+} + 8H^+ \rightarrow Cr_2O_3(s) + 6Fe^{3+} + 4H_2O \quad (18)$$

In basic solutions

$$CrO_4^{2-} + 3Fe^{2+} + 4OH^- + 4H_2O \rightarrow Cr(OH)_3(s) + 3Fe(OH)_3(s) \quad (19)$$

$$4CrO_4^{2-} + 9Fe^{2+} + OH^- + 6HS^{2-} + 10H_2O \rightarrow 2Cr_2S_3(s) + 9Fe(OH)_3(s) \quad (20)$$

Albadarin et al. studied the adsorption of Cr (VI) at carbonate mineral surface. It was concluded that there are two mechanisms in the adsorption of Cr (VI) at carbonate mineral surface [50]. One is physical adsorption caused by attraction of electrical double layer at low pH and the other is chemical adsorption because of ion exchange at dolomite surface. This chemical adsorption process is illustrated by the following equation and can take place even at low pH.

$$MeCO_3 + CrO_4^{2-} \rightarrow MeCrO_4^{2-} + CO_3^{2-} \quad (21)$$

The adsorption of Cr (III) on clay minerals is similar to that of Cu (II), Pb (II), Cd (II) and Hg (II). Cr (III) adsorbs on clay minerals through cation exchange at the planar sites and through the formation of inner-sphere complexes with Si–O– and Al–O– groups at the clay particle edges [31]. In the investigation studying the adsorption of Zn, Cd and Cr on calcite, it was confirmed by X-ray photoelectron

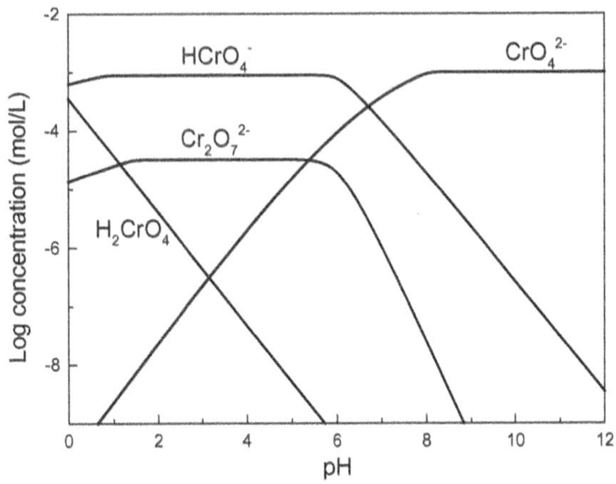

Fig. 10 Speciation diagram for 1×10^{-3} mol/L $HCrO_4^-$

that Cd precipitates on calcite surface via exchange with Ca, Zn adsorbs on calcite surface by precipitation as hydrozincite $Zn_5(OH)_6(CO_3)_2$, while Cr precipitates on calcite as a coating of oxide hydrocarbonate $Cr_2O_3xH_2OyCO_2$ [51]. Yang et al. [52] explored the adsorption of Cr(VI) on iron contaminated graphite surface. It was proved that the adsorption mechanism can be attributed to the formation of ≡Fe–O_4HCr and ≡Fe_2–$(CrO_4)_3$ complexes between the iron on the graphite surface and Cr(VI) in water. The adsorption process is described in the following equation:

$$\equiv Fe{-}OH_2^+ + HCrO_4^- \rightarrow \equiv FeO_4HCr + H_2O \tag{22}$$

$$2\equiv Fe{-}OH^{2+} + 3CrO_4^{2-} + 2H^+ \rightarrow \equiv Fe_2(CrO_4)_3 + 2H_2O \tag{23}$$

The removal of Cr(VI) by Ag^0 decorated molybdenum disulfide (Ag–MoS_2) can be concluded that synergetic effects of photocatalytic reduction and redox reduction. Under dark condition, C(VI) reduces to Cr(III) due to self-oxidation of MoS_2 nanosheets. Under visible light condition, owing to semiconductor properties of MoS_2, photocatalytic reduction started working, in which photoexcited electrons had an effective impact on Cr(VI) reduction, while redox reduction also played a role. The following equation can be used to describe the process.

Chemical reduction process:

$$\text{Anodic: } MoS_2 + 12H_2O \rightarrow MoO_4^{2-} + 2SO_4^{2-} + 24H^+ + 18e^- \tag{24}$$

$$\text{Cathodic: } Cr_2O_7^{2-} + 6e^- + 14H^+ \rightarrow 2Cr^{3+} + 7H_2O \tag{25}$$

$$MoS_2 + 3Cr_2O_7^{2-} + 18H^+ \rightarrow MoO_4^{2-} + 6Cr^{3+} + 2SO_4^{2-} + 9H_2O \tag{26}$$

Photocatalytic reduction:

$$MoS_2 + h\nu \rightarrow h^+ + e^- \quad (27)$$

$$e^- \rightarrow Ag(e^-) \quad (28)$$

$$Ag(e^-) + Cr(VI) \rightarrow Cr(III) + Ag \quad (29)$$

4 Applications of Cation Adsorption on Minerals

4.1 Biological Process

An antimicrobial is an agent that kills microorganisms or stops their growth. Antimicrobial medicines can be grouped according to the microorganisms. Antibacterial can be classified into inorganic antibacterial and organic antibacterial based on their active ingredients. Cations, such as Ag^+, Cu^{2+}, Zn^{2+}, belong to the inorganic antibacterial group. Among various heavy metal ions, Ag^+ is the most commonly used. Specific metal ions are crucial for the structure of cell membranes and DNA; approximately half of all known proteins are predicted to be dependent on metal atoms for their structure and their participation in key cellular processes, such as electron transfer and catalysis [53, 54]. Nonetheless, these essential metals are lethal to all cells when present in excess [55]. Metal ions poison microbial cells through the following mechanisms [55]: (1) Metals can lead to protein dysfunction. (2) They can also lead to the production of reactive oxygen species (ROS) and depletion of antioxidants. (3) Certain metals impair membrane function. (4) Some can interfere with nutrient assimilation. (5) They can also be genotoxic.

The use of metal ions as antimicrobial can date back to long time ago. Pioneering work in the 1880s by French mycologist Pierre-Marie-Alexis Millardet established that the Bordeaux mixture, which is a blend of copper sulphate ($CuSO_4$) and slaked lime ($Ca(OH)_2$), could be used to prevent the growth of downy mildew on grape vines [56]. In 1983, Japan's Shinagawa fuel company started the research using zeolite as antibacterial, which make it an increasing interest for scientist to develop antibacterial using non-metallic minerals as a carrier and metal ions as antibacterial substances. Zeolite X was synthesized and loaded with Ag^+ by ion exchange to be used as antimicrobial [57]. Escherichia coli and Pseudomonas aeruginosa and Staphylococcus aureus suspended in tryptone soya broth were exposed to 0.15, 0.25, 0.5 or 1.0 g/L of silver-loaded zeolite X for a period up to 24 h. No viable cells were detected for any of the three micro-organisms within 1 h. It indicated that silver-loaded zeolite X is a potential antimicrobial materials. Ag^+ is loaded by ion exchange on montmorillonite to be used as antimicrobial [58]. The silver-loaded montmorillonite exhibited strong antimicrobial activity to both Staphylococcus aureus and Escherichia coil, and antifungal activity to *Aspergillus niger*. For practical applications, however, some problems remain to be solved, e.g., how to suppress reduction of Ag^+ into silver particles on standing and how to suppress releasing of Ag^+ into water.

4.2 Ag-PILC Syntheses

Pillared interlayered clays (PILCs) are intercalated porous materials, which can be used as catalysts, catalyst supports and selective adsorbents [59]. The common technique to prepare PILCs is ion exchange of hydrolyzed polyoxocations into clay interlayers followed by calcination that induces thermal stability to the intercalated clays for applications [60, 61]. Polyoxocations were usually prepared by partial hydrolysis of soluble salts in aqueous solutions at optimized pH values. However, only a few of the metals such as aluminum, chromium, titanium and zirconium can be hydrolyzed into polyoxocations [62]. Thus, the ion exchange method is limited to the preparation of some specific PILCs, such as Ti–PILC, Al–PILC, Fe–PILC and Cr–PILC. Rao et al. [63, 64] employed the adsorption of cations on minerals (montmorillonite) in the syntheses of PILCs intercalated by Ag nanoparticles. These Ag–PILCs showed high specific surface area, pore structures and strong stability.

In the first synthesizing route (Fig. 11), aluminum chloride solution was added into the 1% montmorillonite suspension to reach the Al concentration of 5×10^{-3} mol/L at pH 5. Therefore, the montmorillonite layers were positively charged through the specific adsorption of hydrolysed aluminum species on their faces. After that, negatively charged Ag nanoparticles were intercalated into the positively charged montmorillonite interlayers for the pillaring of the interlayers. In the second synthesizing route (Fig. 12), silver ions were adsorbed on montmorillonite layers, and then reduced by sodium citrate into Ag nanoparticles. So the Ag-PILC was formed.

4.3 Porous Hydrogel

Hydrogel is an emerging material with three-dimensional (3D) network-structured construction, which has been extensively studied and designed as adsorbents for wastewater treatment in recent years due to the presence of abundant groups as hydroxyl, carboxyl and amide [65–67]. Normally, hydrogel is prepared based on the crosslinking of polysaccharide chains and chemical bonding, which endows hydrogel the performance of porous networks, swelling and flexibility [68, 69]. Polysaccharides with the properties of rich functional groups, biocompatibility and biodegradability have been widely used in hydrogel. Hydrogels such as chitosan-based polymer [70, 71], cellulose compound [72] and sodium alginate based beads [73] are intensively investigated materials, which has been used to remove organic and inorganic pollutants. However, hydrogels from polysaccharides

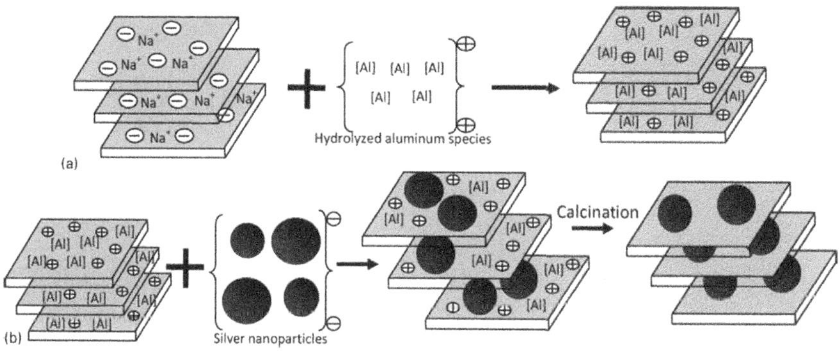

Fig. 11 Ag-PILC synthesis employed the adsorption of montmorillonite to cations: route I

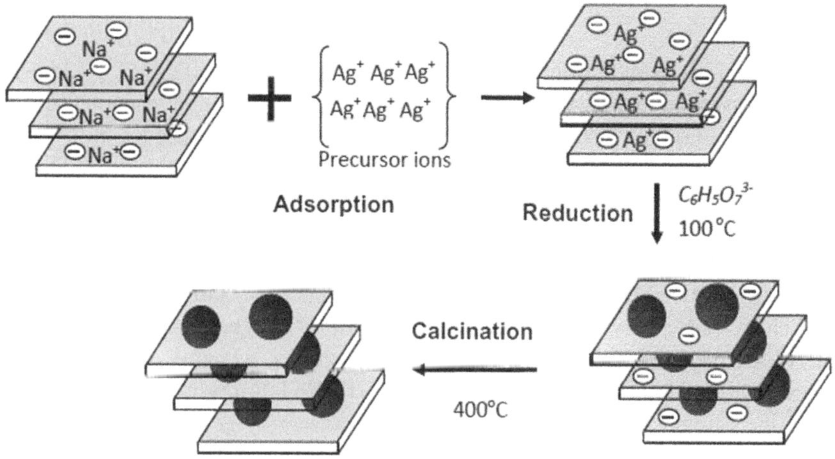

Fig. 12 Ag-PILC synthesis employed the adsorption of montmorillonite to cations: route II

suffer from some drawbacks like low mechanical strength, thermal stability and adsorbability [65], which will eventually affect not only the adsorption efficiency but also long term performance and regeneration. Fortunately, nano-materials like nanoclay, graphene, and carbon nanotubes can significantly improve the strength of hydrogel due to the high surface energy of nano-materials which can generate high constraining forces to the organic chains [74, 75]. Hence, combining the excellent adsorption property of nano-adsorbent and the 3D network-structured construction of hydrogel can fabricate a novel super-adsorbent and give full play to the characteristics of the both.

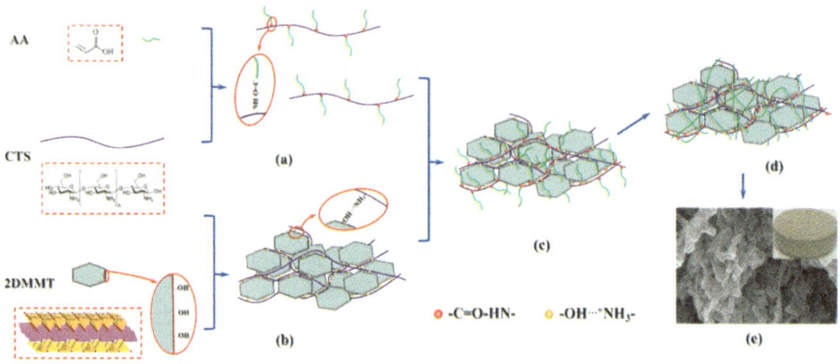

Fig. 13 Formation mechanism schematic of the hydrogel: **a** amidation between AA and CTS; **b** self-assembly of CTS and 2DMMT based on hydrogen bond; **c** amidation and self-assembly among AA, CTS and 2DMMT; **d** polymerization of AA; and **e** hydrogel

Hydrogel prepared from montmorillonite nanosheets (2DMMT), chitosan (CTS) and acrylic acid (AA) was shown as Fig. 13. The amidation reaction (–C=O–HN–) (Fig. 13a) take place between AA and CTS due to the existence of carboxyl (–COOH) and amino (–NH_2) [76]. 2DMMT could combine with CTS via self-assembly based on the hydrogen-bond (–OH…$^{+}NH_3^{-}$) interaction (Fig. 13b) as reported [68]. Hence, the amidation and self-assembly facilitated the combination among AA, CTS and 2DMMT as showed in Fig. 13c. Then the AA monomer molecules polymerized together with the existence of initiator, forming the long-chain poly (acrylic acid) (PAA, Fig. 13d) [77]. The PAA chains would interweave with CTS chains, making the hydrogel a more stable structure (Fig. 13e). The forming process transformed 2DMMT to porous hydrogel, facilitating the entrance and adsorption of metal ions.

Hydrogel prepared from montmorillonite nanosheets (2DMMT), acrylic acid (AA) and acrylamide (AM) was shown as Fig. 14. AA and AM molecules combined together via the amidation [76] (–C=O–HN–, Fig. 14a) and polymerization [78] (P(AA–co–AM), Fig. 14b) reactions. The polymerization could also combine the AA and AA [79], AM and AM [80] to form the PAA and PAM, respectively. Due to the existence of aluminum hydroxyl (Al–OH) on the edge of 2DMMT, AM could connect along the edge via the hydrogen-bond (–OH…NH_2^{-}) interaction [68] as shown in Fig. 14c. Hence, when AA, AM and 2DMMT were mixed together with the initiator of MBAm and $K_2S_2O_8$, amidation, polymerization and hydrogen-bond interactions would synchronize in plane to form the huge lamella as shown in Fig. 14d, e. The prepared porous hydrogel could achieve a 260 mg/g adsorption capacity for Pb(II).

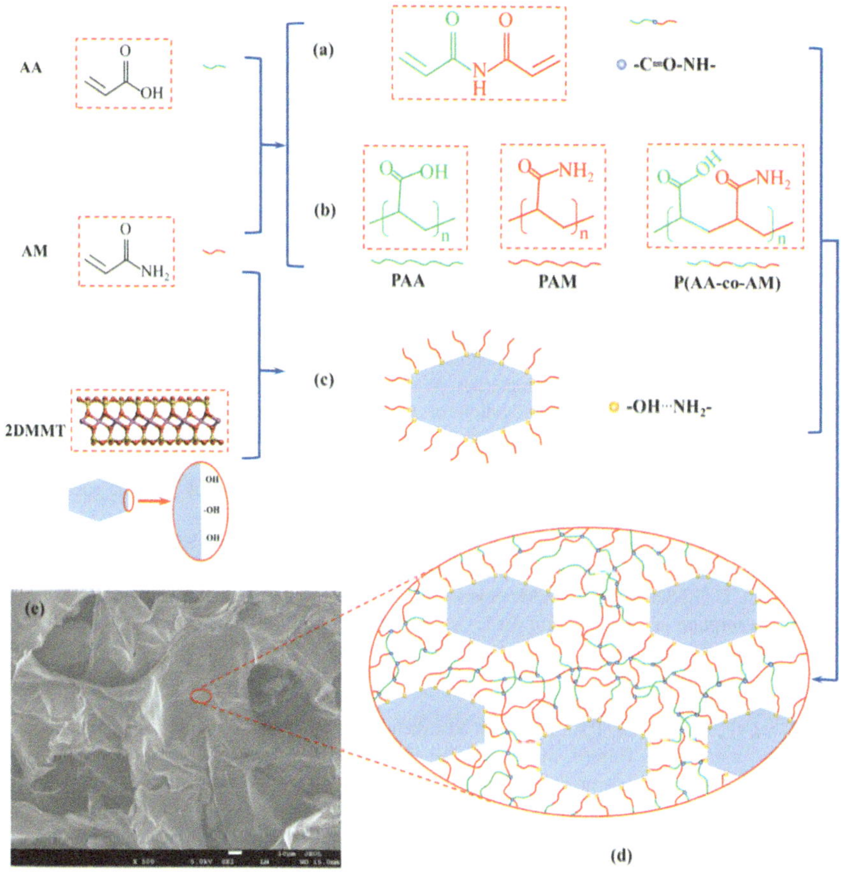

Fig. 14 Schematic diagram of the hydrogel formation process: **a** amidation between AA and AM; **b** polymerization between AA and AA, AM and AM, AA and AM; **c** AM connected around the edge of 2DMMT via hydrogen bond; **d** amidation, polymerization and hydrogen bond interaction among AA, AM and 2DMMT; and **e** hydrogel

References

1. Carroll, D.: Ion exchange in clays and other minerals. Geol. Soc. Am. Bull. **70**, 749–779 (1959)
2. Jiang, M.Q., Jin, X.Y., Lu, X.Q., Chen, Z.L.: Adsorption of Pb (II), Cd (II), Ni (II) and Cu (II) onto natural kaolinite clay. Desalination **252**, 33–39 (2010)
3. Erdem, E., Karapinar, N., Donat, R.: The removal of heavy metal cations by natural zeolites. J. Colloid Interf. Sci. **280**, 309–314 (2004)
4. Davis, J.A., Fuller, C.C., Cook, A.D.: A model for trace metal sorption processes at the calcite surface: adsorption of Cd^{2+} and subsequent solid solution formation. Geochim. Cosmochim. Acta **51**, 1477–1490 (1987)
5. Hayes, K.F., Leckie, J.O.: In: Mechanism of Lead Ion Adsorption at the Goethite-Water Interface. ACS Publications (1986)

6. Tahir, S., Rauf, N.: Thermodynamic studies of Ni (II) adsorption onto bentonite from aqueous solution. J. Chem. Thermodyn. **35** (2003)
7. Adebowale, K.O., Unuabonah, I.E., Olu-Owolabi, B.I.: The effect of some operating variables on the adsorption of lead and cadmium ions on kaolinite clay. J. Hazard. Mater. **134**, 130–139 (2006)
8. Zhao, D., Yang, X., Zhang, H., Chen, C., Wang, X.: Effect of environmental conditions on Pb (II) adsorption on β-MnO_2. Chem. Eng. J. **164**, 49–55 (2010)
9. Sheng, G., Wang, S., Hu, J., Lu, Y., Li, J., Dong, Y., Wang, X.: Adsorption of Pb (II) on diatomite as affected via aqueous solution chemistry and temperature. Colloid Surf. A **339**, 159–166 (2009)
10. Liang, P., Qin, Y., Hu, B., Li, C., Peng, T., Jiang, Z.: Study of the adsorption behavior of heavy metal ions on nanometer-size titanium dioxide with ICP-AES. Fresen J. Anal. Chem. **368**, 638–640 (2000)
11. Dong, D., Derry, L.A., Lion, L.W.: Pb scavenging from a freshwater lake by Mn oxides in heterogeneous surface coating materials. Water Res. **37**, 1662–1666 (2003)
12. Eren, E., Afsin, B.: An investigation of Cu (II) adsorption by raw and acid-activated bentonite: a combined potentiometric, thermodynamic, XRD, IR, DTA study. J. Hazard. Mater. **151**, 682–691 (2008)
13. Abdel-Samad, H., Watson, P.R.: An XPS study of the adsorption of lead on goethite (α–FeOOH). Appl. Surf. Sci. **136**, 46–54 (1998)
14. Li, H., Mu, S., Weng, X., Zhao, Y., Song, S.: Rutile flotation with Pb^{2+} ions as activator: adsorption of Pb^{2+} at rutile/water interface. Colloid Surf. A **506**, 431–437 (2016)
15. Strawn, D.G., Scheidegger, A.M., Sparks, D.L.: Kinetics and mechanisms of Pb (II) sorption and desorption at the aluminum oxide–water interface. Environ. Sci. Technol. **32**, 2596–2601 (1998)
16. Fu, F., Wang, Q.: Removal of heavy metal ions from wastewaters: a review. J. Environ. Manag. **92**, 407–418 (2011)
17. Babel, S., Kurniawan, T.A.: Low-cost adsorbents for heavy metals uptake from contaminated water: a review. J. Hazard. Mater. **97**, 219–243 (2003)
18. Uddin, M.K.: A review on the adsorption of heavy metals by clay minerals, with special focus on the past decade. Chem. Eng. J. **308**, 438–462 (2017)
19. Li, Y.H., Wang, S., Wei, J., Zhang, X., Xu, C., Luan, Z., Wu, D., Wei, B.: Lead adsorption on carbon nanotubes. Chem. Phys. Lett. **357**, 263–266 (2002)
20. Sekar, M., Sakthi, V., Rengaraj, S.: Kinetics and equilibrium adsorption study of lead (II) onto activated carbon prepared from coconut shell. J. Colloid Interf. Sci. **279**, 307–313 (2004)
21. Bhattacharyya, K.G., Gupta, S.S.: Kaolinite, montmorillonite, and their modified derivatives as adsorbents for removal of Cu (II) from aqueous solution. Sep. Purif. Technol. **50**, 388–397 (2006)
22. Li, N., Bai, R.: Copper adsorption on chitosan–cellulose hydrogel beads: behaviors and mechanisms. Sep. Purif. Technol. **42**, 237–247 (2005)
23. Tan, G., Xiao, D.: Adsorption of cadmium ion from aqueous solution by ground wheat stems. J. Hazard. Mater. **164**, 1359–1363 (2009)
24. Boparai, H.K., Joseph, M., O'Carroll, D.M.: Kinetics and thermodynamics of cadmium ion removal by adsorption onto nano zerovalent iron particles. J. Hazard. Mater. **186**, 458–465 (2011)
25. Nabais, J.V., Carrott, P., Carrott, M.R., Belchior, M., Boavida, D., Diall, T., Gulyurtlu, I.: Mercury removal from aqueous solution and flue gas by adsorption on activated carbon fibres. Appl. Surf. Sci. **252**, 6046–6052 (2006)
26. Owlad, M., Aroua, M.K., Daud, W.A.W., Baroutian, S.: Removal of hexavalent chromium-contaminated water and wastewater: a review. Water Air Soil Poll. **200**, 59–77 (2009)
27. Schneider, R., Cavalin, C., Barros, M., Tavares, C.: Adsorption of chromium ions in activated carbon. Chem. Eng. J. **132**, 355–362 (2007)

28. Strawn, D.G., Sparks, D.L.: The use of XAFS to distinguish between inner-and outer-sphere lead adsorption complexes on montmorillonite. J. Colloid Interf. Sci. **216**, 257–269 (1999)
29. Oubagaranadin, J.U.K., Murthy, Z.: Adsorption of divalent lead on a montmorillonite–illite type of clay. Ind. Eng. Chem. Res. **48**, 10627–10636 (2009)
30. Weaver, C.E., Pollard, L.D.: The Chemistry of Clay Minerals. Elsevier (2011)
31. El-Bayaa, A., Badawy, N., AlKhalik, E.A.: Effect of ionic strength on the adsorption of copper and chromium ions by vermiculite pure clay mineral. J. Hazard. Mater. **170**, 1204–1209 (2009)
32. Coles, C.A., Yong, R.N.: Aspects of kaolinite characterization and retention of Pb and Cd. Appl. Clay Sci. **22**, 39–45 (2002)
33. Gu, X., Evans, L.J.: Surface complexation modelling of Cd (II), Cu (II), Ni (II), Pb (II) and Zn (II) adsorption onto kaolinite. Geochim. Cosmochim. Acta **72**, 267–276 (2008)
34. Zhao, D., Chen, S., Yang, S., Yang, X., Yang, S.: Investigation of the sorption behavior of Cd (II) on GMZ bentonite as affected by solution chemistry. Chem. Eng. J. **166**, 1010–1016 (2011)
35. Jia, F., Wang, Q., Wu, J., Li, Y.: Two-dimensional molybdenum disulfide as a superb adsorbent for removing Hg^{2+} from water. Acs. Sustain. Chem. Eng. **5**, 7410–7419 (2017)
36. Liu, C., Jia, F., Wang, Q., Yang, B., Song, S.: Two-dimensional molydbenum disulfide as adsorbent for high-efficient Pb(II) removal from water. Appl. Mater. Today **9**, 220–228 (2017)
37. Jia, F., Zhang, X., Song, S.: AFM study on the adsorption of Hg^{2+} on natural molybdenum disulfide in aqueous solutions. Phys. Chem. Chem. Phys. **19**, 3837–3844 (2017)
38. Jia, F., Liu, C., Yang, B., Zhang, X., Yi, H., Ni, J., Song, S.: Thermal modification of the molybdenum disulfide surface for tremendous improvement of Hg^{2+} adsorption from aqueous solution. Acs. Sustain. Chem. Eng. **6**, 9065–9073 (2018)
39. Jean, G.E., Bancroft, G.M.: Heavy metal adsorption by sulphide mineral surfaces. Geochim. Cosmochim. Acta **50**, 1455–1463 (1986)
40. McKenzie, R.: The adsorption of lead and other heavy metals on oxides of manganese and iron. Soil Res. **18**, 61–73 (1980)
41. McBride, M.: Environmental Chemistry of Soils. Oxford University Press, New York (1994)
42. Dzombak, D.A.: Surface Complexation Modeling: Hydrous Ferric Oxide. Wiley (1990)
43. Sturchio, N.C., Chiarello, R.P., Cheng, L., Lyman, P.F., Bedzyk, M.J., Qian, Y., You, H., Yee, D., Geissbuhler, P., Sorensen, L.B.: Lead adsorption at the calcite-water interface: synchrotron X-ray standing wave and X-ray reflectivity studies. Geochim. Cosmochim. Acta **61**, 251–263 (1997)
44. Qian, Y., Sturchio, N.C., Chiarello, R.P., Lyman, P.F., Lee, T.L., Bedzyk, M.J.: Lattice location of trace elements within minerals and at their surfaces with X-ray standing waves. Science-New York Then Washington, 1555–1555 (1994)
45. Brown, J.R., Bancroft, G.M., Fyfe, W.S., McLean, R.A.: Mercury removal from water by iron sulfide minerals. An electron spectroscopy for chemical analysis (ESCA) study. Environ. Sci. Technol. **13**, 1142–1144 (1979)
46. Özverdi, A., Erdem, M.: Cu^{2+}, Cd^{2+} and Pb^{2+} adsorption from aqueous solutions by pyrite and synthetic iron sulphide. J. Hazard. Mater. **137**, 626–632 (2006)
47. Richard, F.C., Bourg, A.C.: Aqueous geochemistry of chromium: a review. Water Res. **25**, 807–816 (1991)
48. Aoki, T., Munemori, M.: Recovery of chromium (VI) from wastewaters with iron (III) hydroxide—I: adsorption mechanism of chromium (VI) on iron (III) hydroxide. Water Res. **16**, 793–796 (1982)
49. Lu, A., Zhong, S., Chen, J., Shi, J., Tang, J., Lu, X.: Removal of Cr (VI) and Cr (III) from aqueous solutions and industrial wastewaters by natural clino-pyrrhotite. Environ. Sci. Technol. **40**, 3064–3069 (2006)
50. Albadarin, A.B., Mangwandi, C., Ala'a, H., Walker, G.M., Allen, S.J., Ahmad, M.N.: Kinetic and thermodynamics of chromium ions adsorption onto low-cost dolomite adsorbent. Chem. Eng. J. **179**, 193–202 (2012)

51. Garcıa-Sánchez, A., Alvarez-Ayuso, E.: Sorption of Zn, Cd and Cr on calcite. Application to purification of industrial wastewaters. Miner. Eng. **15**, 539–547 (2002)
52. Yang, L., Yang, Q., Jia, F. et al.: Cr(VI) Removal from water with amorphous graphite concentrate contaminated by iron. Min. Proc. Ext. Met. Rev. **38**(2) (2017)
53. Waldron, K.J., Robinson, N.J.: How do bacterial cells ensure that metalloproteins get the correct metal? Nat. Rev. Microbiol. **7**, 25–35 (2009)
54. Andreini, C., Bertini, I., Rosato, A.: A hint to search for metalloproteins in gene banks. Bioinformatics **20**, 1373–1380 (2004)
55. Lemire, J.A., Harrison, J.J., Turner, R.J.: Antimicrobial activity of metals: mechanisms, molecular targets and applications. Nat. Rev. Microbiol. **11**, 371–384 (2013)
56. Ayres, P.G., Millardet, A.: France's forgotten mycologist. Mycologist **18**, 23–26 (2004)
57. Kwakye-Awuah, B., Williams, C., Kenward, M., Radecka, I.: Antimicrobial action and efficiency of silver-loaded zeolite X. J. Appl. Microbiol. **104**, 1516–1524 (2008)
58. Ōya, A., Banse, T., Ohashi, F., Ōtani, S.: An antimicrobial and antifungal agent derived from montmorillonite. Appl. Clay Sci. **6**, 135–142 (1991)
59. Garrido-Ramirez, E.G., Theng, B.K.G., Mora, M.L.: Appl. Clay Sci. **47**, 182 (2010)
60. Gil, A., Gandia, L.M., Vicente, M.A.: Catal Rev Sci Eng **42**, 145 (2000)
61. Zhang, Y., Song, S., Zhang, M., Tuo, B.: Surf. Rev. Lett. **15**, 329 (2008)
62. Li, J., Mu, Z., Xu, X., Tian, H., Duan, M., Li, L., Hao, Z., Qiao, S., Lu, G.: Micropor. Mesopor. Mat. **114**, 214 (2008)
63. Rao, F., Song, S., Lopez-valdivieso, A.: Synthesis and characterization of Ag-PILC through the formation of Ag@Montmorillonite nanocomposite. NANO: Brief Reports and Reviews **10**, 1–9 (2015)
64. Rao, F., Song, S., Lopez-valdivieso, A.: Synthesis of Ag-PILC through direct insertion of silver nanoparticles into smectite interlayers. Mater. Technol. **27**, 186–190 (2012)
65. Mohammadzadeh Pakdel, P., Peighambardoust, S.J.: Review on recent progress in chitosan-based hydrogels for wastewater treatment application. Carbohydr. Polym. **201**, 264–279 (2018)
66. Han, H., Lu, X., Liu, L., Yang, W., Shi, W., He, X.: Bioinspired and microgel-tackified adhesive hydrogel with rapid self-healing and high stretchability. Macromolecules **52**, 72–80 (2018)
67. Yang, X., Qin, Z., Wu, H., Lv, H., Yu, X.: Tuning hydrogel mechanics by kinetically dependent cross-linking. Macromolecules **52**, 1249–1256 (2019)
68. Wang, W., Zhao, Y., Yi, H., Chen, T., Kang, S., Li, H., Song, S.: Preparation and characterization of self- assembly hydrogels with exfoliated montmorillonite nanosheets and chitosan. Nanotechnology **29**, 025605 (2018)
69. Feng, Q., Wei, K., Lin, S., Xu, Z., Sun, Y., Shi, P., Li, G., Bian, L.: Mechanically resilient, injectable, and bioadhesive supramolecular gelatin hydrogels crosslinked by weak host-guest interactions assist cell infiltration and in situ tissue regeneration. Biomaterials **101**, 217–228 (2016)
70. Boardman, S., Lad, R., Green, D., Thornton, P.: Chitosan hydrogels for targeted dye and protein adsorption. J. Appl. Polym. Sci. **134**, 1–10 (2017)
71. Salzano de Luna, M., Castaldo, R., Altobelli, R., Gioiella, L., Filippone, G., Gentile, G., Ambrogi, V.: Chitosan hydrogels embedding hyper-crosslinked polymer particles as reusable broad-spectrum adsorbents for dye removal. Carbohydr. Polym. **177**, 347–354 (2017)
72. Maity, J., Ray, S.: Removal of Cu (II) ion from water using sugar cane bagasse cellulose and gelatin based composite hydrogels. Int. J. Biol. Macromol. **97**, 238–248 (2017)
73. Hu, Z., Omer, A., Ouyang, X., Yu, D.: Fabrication of carboxylated cellulose nanocrystal/sodium alginate hydrogel beads for adsorption of Pb(II) from aqueous solution. Int. J. Biol. Macromol. **108**, 149–157 (2018)
74. Heng, L., Guo, X., Guo, T., Wang, B., Jiang, L.: Strengthening of polymer ordered porous materials based on a layered nanocomposite internal structure. Nanoscale **8**, 13507–13512 (2016)

75. Liu, C., Liu, H., Zhang, K., Dou, M., Pan, B., He, X., Lu, C.: Partly reduced graphene oxide aerogels induced by proanthocyanidins for efficient dye removal. Bioresour. Technol. **282**, 148–155 (2019)
76. Kumararaja, K., Manjaiah, P., Datta, S., Shabeer, T., Sarkar, B.: Chitosan-g-poly (acrylic acid) -bentonite composite: a potential immobilizing agent of heavy metals in soil. Cellulose **25**, 3985–3999 (2018)
77. Ma, Y., Lv, L., Guo, Y., Fu, Y., Shao, Q., Wu, T., Guo, S., Sun, K., Guo, X., Wujcik, E.K., Guo, Z.: Porous lignin based poly (acrylic acid)/organo-montmorillonite nanocomposites: swelling behaviors and rapid removal of Pb (II) ions. Polymer **128**, 12–23 (2017)
78. Ji, J., Zeng, C., Ke, Y., Pei, Y.: Preparation of poly(acrylamide-co-acrylic acid)/silica nanocomposite microspheres and their performance as a plugging material for deep profile control. J. Appl. Polym. Sci. **134**, 1–11 (2017)
79. Liang, T., Yan, C., Li, X., Zhou, S., Wang, H.: Polyacrylic acid grafted silica fume as an excellent adsorbent for dysprosium(III) removal from industrial wastewater. Water Sci. Technol. **77**, 1570–1580 (2018)
80. Wan, Y., Wang, T., Lu, H., Xu, X., Zuo, C., Wang, Y., Teng, C.: Design and synthesis of graphene/SnO_2/polyacrylamide nanocomposites as anode material for lithium-ion batteries. Rsc. Adv. **8**, 11744–11748 (2018)

Adsorption of Organic Compounds on Minerals

Jing Su

Abstract Organic compounds have attracted extensive concerns in water treatment, agriculture and ecological engineering due to most of which are poorly-biodegradable and highly-toxic to aquatic ecosystem as well as the human beings. Therefore, developing efficient and economical technologies to remove organic contaminants from aqueous system is highly essential for ensuring the safty of drinking water for human beings. In this chapter, the adsorption of organic compounds on minerals, which is one of the most promising technologies for removing organic compounds, will be discussed. Four kinds of typical organic compounds, i.e. dyes, phenolic compounds, antibiotics and perfluorooctane sulphonate were discussed in detail. The physical and chemical properties of each organic compounds were firstly introduced. Then the structure and properties of minerals which were appropriate to immobilize organic compounds were reviewed. In the meanwhile, in each section, the adsorption process of those organic compounds on minerals and the related adsorption mechanism were also discussed. This chapter provided a comprehensive insight into the role of natural or modified minerals in organic compounds adsorption, which probably pave the way for organic wastewater treatment.

Keywords Organic compounds · Minerals · Hydrophobic interaction · Ion exchange · Adsorption

---

J. Su (✉)
Key Laboratory of Environmental Nanotechnology and Health Effects, Research Center for Eco-environmental Sciences, Chinese Academy of Sciences, Beijing 100085, People's Republic of China
e-mail: 1062908702@qq.com

S. Song and B. Li (eds.), *Adsorption at Natural Minerals/Water Interfaces*, Engineering Materials,
https://doi.org/10.1007/978-3-030-54451-5_6

1 Introduction

Organic compounds in water treatment, attracted extensive concerns because most of which are poorly–biodegradable and do great harm to aquatic ecosystem as well as the health of human beings. Compared with inorganic ions, the larger size, more complicated structure and higher toxicity have made the removal of organic compounds more challengeable. In this chapter, the adsorption of organic compounds on minerals, which is one of the most promising technologies for removing organic compounds, will be discussed. The development of organics adsorption by minerals as well as the related physical and chemical adsorption principles will be introduced. Detailed discussion will be given on the adsorption of dyes, phenolic compounds, antibiotics and perfluorooctane sulphonate. The physical and chemical properties of each organic compounds, the type and structure of mineral adsorbents, and the probable adsorption mechanisms will also be described.

1.1 Organic Compounds

Organic compounds, especially organic contaminants, deriving from industrial production, waste water disposal and agricultural fertilization, have attracted highly-intensive attention of the government, public and environmental scientists in past few decades (Fig. 1). Most of organic compounds are easily leaked into aqueous ecosystem if no proper control or management was taken [1–3]. In general, those organic compounds contained one or more aromatic rings were stable in aqueous system and hard to be biodegraded, which would lead to high toxicity and biocumulativity in ecosystem [4]. It was reported that many emerging contaminants were supposed of high correlation with genetic or malignant disease such as breast cancer, reproductive multifunction, kidney damage, liver damage, and renal disorders [5]. For example, antibiotics, as kind of common PPCPs (shorted from pharmaceutical and Personal Care Products, broadly referring to any product with healthcare or medical purposes for humans and animals), not only do harm to metabolism of microbials but also cause resistance gene pollution in underground water, drinking water and soil, which lead to high risk of ecological security [6, 7]. Perfluorooctane sulfonate (PFOs), a kind of organics composed of carbon chains without any aromatic rings, also show high environmental risk [8–10]. They were used widely in industrial and commercial applications because of their water–resistant properties [3] and its derivative has been recently recognized as persistent organic pollutants (Pops) due to their endocrine disrupting effects [11]. It should be noted that extensive researches have focused on the environmental behavior of those organic compounds, including their origination, transportation, transformation and fate. However, for environmental engineers and water chemistry scientists, how to remove those emerging contaminant to satisfy the environmental standard and minimize the toxicity of aqueous system is an urgent problem to solve out.

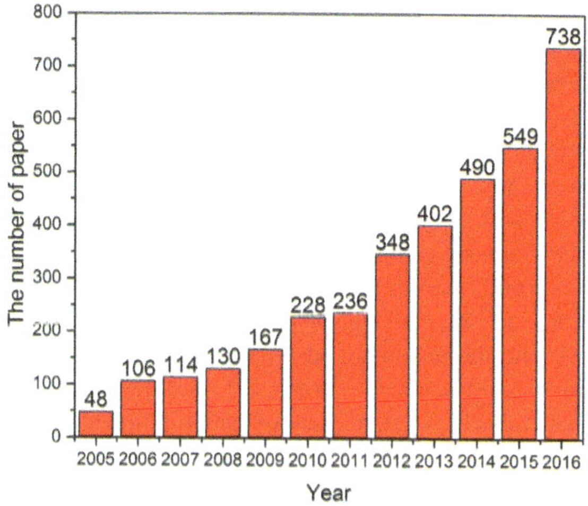

Fig. 1 The number of published paper about "emerging contaminants" during 2005–2016. Cited from Web of Science

In order to control the concentration of organic pollutants draining into the water environment, much effort have been made during sewage treatment. In general, the technologies for organic removal include adsorption, coagulation, membrane process, advanced oxidation process and biofilm treatment. Flocculating agent is indispensable during coagulation, thus may deteriorate the water quality and lead to relatively low efficiency [12]. Membrane fouling would happen during membrane process and that also cause low removal efficiency [13]. As a comparison, advanced oxidation process (AOPs) has attracted great interests due to its high efficiency. The toxic pollutants can be degraded to low molecular organic matter which was less harmful during AOPs. However, it may also cause secondary pollution because of the production of by–product [14]. Qiu et al. [15] reported that bisphenol A (BPA) can be photodegraded thoroughly by oxygen doped g–C_3N_4 although its mineralization rate were just 56%, which means bisphenol A cannot be converted completely to CO_2 and H_2O. The remnant organics still have negative effects on water quality. Biofilm process is an promising method which is environmentally friendly without secondary pollution production compared to coagulation and AOPs, but it also suffers from complicated equipment operation and expensive maintenance. Therefore, considering the comprehensive facts of low cost, high efficiency and non-secondary pollution, adsorption has shown a great potential in practical organic contaminants remediation among above technologies and methods.

1.2 Adsorption of Organic Compounds

Various adsorbents have been developed to decontaminate organic waste water, such as carbon materials [16] (graphene, carbon nanotube, activated carbon, biochars, etc.), polymers [17, 18] (polymer resins, natural polymers and synthesized polymers, etc.), minerals and their modified compounds [19]. It was known that the enhanced removal efficiency could be acquired by various modification methods. Yang et al. prepared multifunctional magnetic N-doped graphene composite for Rhodamine 6G detection, adsorption separation and photocatalytic degradation. It was showed that modified graphene exhibited improved removal ability towards R6G with an adsorption capacity of 44.0 mg g^{-1} [20]. Xu et al. [21] reported that the graphene composites exhibited high removal efficiency of bisphenol A with an adsorption capacity of 181.8 mg g^{-1}. Sadeghi-Kiakhani et al. [22] founded that ethyl acrylate grafting played excellent roles on removing basic blue by chitosan 41, and the maximum adsorption capacity of the modified chitosan towards basic blue could reach 217.39 mg g^{-1}. Besides those synthetic nanomaterials, naturals minerals also showed great potential in organic pollutants adsorption attributing to their intrinsic large ion exchange capacity, high charge density and surface area, abundant content in earth as well as chemical stability [23]. However, most of natural clay minerals are hydrophilic and they are not suitable for removing organic compounds that are hydrophobic. Thus it is essential to graft appropriate functional groups on natural clay minerals to change the surface properties and increase the number of adsorption sites of minerals.

In general, chemical modification includes surfactant modification, amino/ oxygen containing groups grafting, transitional metals doping and other adsorbents recombination. The detailed examples will be given as follows:

(1) Modifying surfactant on minerals. The surface of minerals can be converted totally from hydrophilic to hydrophobic by surfactant modification. Surfactant is a kind of amphipathic molecule of which hydrophobic end has a great affinity to organic compounds via hydrophobic interaction. It has been found that the adsorption of BPA on montmorillonite modified with dodecyl dimethyl betaine was much more efficient [24].

(2) Grafting amino or oxygen-containing groups. The hydrogen bonds between organic compounds and active groups such as –OH, $-NH_2$, –COOH, etc. could be an effective interaction for organic compounds immobilization [25]. Additionally, carboxyl and carbonyl with abundant electron groups can alter surface electrical properties of minerals, thus in favor of electrostatic attraction between organic compounds and adsorbents. Take hydrotalcite as an example, several studies revealed that organic pollutants can be effectively adsorbed on the interlayer of the anionic surfactant-layered double hydroxides (LDH) due to the partitioning mechanism. However, Li et al. [26] grafted rhamnolipid with abundant carboxyl and carbonyl on LDH and the composite exhibited higher adsorption towards pcresol than that of presitine LDH, and found that electrostatic attraction and hydrophobic interaction were the main adsorption mechanisms. Furthermore, Munagapat et al. [27] proved that nano goethite modified with calcium alginate beads exhibited great efficiency on congo red adsorption, which mainly attributed to the strong electrostatic interaction between the $-OH_2^+$ of calcium alginate beads and dye anions in acid condition. The maximum monolayer adsorption capacity was 181.1 mg/g at pH 3.

(3) Doping transition metal or their oxides. Gamba et al. [28] found that Cu^{2+}–loaded organo–montmorillonites were efficient for thiabendazole adsorption due to the formation of Cu^{2+}–thiabendazole complexes during adsorption process. Additionally, the surface area and active sites were also increased by transition metal and oxides doping, further improving the adsorption capacity and adsorption kinetics. Functional bentonite supporting with Fe/Pd nanoparticles (B–Fe/Pd) was used to remove microcystin–LR (MC–LR) in Wang's work and the removal efficiency for MC–LR could reach 96.86% based on adsorption and degradation [29]. Iron nanoparticles could contribute more adsorption sites for MC–LR removal and excellent degradation was achieved attributing to higher adsorption capacity. In the meanwhile, the ferric oxides or copper oxides could also act as Fenton or Fenton—like heterogeneous catalysts, which not only adsorb organic compounds effectively but also catalyze degradation. Indeed, pyrite and iron sulfide have been explored to boost the degradation of recalcitrant organic pollutants such as chlorinated organic compounds, benzene, and PAHs, etc. in recent years [30].

(4) Compounding with other adsorbents. Compounding with Nano materials was also an effective method to improve the adsorption efficiency of minerals torwards organic contaminants since nano composites was of large surface area and enormous active sites. For example, many studies have proven that clay–nZVI composites were efficient remediation materials for organic waste water treatment because they exhibited better adsorption and catalytic performance than sole nZVI and clay. In the hybrid composites, clay minerals inhibited agglomeration of nZVI particles, which might expose more active sites to contaminants, thereby showing a synergistic effect for the removal of phenolic compounds, dyes and poly brominated diphenyl ethers (Fig. 2) [31]. Besides iron modified composites, there are other nano materials modified minerals appropriate to adsorb organic compounds. Liu et al. [32] combined reduced graphene oxide with halloysite nanotube (HNT) via an electrostatic self-assembly process and this composite showed moderate adsorption capacity for rhodamine B (RhB). Wang et al. synthesized a spherical zeolite/reduced graphene oxide composite by blending HNT–derived zeolite with graphene oxide nanosheets. The synthesized composite showed an adsorption capacity of 53.3 mg g^{-1} for methylene blue and a smaller value of 48.6 mg g^{-1} for malachite green [33]. It can be concluded that the synergistic interaction between nanomaterials and minerals could result in the improvement of the adsorption efficiency. To sum up, the relative high capacity and fast adsorption behavior of organic compounds on minerals can be achieved by surface modification or activate composite combination.

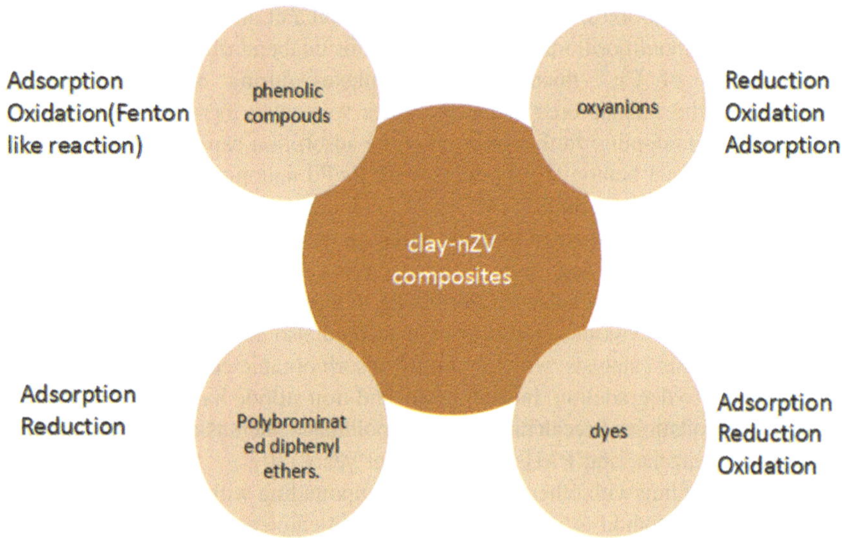

Fig. 2 Schematic diagram of reaction mechanism between clay–nZVI composite with different aqueous contaminants [31]

1.3 The Adsorption Mechanisms on Minerals

1.3.1 Electrostatic Attraction

Electrostatic interaction is a kind of physical interaction. The strength of electrostatic interaction depends on the charge properties of organic compounds and minerals. The electrostatic attraction present when the adsorbents and target pollutants were charged oppositely in water solution, otherwise the electrostatic repulsion would predominate. It should be noted that the electrostatic interaction were easily affected by the solution acidity condition. Therefore, pH is an important factor to affect the charge of both adsorbents and adsorbates and then further influence the adsorption efficiency. It has been reported that the adsorption efficiency for natural organic matter (NOM) and phenolic compounds on nanomaterials decreased with increasing pH. Because NOM and phenolic compounds were easily hydrolyzed into anions when pH increased, both target pollutants and the surface of nanomaterials were electronegative under alkaline condition, which cause a strong electrostatic repulsion and thus reducing the adsorption efficiency when pH was high. Similar to nanomaterials, the clay minerals were also electronegative over a wide environmental pH range attributing to their low isoelectric point, thus might caused that only cationic organics can be adsorbed effectively. For instance, cation dyes, such as methylene blue which is positively charged in neutral pH condition was always remarkably adsorbed on the surface of the minerals which were electronegative. In contrast, the surface of LDH was positively charged, which showed a great removal efficiency towards phenolic compounds that was usually negatively charged in solution.

1.3.2 Hydrophobic Interaction

Hydrophobic interaction would occur when the surface of minerals tend to be hydrophobic and the adsorbates contain aromatic structures. The hydrophobic interaction can be regarded as a kind of Van der Waal's force. Although most of natural clays are hydrophilic, modifying with hydrophobic groups or composites can reverse the hydrophilicity of minerals. For example, recombination of minerals with carbonaceous materials like graphene or activated carbon often showed enhanced adsorption capacity towards organic compounds attributing to the strong hydrophobic interaction between carbon composites and hydrophobic compounds. Besides carbon based minerals, modifying with surfactant on clay minerals was also a promising method to enhance the hydrophobicity of adsorbents. Montmorillonite were usually intercalated or functionalized with surfactant to form organoclay so that were in favor of PHAs adsorption due to the swelling property of montmorillonite and strong affinity between surfactant and pollutants [34]. Many investigations have proven that the strength of hydrophobicity interaction showed an inverse relationship with the oxygen content of the adsorbents and the solubility of pollutants [35, 36].

1.3.3 Ion Exchange

Ion exchange, especially cation exchange, was the dominated process during organics adsorption on minerals. Many researchers have found that montmorillonite or smectites could be intercalated by ionic surfactants to form organoclays through ion exchange. Some ionic organic compounds, especially cation organics similar to those surfactant, also show the same adsorption behavior through intercalation or ion exchange. Because various alkali metals and alkaline earth metals exist in the silicate interlayer of the clay to balance the surface charge, most of them are readily to be exchanged out with other cations or cation organic compounds. Those exchanged organic cations in interlayers can further interact with the silicate layer through electrostatic attraction, or form coordinate bonds, making organic compounds stay in the interlayer stablely and giving rise to larger space distance of clay minerals. The montmorillonite basal spacing could enlarge to 1.5–2.3 nm after organic surfactant intercalating into the interlayer, indicating of larger specific surface area and probable higher adsorption capacity towards organic compounds [34]. It should be noted that the orientation of surfactant in the galleries of the silicate layers could change the layer space of minerals conspicuously. Take alkylammonium ions as an example, when the intercalated orientation of alkylammonium ions gradually changed from horizontal to vertical, the interlamellar spacing of clay minerals would increase, and thus might further facilitate the process of ion exchange or adsorption (Fig. 3).

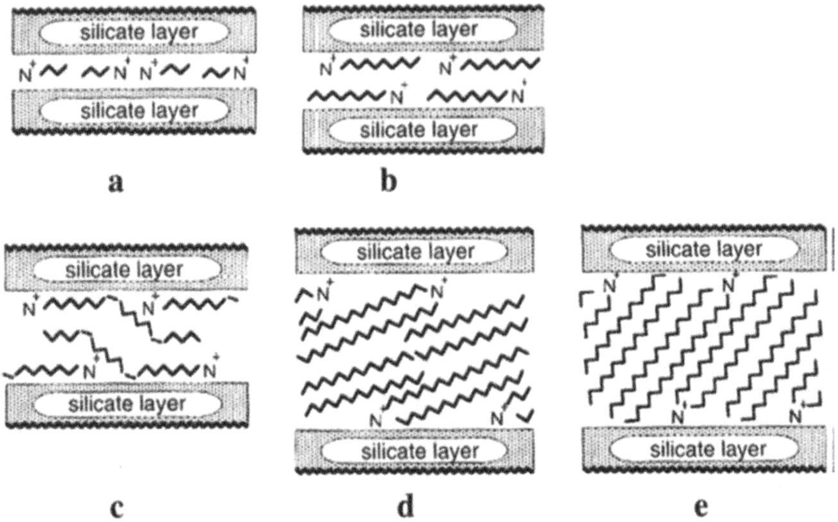

Fig. 3 Orientations of alkylammonium ions in the galleries of layered silicates: **a** Monolayer, **b** bilayers, **c** pseudotrimolecular layers, and **d**, **e** paraffin–type arrangements of alkylammonium ions with different tilting angles of the alkyl chains [34]

Moreover, exchanging with larger molecules could increase the hydrophobicity of clay and that would further improve the removal of apolar non-ionic organic compounds. Yu et al. prepared an organo–montmorillonite using alkylphosponium cations of (4-carboxybutyl)–triphenylphosphonium bromide as intercalation agents through cation exchanging [37]. Study about adsorption of methylene blue on bentonite and sepiolite samples has been carried out by Bilgiç et al. and the cation exchange has proved to be the main mechanism during adsorption [38]. Besides cation exchanging, anion exchanging is also a common pathway for organic compounds removal. In this case, the organic compounds should be anion organics which could occupy electronegative exchange sites to balance the minerals surface charge. For example, Yao et al. [39] found glycerol-modified nanocrystallined Mg/Al layered double hydroxides could adsorb methyl orange effectively and anion exchange was one of the main adsorption mechanism.

1.3.4 Hydrogen Bonding

Forming hydrogen bonding is an important pathway for minerals to immobilize organic compounds when adsorbents contained functional groups such as–OH, –COOH, $-NH_2$. Some ferric minerals, such as magnetite, goethite and hematite usually exist in the form of hydrate with abundant surface hydroxyl. Those hydroxyl could contribute electrons to aromatic compounds and capture them firmly through hydrogen bonding. Filius et al. confirmed that H–bonding is an important sorption mechanism for humic substances removal by goethite. The detailed adsorption mechanism could be identified by NMR, IR or modeling. Solution pH and ion strength could affect the adsorption efficiency of fulvic acids on goethite remarkably, indicating that the strength of H bonding could be influenced by environmental conditions [40]. It has been demonstrated that the adsorbed amount of dyes which contained hydroxyl (–OH) groups (EBT, BPB, BCG, and FLU) was much higher than those of non –OH groups by clay minerals. Therefore, the hydrogen bonding played an important role in dye adsorption on magnetite. However, sometimes the water molecules would compete the surface site of mineral with organic compounds due to stronger H bonding interactions exist between H_2O molecules and minerals surface.

1.3.5 Ligand Exchange/Coordination

Ligand exchange refers to one or more kinds of ligands of adsorbents that substituted by other ligands of adsorbates. It would occur in soil where abundant humic substances and clay minerals coexist and both of them contain carboxyl and carbonyl groups which are exchangeable. It has been found that the exchange of organic ligands with –OH groups may help stabilize organics on the surface of hematite [41]. In addition, organic compounds tend coordinate with iron containing minerals because the orbital of Fe was available to accept electrons from –OH, –NH_2 or–SH. Chen et al. [42] reported that the high exothermic molar enthalpy of humic acid binding to the clays occurred was probably ascribed to ligand exchange and electrostatic binding, meaning that ligand exchange played an important role in adsorption process between DOM and clay.

To sum up, the adsorption mechanisms of organic compounds by minerals depend on the properties of mineral and organic compounds as well as the solution conditions, such as pH, ion strength and coexistent substances. In the following chapters, some typical organic compounds would be given as examples and their adsorption behavior on minerals will also be discussed in detail.

2 Adsorption of Dye on Minerals

2.1 General Introduction of Dyes

Dyes mainly come from food processing, medication, printing and maquillage, etc. It was reported that the annual global production of dyes was nearly 7–9 million tons, and 10–15% of them were discharged into water ecosystem without appropriate disposals. They can be classified as sulphur dyes, vat dyes, oxidation dyes, acid dyes, insoluble azo dye, disperse dyes and cation dyes according to their chemical structures and properties. Most dyes contained one or more aromatic rings which derived from benzene, naphthalene, anthracene etc., suggesting of chemical stability and high toxicity of them. Moreover, high concentrations and high colority made dyes difficult to be removed. According to the practical cases, the concentration of colored suspended matter in dyes waste water could reach as high as 100–150 mg/L and the COD (Chemical Oxygen Demands) were almost 3000–16,000 mg/L. Additionally, the large amount of acid and base were used during dyes production, which induced obvious pH fluctuation in wastewater treatment and that further increase the treatment cost. Therefore, it was clear that dyes were hard biodegradable and its wastewater is pretty complicated. In the aspect of eco toxicity, a few reports have confirmed the adverse effects of dyes on the health of human beings. For example, azo dyes were commonly used in industrial manufacture such as textile coloring and oil painting, however, malignant tumor like bladder cancer, spleen tumor and liver cancer could be induced by anilines coming from incompletely degradation of azo dyes. A few investigations suggested that Brown FK, D & C Red No. 9, Diamond Green B and Direct Black 38 had direct mutation on Clostridium species, bacterial [43], Ames typhimurium Salmonella [44] and S. typhimurium TA 98 [45], respectively. Oliveria et al. [46] also proved that Disperse Red 1, Disperse Red 13 and Disperse Orange 1 had significant impaired effect on DNA of HepG2 cells. Besides azo dyes, anthraquinone dyes and triphenylmethane dyes also exhibited high risk in eco environmental security. Marrs et al. [47] found that anthraquinone dyes like Disperse Blue 180 had relatively close relationship to cancer incidence of workers in the dyes plants. Malachite Green, one kind of triphenylmethane dyes, has been listed as prior chemical in carcinogenetic test due to its hard-biodegradation, toxicity and fungus inhibition [48]. With more and more investigations and reports discovering physio-toxicity of triphenylmethane dyes on living organisms, many countries and regions in world have inhibited the production and application of those highly toxic dyes. Dyes with aromatic rings showed persistence and high ecotoxicity, therefore, developing efficient methods to reduce dyes concentration in practical water system is of highly urgent.

There were many methods and technologies for dyes waste water disposal, including adsorption, membrane process, flocculation, electrodegradation, photodegradation, Fenton process and biotreatment. Advanced oxidation process showed a high decolorization efficiency for dyes but still suffered from expensive costs, complicated operation and secondary pollution. Flocculation is more appropriate for hydrophobic dye removal and the decolorization rate could reach as high as 92%. Nevertheless, the removing efficiency for hydrophilic dyes was pretty low. Biological process showed feasibility in actual dyes water treatment due to its low cost and high BOD removal efficiency, however, it also suffered from low decolorization rate and putrid odor, suggesting that further operations or technologies were needed.

Adsorption, compared with other techniques, showed special advantages due to its low cost, non-secondary pollution and high removal efficiency. Although the aromatic rings of dyes could not be destroyed by adsorption, the concentration of dyes in water could be reduced under the permissive value for drinking water via strong affinity between active groups of dyes and adsorbents. It was believed that high removal efficiency for dyes could be achieved as long as proper adsorbents were selected.

In this chapter, we will focus on the mineral and their modifications in adsorbing dyes. The detailed discussion about adsorption behaviors and mechanisms would be given. Additionally, the application of minerals for dyed immobilization in practical case would be given as examples and related prospectives will also be shown.

2.2 *Adsorption of Dye on Minerals*

(1) Clay minerals

Previous studies showed a promising potential of clay minerals and their modifications in dyes adsorption. Silicate minerals such as rectorite, bentonite, kaolin and zeolite, carbonaceous minerals such as graphene and other nonmetallic minerals like diatomite were usually chosen as the appropriate dyes adsorbents in actual wastewater treatment. The adsorption capacity of several common clay minerals towards ionic dyes has been listed in Table 1.

Rectorite was a kind of silicate minerals which was stacked up by one layer of mica and one layer of montmorillonite, thus indicating both of the properties of mica and montmorillonite are involved in rectorite. Because the balanced cations (i.e. Ca^{2+}, Mg^{2+}, Na^{+}) in the interlayer of montmorillonite were exchangeable, thus rectorite showing a high exchangeable capacity for cations. In addition, the swelling properties, tailored channels and large surface area also made rectorite a good adsorbent for dye removal.

Better adsorption efficiency for dye has been achieved by rectorite and their modifications. He [49] investigated the best conditions for safranine adsorption on

Table 1 Adsorption capacity q_m(mg/g) of natural inorganic materials to dye [62]

Material	Dye	q_m (mg/g)	Reference
Modified ball clay	Methylene blue	100	[124]
Modified montmorillonite	Methylene blue	322.6	[125]
Clay	Methylene blue	58.2	[126]
Kaolin	Congo red	1.98	[127]
Na–bentonite	Congo red	35.84	[127]
Bentonite	Methylene blue	151/175	[128]
Pyrophyllite	Methylene blue	70.42	[129]
Spent active clay	Methylene blue	127.5	[130]
Charred dolomite	Dye E–4BA	950	[131]
Zeolite	Basic dye	55.86	[132]
Modified silica	Acid Blue 25	45.8	[133]
Activated clay	Basic Red 18	157	[134]

rectorites and they found when the initial concentration of dyes was 50 mg/L, rectorite dosage was 2 g/L and pH was 6, the adsorption efficiency could reach 95%. The main mechanism for safranine removal on rectorites could attribute to cation exchange. Chen et al. [50] found rectorite after calcination showed a high adsorption capacity for methylene blue due to strong electrostatic attraction between cation type methylene blue and electronegative minerals surface, and the adsorption behavior could be better described by Langmuir isotherm and quasi–secondary dynamics model. Zheng et al. [51] prepared three dimensional rectorite gels and found that the porous rectorite exhibited good ability to adsorb methylene blue. The maximum adsorption capacity for MB on rectorite gel networks could reach 95.24 mg/g. During the synthesis of rectorite gels, cationic guar gum could intercalate into the rectorite layers or attach on the rectorite surface, enlarging the interlayer spacing of rectorite and inducing higher surface area as well as more sites for dye adsorption.

Bentonite is a phyllosilicate clay mineral mainly consists of montmorillonite. Many studies showed that bentonite and their organic modifications had a promising effect on dye adsorption, as compared to active carbon. Some dyes in water solution could hydrolyze into anions or cations, which could be adsorbed on the surface of bentonite firmly via electrostatic attraction and cation exchanging. Zhao et al. [52] found that sodium bentonite exhibited faster adsorption for methylene blue than that of calcium bentonite. The maximum adsorption capacity on sodium bentonite and calcium bentonite could reach 86.51 mg/g and 66.67 mg/g, respectively. The main mechanism was probably ascribed to cation exchange. Ihssane Belbachir et al. [53] found that sodic bentonite was also a suitable sorbent for Bezathren dye adsorption and maximum adsorption capacity could reach 35.08 mg/g, 32.88 mg/g and 48.52 mg/g for Bezathren–Blue, Bezathren–Green and Bezathren–Red, respectively. Chakraborty et al. [54] fabricated layer by layer film with montmorillonite and methylene blue. The study showed cation methylene blue could be adsorbed on MMt organizedly through electrostatic interaction, which identified by Fourier Transform infrared (in ATR mode) spectroscopic. The Atomic Force microscopic image of hybrid films revealed the relative height of the organic–inorganic hybrid molecular assemblies increased from 3.672–7.658 nm, indicating that the interlayer spacing enlarged sharply after methylene blue incorporation. Bentonite modified with surfactant to form organoclays also showed improved adsorption efficiency for dyes. As mentioned in Sect. 1.3.3, intercalating surfactant into phyllosilicate could enlarge the interlamellar spacing of minerals, which further favor of dye adsorption. Furthermore, functionalizing organic surfactants on bentonite surface alter the hydrophilic surface to hydrophobic, strengthening the affinity of clay surface to organic compounds. Wang et al. [55] prepared lignocellulose/organic montmorillonite (LNC/OMMT) composite by intercalation interaction. The adsorption capacity of LNC/OMMT towards Congo Red was investigated and the results showed the maximum capacity of LNC/OMMT towards Congo red was 102.32 mg/g, which was much higher than the capacity of LNC/MMT composite of 51.16 mg/g. Kang [56] found that MB could be effectively removed by exfoliated MMT/chitosan when the concentration of MMT was lower. The adsorption capacity of MB on the composite with MMTNS concentration of

15 g/L could reach 538 mg/g at 318 K. Wang et al. [57] synthesized MMT-based hydrogel beads by using poly (vinyl alcohol)–sodium alginate and chitosan to remove MB. The adsorption capacity of MB reached 137 mg/g via hydrogen bonding as well as ion–exchange. Tong et al. modified montmorillonite with octadecyltrimethylammonium (ODTMA) for textile dye (acid red G (ARG), $C_{18}H_{13}N_3Na_2O_8S_2$) sorption from aqueous solution. The modified composite resulted in an increase of basal spacing of montmorillonite (Mmt) from 1.48 to 2.35 nm, which facilitated ARG adsorption. Both physiosorption and chemisorption phenomena were observed during sorption process and the maximum adsorption capacity of ARG on ODTMA–Mmt was 149.3 mg/g at 30 °C. Besides surfactant intercalation, grafting or functionalizing organic monomers or polymers on bentonite would also improve dyes adsorption efficiency. For example, Du et al. [58] grafted polyethyleneimine on bentonite and the modified adsorbents showed high adsorption capacity towards amino black 10B which could reach 327.7 mg/g. The adsorption mechanism was depended on pH. When pH < 7, the amino groups of polyethyleneimine were protonated to $-NH_3^+$. $-NH_3^+$ could interact with sulfonic acid groups in amino black 10B by electrostatic attraction. While pH > 7, the amino groups of polyethyleneimine was in the form of $-NH_2$ and–NH–, which can interact with amino black 10B molecules through hydrogen bonding. Elsherbiny et al. [59] prepared Na^+–Mt grafting with polyaspartate composite to adsorb methylene blue from aqueous solution. The results showed that the maximum adsorption capacity of modified clay composite could reach 9.95 mmol/g based on electrostatic interaction between cationic active sites of adsorbents and anionic MB dyes. Aguiar et al. [60] prepared bentonite heterostructures that contained silica or silica—zirconium pillars, and this heterostructure showed significant improvement in Violet 5R–RV5R and Acid Blue 25–AB25 adsorption. The maximum capacity could reach 265.9 mg/g for AB25. FT–IR and XPS were used to evaluated the adsorption mechanism and the electrostatic interactions between the silanol groups of the adsorbents and the complexation with functional groups of the dyes, such as amine or hydroxyl, were considered to dominate the whole adsorption process.

As a kind of phyllosilicate, kaolin was in 1:1 structure composing of silicon oxygen tetrahedron and hydroxide octahedron. It was widely used in paper, ceramics, refractory, painting, rubber filler and enamel glaze etc. Additionally, kaolin was also used as an efficient adsorbent for dye removal due to its high ion exchange capacity and porous structure. Abidi et al. [61] investigated adsorption behavior of anionic dyes RR120 on kaolinite and the results showed standard kaolin KGa–2 had much higher adsorption capacity than natural clay TBK.

Other minerals, such as serpentine, pyrophyllite (talc), vermiculite and sepiolite etc. were also appropriate adsorbents for dyes removal. Tang et al. [62] prepared magnetic attapulgites incorporated with carbon nanocomposites. The synthesized clay minerals material showed high adsorption capacity towards MB of 254.83 mg/g. Adsorption mechanism study showed that deposition of carboxyl groups on the surface of synthesized materials may provide more binding sites. The FTIR results showed that those abundant carboxyl groups could strengthen interaction between methylene blue and clay surface by forming cation–π bonding, hydrogen bonding and electrostatic attraction. The adsorption behavior between dyes and materials could be better described by Langmuir and pseudo–second–order kinetic model. Among those results, it can be concluded that clay minerals showed great prospect in adsorbing ionic dyes (Table 1).

(2) Molybdenite

Molybdenite (MoS_2), an emerging mineral with a layered structure consisting of covalently bonded S–Mo–S, exhibited a great potential in the field of adsorption, due to its high specific surface area [63], suitable active edge sites for adsorption [64], and economic feasibility (earth–abundant material) [65]. In addition, the morphologies and number of layers of MoS_2 is tunable via controlling the preparation methods, which may expose more active edge sites for adsorption. Li et al. [66] synthesized MoS_2 with different S sources via hydrothermal method, and the adsorption performance of these samples was investigated by adsorbing Rhodamine B (RhB). It was found that the hierarchical MoS_2 microspheres prepared by CH_4N_2S exhibited the high adsorption capacity (136.99 mg/g). Moreover, the intra-particle diffusion model was also employed to analyze the adsorption rate, which demonstrated that the intra-article diffusion played an important role in the overall adsorption process. The flower–like MoS_2 nanosheets was synthesized by Han et al. [67], which had superior ability to adsorb various dyes and organic pollutants, especially cationic dyes. The surface of obtained MoS_2 samples was electronegative potential, resulting in superior adsorption of cationic dye rather than anodic dye, indicating the adsorption performance of dyes on MoS_2 strongly depends on their surface charge.

Although MoS_2 nanopowders exhibited excellent adsorption performance toward RhB dyes, the difficulty of the separation process of powder adsorbents was still a challenge. Fang et al. [68] introduced MoS_2 nanoflowers on a melamine–formaldehyde sponge through a facile dip–coating method. The MoS_2 sponges exhibited highly–efficient adsorption performance to cationic dye, i.e. RhB, with a maximum adsorption capacity of 104.78 mg/g within 60 min, which was primarily attributed to the strong electrostatic interactions between electronegative MoS_2 nanoflowers and positively charged RhB molecules. More remarkably, the as–prepared sponge adsorbent can be separated from the dye solution easily through a simple salvage instead of any tedious centrifugation or filtration processes.

In addition, MoS_2 is considered to be a promising photocatalyst in the field of degradation of organic compounds, due to its inherent narrow band gap (1.23–1.9 eV) accompanied with good visible–light absorption. Sadhanala et al. [69] prepared MoS_2 nanoflowers (NFs) with an optical bandgap of 1.55 eV. It was found that the MoS_2 NFs showed excellent photocatalytic activity for methylene blue degradation with almost 100% degradation efficiency under vigorous stirring and natural sun light. In order to inhibit the recombination of photogenerated electron–hole pairs, Li et al. [70] synthesized MoS_2/CdS heterostructures in situ where interface between MoS_2 and CdS facilitated the transportation and separation of photogenerated charge. Besides, the dispersion of MoS_2 nanodots on CdS nanorods provided more available active sites. As a result, the optimized MoS_2/CdS hybrid shows an excellent photodegradation performance for RhB with the efficiency of 99.11% in 45 min.

(3) Other minerals

Zeolite was aluminosilicate mineral containing alkali metals and a small amount of water. It was produced from fissure of volcanic rocks. The morphology of zeolite crystals varied from fibrous, hairy, columnar and minority of those were clintheriform. Because of its abundant of pores and channels, zeolite was often used as molecular sieves and efficient adsorbents to remove trace contaminants, such as heavy metal ions and organic compounds. Several studies have been conducted on the zeolites adsorption behavior. However, raw clinoptilolite was not suitable for dyes adsorption due to realtively small amunt od active sites or groups. Therefore, modification with functional groups on clinoptilolite was essential for dyes efficient removal. Wei et al. [71] prepared natural clinoptilolite modified with cetyl tri–methyl ammonium bromide (CTMAB) and investigated the adsorption behavior of modified zeolite for acid carmoisine B. It was found that the adsorption capacity was improved 20 times after modification and the decoloration rate of acid carmoisine B could reach 95% with the adsorption capacity of 38 mg/g. The sorption mechanism of dye on zeolite particles is complex because of their porous structure and variable surface character. However, it has been recognized that the adsorption capability of zeolites mainly originated from their ion–exchange properties. Although the removal efficiency of zeolites for dyes may not be as good as that of clay materials, their availability and low cost may compensate the associated drawbacks.

Graphite was an allotrope of carbon and it was in the honeycomb like structure. Graphite could be classified into crystalline graphite and cryptocrystalline graphite according to their crystal morphology. Crystalline graphite, especially flaky graphite, showed perfect crystallinity, plasticity and unctuosity. The Flaky graphite could be intercalated by organic compounds or composites and then expanded at higher temperature. Those expanded graphite exhibit high surface area and large pore volume due to its wormlike morphology. In recent years, researchers found it exhibited a promising potential as adsorbent for organic pollutants removal. Excellent adsorption capability for dyes could be observed on expanded graphite under the conexistence of other organic compounds in wastewater, suggesting of the selective adsorption of expanded graphite. Besides intercalation, changing graphite morphology and incorporating with other functional groups on graphite may also be promising methods to improve dyes removal. Muthoosamy et al. [72] summarized the recent advances in graphite synthesis. Ultrathin graphene nanosheets could be produced from graphite by ultrasound and thus increasing surface area and the number of active sorption sites. It was reported that sonication is also a useful method to facilitate the formation of reduced graphene oxide (RGO), which is a good adsorbents and catalyst for organic compounds removal. Exfoliation in different solvent may also change the surface structure of graphite and further affect the adsorption capacity.

Both RGO and GO were good adsorbents and catalysts for dyes removal due to the surface π electrons structure and abundant oxygen–containing functional groups. RGO was rich in π electrons although there were less carboxyl and carbonyl groups in it. Those π electrons could transfer around the entire carbon plane and be in favor of dye adsorption through π–π interaction. As a comparison, the amount of surface π electrons in GO was not as much as that in RGO but abundant oxygen containing groups were present on the surface of GO, which could provide more adsorption sites for dyes removal through H bonding. Cheng et al. [73] fabricated graphene oxide/silicalite–1 composites with a hierarchical porous structure and found that the composite showed a good adsorption capacity for rhodamine B removal. The adsorption capacity could reach 56.55 mg/g and adsorption behavior could better follow Langmuir isotherm and pseudo–second–order models. The batch experiment and FTIR results suggested that besides electrostatic interaction, hydrogen bonding interactions also existed between the carboxylic acid groups in GO and RB molecular. Additionally, at lower pH, hydrophobic interaction played an important role for RB removal, which derived from the π–π stacking of GO.

Diatomite was porous rocks which composed more than 80% amorphous SiO_2 and a small amount of Fe_2O_3, Al_2O_3, MgO, CaO and organic matter etc. In these years, it was often used as filter or adsorbents due to its wide variety of morphologies, high surface area and a high porosity which could up to 80% [74]. However, the adsorption efficiency of natural diatomite towards dyes was not satisfactory due to the limited active sites and hydrophilicity properties of the presitine diatomite. Therefore, further modifications were needed to improve the adsorption capabilities of diatomite. Researchers had developed some modification methods, e.g., compounding with manganese–oxides, microemulsion and calcination to improve the performance of diatomite [75]. Mohammad et al. [76] impregnated manganese oxides onto diatomite and found adsorption efficiency of manganese oxides–modified diatomite (MOMD) towards basic dyes had been improved. The adsorption capacity of MOMD for methylene blue (MB), hydrolysed reactive black (RB) and hydrolysed reactive yellow (RY) could reach 320, 419, and 204 mg/g, respectively. A slight shift to smaller 2 theta value after MB adsorbing onto modified diatomite was observed from XRD pattern, which indicated that MB was intercalated into the octahedral layers of diatomite. FTIR results showed that dyes could attach to the oxygen atoms near the octahedral vacancy through coordination interaction and MB, RY, and RB interacting with MOMD may form monodentate, bidentate, and tridentate coordination products, respectively. Zhang et al. [75] prepared chitosan modified diatomite and the adsorption capacity for Reactive Red M–8B (RR) and Direct Green B(DG) were 94.46 and 137.0 mg/g, respectively. Both of two dyes are anion dyes, and chitosan under acid condition was positively charged, indicating that electrostatic attraction was the main adsorption mechanism. Zhang et al. [72] used sodium hydroxide to modify diatomite and the adsorption performance for MB had improved significantly. The adsorption capacity for MB was increased from 1.72 to 18.15 mg/g and removal efficiency from 8.60 to 90.75% under the initial concentration of 100 mg/L. SEM analysis suggested well–developed porous structure was formed after sodium hydroxide treatment and the surface area increased according to BET results, indicating more adsorption sites were exposed by NaOH treatment. The adsorption process could be described by Langmuir isotherm and pseudo–second–order model.

Compound minerals have caused much attention because they combined advantages of each composition and may exhibit synergistic effect for dyes adsorption. For example, Liu et al. [77] investigated the removal behavior of methyl orange (MO) and methylene blue (MB) by montmorillonite–pillared graphene oxide (MGO). The batch experiment results showed MGO possessed superior adsorption capability for both anion dye (MO) and cation dye (MB) with adsorption capacity of 250 and 144.9 mg/g, respectively. In addition, the uptake amount of MB and MO increased in the binary system. The higher adsorbed amount of MO could be explained by π–π stacking interaction between columnar structure of MGO and the aromatic ring of MO through FTIR, zeta potential analysis. However, the main adsorption mechanism attributed to electrostatic interaction was observed in MB adsorption. From BET and TEM analysis, it could be seen that the surface area of adsorbents was increased after compounding two minerals, which indicated high surface area and more active sites of complex minerals facilitate dyes adsorption.

In summary, minerals or modified minerals could be applied as excellent adsorbents for dyes removal. However, several aspects are still needed to be focus on (1) Higher adsorption capacity. With the development of nanotechnology, novel adsorbents compounding minerals with nanomaterials need to be developed to achieve higher adsorption capacity and reusability. (2) Selective adsorption. In practical application, there were various pollutants in waste water and adsorbing dyes specifically was essenitial for the longevity of adsorbents. (3) Stability of the adsorbents. Adsorbent which is thermodynamically stable is need to be developed because the water condition in practice was complicated and those adsorbents which were acid–proof, alkali–proof and corrosion resistance would be in favor of recycling and showed promising potential in real dyes treatment.

3 Adsorption of Phenolic Compounds on Minerals

3.1 General Introduction of Phenolic Compounds

Phenolic compound is a kind of polycyclic aromatic hydrocarbons (PAHs) of which hydrogen atoms were partially substituted by one or more hydroxyl groups. From their chemical structure, phenolic compounds contained at least one aromatic rings, suggesting of an unique π electron structure and hydrophobicity in solution [78]. Phenolic compounds in natural environment mainly come from plants and that's called endogenous phenol. Endogenous phenol was biologically active and showed antioxidant activity, which is good to humans health. Beside natural phenols, a large amount of phenolic compounds come from human activities called exogenous phenols. Unlike endogenous phenols, exogenous phenols are considered as priority pollutant because most of them showed harm effects on the health of animals and human beings [79]. It was reported that the total production of phenols around world reached 7.8 million tons in 2001 [80]. With the increasing global production of phenolic compounds these years, the concentration of phenolic compounds in water, soil and air had been detected beyond limits value in several regions or countries, which indicates that further strict management and control strategies are urgently needed. It has been proven that the phenols in water over 2 mg/L are toxic to fish and concentrations between 10 and 100 mg/L would result in death of aquatic life within 96 h [81]. US Environmental Protection Agency (EPA) limited that the permissive concentration of phenols in aqueous system is 1 mg/L [82].

Some phenolic compounds are volatile, such as p–nitrophenol (PNP) and p–chlorophenol (PCP), in para position of which the hydrogen atom were substituted by nitryl or chlorine atoms. Those phenolics which can inflame the respiratory tract are odorous and toxic, might further causing chronic disease like bronchitis. Another example, bisphenol A (BPA), as a derivative of phenol, which also belongs to PHAs, is the raw material used in polycarbonate (PC) and epoxy resin industry. It is known that the production amount of plastics in the world was 27 million every year and most of them contain BPA. The chemical structure of BPA can be noted that two hydrogen atoms of methane (CH_4) were replaced by a methyl and the rest two hydrogen atoms were replaced by phenols. Because of the presence of two aromatic rings, BPA shows strong hydrophobicity in water solution, which means it is likely to be removed via hydrophobic interaction according to the probable adsorption mechanisms mentioned in Sect. 1.3.2. In natural environment, BPA was regarded as an endocrine disrupter which could damage the endocrine system of animals and humans. In the past thirty years, many investigations have confirmed that BPA had adverse effects on physiological metabolism of animals and the malformation would occur in the next generation. Liu et al. [83] reported that BPA could remarkably retard the growth of zebrafish embryos, which may induce embryo abnormality and even death. The EC_{50} for zebrafish embryos was 2.9 mg/L. There is a good affinity between BPA and estrogen receptors of human beings, which induced unwished expression of progesterone receptor in breast cancer cell MCF–7 and breast cancer cell proliferation [84]. In addition, BPA tend to react with chlorine compounds when BPA contacted with household cleaning products to form more poisonous products Clx–BPA [85]. It was reported that chlorinated BPA showed higher estrogenic activity (10–40 times) than BPA [86] and can increase the incidence of breast cancer and uterine endometrium lesion [87, 88]. Because of its hydrophobicity and lipotropy, BPA were more likely to be bioaccumulation and difficult to biodegrade. Some measures have been adopt. In March 2nd, 2011, the production of feeding–bottle containing chemical bisphenol A(BPA) has been banned by European Union. Further strategies, such as effectively remove or immobilize BPA need to be further focus on in order to reduce the risk of BPA in the environment.

3.2 Adsorption of Phenolic Compounds on Minerals

There is a lot of methods including adsorption, membrane process, biodegradation and advanced oxidation process such as photodegradation, electrochemical degradation, Fenton process etc. for phenolic compounds removal. Adsorption was one of the most extensively employed technique and promising method for phenolic compounds effective immobilization. Moreover, minerals and their modifications being used as good adsorbents for phenolic compounds have been investigated for many years because their availability, low cost, and ion exchange properties made them good adsorbents in practical. It is known that natural clay minerals are in layered structure, in which alkaline earth metals as well as transition metals are involved [81], thus the present of hydroxyl groups on the minerals surface would give rise to the electronegativity of the surface of minerals. Those properties made clay minerals were not so suitable for phenolic compound removal if without any modifications [78].

Grafting hydrophobic groups or functionalizing organic compounds on clay minerals, like surfactants through cation exchanging to form organoclay would increase surface hydrophobicity, surface areas and electric charge densities of minerals and that would be a promising strategy to enhance the adsorption efficiency of natural clay minerals towards phenols. Various kinds of modification composites based on natural minerals have shown remarkable improvement for phenolic compounds adsorption in lab scale. Ake et al. [89] found that cetylpyridinium–exchanged montmorillonite clay could reduce PNP load from the oil water system by more than 99%, which attributed to good hydrophobic interaction of cetylpyridinium with PNP. Mehmet and Akcay [90] modified bentonite with dodecylammonium to synthesize organic clay which showed greater cation exchange capacity and enhanced hydrophobicity than bentonite without any modification. The dodecylammonium bentonite, DDAB had a positive effect on p–chlorophenol (p–CP) and p–nitrophenol (p–NP) adsorption and the maximum adsorption capacity for p–CP and p–NP was 629.24 and 771.10 mmol/kg, respectively. It was found that DDAB prefers to interacted with p–NP rather than p–CP due to donor–acceptor interaction and hydrogen bonding. In recently, Zhu et al. [91] modified montmorillonite with hexadecyltrimethylammonium (HDTMA) and chitosan to form hydrophobic organo phases, which could decontaminate complicated wastewater. Both HDTMA and chitosan changed the surface properties of montmorillonite and make it more hydrophobic and positively charged, which benefited phenol removal. Wang et al. [92] prepared dual–cation organomontmorillonites by cation surfactant (trimethyltetradecyl ammonium chloride, hexadecyltrimethyl ammonium chloride or octadecyltrimethyl ammonium chloride and cysteamine hydrochloride) and the modified montmorillonite had positive effect on PNP adsorption with increasing pH. The partitioning interaction was considered to be the main adsorption mechanism.

The modified minerals with unique surface properties can potentially immobilize phenolic compounds and have a great prospect in environmental remediation. Next we will focus on two phenolic compounds, e.g., p–nitrophenol and BPA, where detailed adsorption behavior and related adsorption mechanisms on minerals will be given.

3.3 Adsorption of p–Nitrophenol(PNP) on Minerals

As a derivative of phenol, p–nitrophenol is an important intermediate in the production of pharmaceuticals, pesticides, dyes, artificial resins, explosives, and other valuable fine chemicals [93]. It was known that p–nitrophenol was persistent, bio–accumulative and highly toxic even at very low concentration. Frequent exposure to p–nitrophenol can cause methemoglobin inhibition, anemia, liver and kidney damage, eye and skin irritation, tumor formation, cancer, and systemic poisoning. Nowadays, p–nitrophenol has become a commonly encountered pollutants in water pollution. As a result, p–nitrophenol has been categorized as one of the 126 priority pollutants by the U.S. Environmental Protection Agency (U.S.EPA). Therefore, fast and effective removal of p–nitrophenol from industrial wastewater is becoming increasingly urgent.

Many studies have been carried out aiming at p–nitrophenol adsorption on minerals these years due to the low cost, extensive sources and relatively high adsorption capabilities of minerals. However, considering the similar structure and properties of PNP with other phenolic compounds, natural clay minerals were not suitable for PNP removal. Surfactant modification or preparing organoclay was the common resolution in improving the affinity between organic compounds and the mineral [94]. It was reported that clays modified by quaternary ammonium cations were effective in adsorbing phenol, p–chlorophenol, benzene and toluene [92]. To date, the most commonly used surfactants in fabricating organoclays are cation surfactants because they can be easily intercalated into the layers of clay via cation exchange. However, anionic, nonionic and amphoteric surfactants could also be modified on clay minerals and those modifications have also been proven to have positive effect on PNP removal through electrostatic interaction and hydrophobic interaction. Zhu et al. [95] modified montmorillonite with zwitterionic surfactants, namely sulphobetaine to enhance the adsorption efficiency of PNP on montmorillonite, and the results showed that under the same condition, the sorption capacities of modified montmorillonite towards p–nitrophenol increased with alkyl chain length increasing, indicating a stronger hydrophobic interaction occurred when loner alkyl chain being intercalated into the silicate layers. The sorption isotherms were linear, suggesting of the partition process. Beside montmorillonite, palygorskites could also be intercalated with organic surfactant and be used as superior adsorbents for PNP removal. Palygorskites is a 2:1 type clay mineral that was highly negatively charged due to the considerable substitution of Al^{3+} by Mg^{2+} and Fe^{2+} in the octahedral sheet, which is similar to montmorillonite. The morphology of palygorskites is fibrous like with lots of internal channels, thus inducing high surface area, making palygorskite a promising candidate for environmental remediation and organic contaminants removal. Sarkar et al. [96] prepared novel organopalygorskites by using dimethyldioctadecylammonium bromide (DMDOA) and cetylpyridinium chloride (CP) as intercalators and the organopalygorskites showed enhanced capability for PNP adsorption compared with presitine palygorskites. According to bath experiment data, the adsorption capacity varies with pH value and the main adsorption mechanism was considered to be electrostatic bonding between PNP anions and surfactant cations. The partition of the undissociated PNP molecules on organopalygorskites through hydrophobic interaction was another reason for PNP adsorption. The adsorption behavior could be described by Freundlich model, indicating of multiple layer adsorption. Except for organoclays, there were many other minerals used for PNP adsorption. For example, graphene, the exfoliation product of graphite, has been proved to be a good adsorbent for PNP. Shah et al. [97] synthesized graphene magnetic nanocomposites and used it to extract PNP from aqueous samples. The batch experiment showed that the maximum adsorption capacity could reach as high as 500 mg/g, fitting well with Langmuir isotherm model and pseudo–second–order kinetic model. The sorption of p–nitrophenol was found exothermic in nature condition and the physisorption process was involved during adsorption. It was considered that hydrogen bonding and electron transfer from donor to accepter was the main adsorption mechanism during the process. Ji et al. [96] found Al–doped graphene could be an effective adsorbent for PNP through density functional theory calculation and the results suggested that chemisorption dominate the adsorption process. In conclusion, most of organoclays could significantly improve adsorption capacity for PNP due to large interlamellar spacing and hydrophobic interaction between adsorbates and adsorbents. Functionalizing active groups or sites on graphite also improved the adsorption effectiveness while hydrogen bonding would be the main sorption mechanism.

3.4 Adsorption of Bisphenol A on Minerals

As it mentioned above, BPA was a kind of endocrine disrupter and defined as emerging contaminant by EPA, attracting a great attention of the public and environmentalists. Therefore, developing appropriate adsorbents for BPA removal has become increasing urgent. It is known that BPA was poorly soluble in water because it contains two aromatic rings, thus the main pathway of its removal should be ascribed to hydrophobic interaction and π–π interaction between adsorbents and BPA. From the previous researches, the organoclays were the common adsorbents for phenolic compounds removal, including BPA. Intercalation of surfactant into the montmorillonite layers altered the clay surface from hydrophilicity to hydrophobicity, which strengthen the affinity between BPA and the surface of mineral. Liu et al. [23] modified vermiculites with amphoteric surfactants with different negatively charged groups (carboxylate, sulfonate and phosphate) to fabricate dodecyl dimethyl (3–sulphonate) ammonium (SB), dodecyl dimethyl(N–

carboxylate) ammonium (BS) and dodecyl dimethyl (N–phosphate) ammonium (PBS), respectively to adsorb BPA and tetrabromobisphenol A (TBBPA). The results showed vermiculites modified with PBS exhibited the highest adsorption capacity for BPA and TBBPA, which could reach 92.67 and 88.87 mg/g, respectively. A series of characterizations suggested PBS not only adsorbed on the surface of vermiculites but also intercalated into the layers, resulting in increased number of adsorption sites exposed. It was also found that the loading capacity of surfactant greatly affect the adsorption capacity, suggesting the main sorption mechanism was hydrophobic interaction. It was worthy to notice that Ca^{2+} could inhibit the sorption of BPA and TBBPA significantly, mainly because Ca^{2+} impaired the hydrophobic interaction between the modified VER and pollutants as well as the competition effect of Ca^{2+} on the adsorption sites. Wang et al. [98] fabricated montmorillonite intercalated with alkyltrimethylammonium ions to remove the trace amount of BPA in aqueous solution. The result showed that the organo–montmorillonite could completely removed BPA. Similar to PNP, the enhanced adsorption capacity could be achieved when surfactant with longer chain was intercalated into the silicate layers. The study also found the coexistence of matrix organic substances could increase the uptake amount of BPA, suggesting modified with long chain surfactant was an effective way not only to remove trace BPA completely and but alos achieve the selective adsorption performance by montmorillonite. The same results were also showed in Xie's [99] work that the modification of zeolite with hexadecyltrimethylammonium (HDTMA) has exhibited enhanced capability for BPA removal due to stronger hydrophobic interaction after loner chain intercalation. Besides clay minerals and silicate minerals modification, graphite and its derivative also showed superior performance on BPA adsorption. Xu et al. [21] found that the maximum adsorption capacity of graphene for BPA could reach 182 mg/g at 302.15 K and the main adsorption mechanism would be ascribed to π–π interaction and H bonding, which was identified by FTIR. Some other studies also showed that graphene exhibited high selectivity towards organic compounds due to its unique sp^2–hybridized layer structure.

Both PNP and BPA were hydrophobic compounds. In order to improve the removal efficiency of those phenol compounds, it was essential to alter the surface properties of minerals. Organoclays, silicate minerals modified with functional groups and graphite were suitable to be chosen as adsorbents for phenolic compounds removing, which based on the hydrophobic interaction. Additionally, it also need to be noticed that compounding various minerals together or hybridizing different nanomaterials could also increase the adsorption sites and surface area of adsorbents and then, more effective adsorption performance could be achieved. Further works aiming at developing more effective, reusable and selective minerals composite are indispensable. The detailed adsorption mechanisms are also needed to investigate to help us know more about the interaction between adsorbates and adsorbents.

4 Adsorption of Antibiotic on Minerals

4.1 General Introduction of Antibiotics

In recent years, pharmaceutical and personal care products (PPCPs) have caused extensively attentions. PPCPs include antibiotics, imaging agents, painkillers, etc. They were widely used in medical care and agriculture. Antibiotics, as one kind of pharmaceutical products, were widely used in disease prevention and therapy for both animals and humans. Also, antibiotics are used in agriculture to promote the growth of living organisms and enhance their economic efficiency. Accompanied with the large production of antibiotic, the antibiotic abuse has been a serious problem in China these years. It was reported that the total production of penicillin and oxytetracycline in china in 2003 was 2.8 million tons and 1 million tons, respectively, accounting for more than 60% of all penicillion production globally. More than 70% of antibiotics in China were used for inpatient every year and the utilization rate was relatively high compared to WHO permissive value that is 30%. The harmful effects of antibiotics on eco–system and human beings have been gradually noticed by researchers. It was proven that antibiotics couldn't be completely absorbed by living organism. For example, Trimethoprim (TMP), an antibacterial drug used for infections therapy, can inhibit the growth of gram–positive and gram–negative bacteria, thus being currently used as an antibacterial agent. However, TMP is incompletely metabolised by human beings during the therapeutic process, and about 80% is excreted. Another example, tetracyclines (TCs), most widely used in livestock due to its desirable antimicrobial activity, has been found in high concentration of swine water because it was hard to be adsorbed or disintegraded completely by animals and microbials. Some reports showed that the maximum concentration of tetracyclines could reach 685 μg/L in surface water. One of the largest concerns related to this fact is the production of resistant organisms and the pollution of resistence genes. The emergency and wide spread of drug–resistant Gram–positive pathogens including MRSA and MRSE has caused great concern throughout the world. Zhang et al. [100] found that the resistance of Gram–positive organisms to 4–quinolones become more prominent these years, making antibiotics less effective. As a result, longterm exposure to high concentration of antibiotics could pose serious health problems. In fact, the surface water in many other countries have also been under a high risk of antibiotic contamination (Table 2), therefore, it was necessary for researchers and scientists come up with more favorable methods to dispose the antibiotic wastewater.

4.2 Minerals for Antibiotic Adsorption

Clay minerals were the common adsorbents for antibiotic removal due to its high capacity of cation exchange and layered structure which was in favor of organic

Table 2 The concentration of antibiotics in global surface water (1998–2010)

Year	Place	Antibiotics	Concentration /ng/L	Reference
1998	Germany, Elbe River	Sulfamethoxazole	30–70	[135]
		Erythrocin	30–40	
1999–2000	America	Benzylidine triamine	150	[136]
		Sulfamethoxazole	150	
		Dehydrated erythromycin	100	
2001	Italy, River Po	Lincomycin	3.13–248.9	[137]
2003–2005	America	Sulfamethoxazole	40–320	[138]
		Dehydrated erythromycin	20–450	
2005	Japan, Tama River	Azithromycin	43–448	[139]
		Clarithromycin	55–254	
2005	China, Pearl River	Dehydrated erythromycin	460	[140]
		Sulfadiazine	209	
2006	China, Yellow River	Ofloxacin	<LOQ[b]–264	[141]
2006	China, Victoria Harbour	Dehydrated erythromycin	9.5–486	[142]
2006	France, Seine	Sulfamethoxazole	23–69	[143]
		Erythrocin		
2008	China, Bohai bay	Norfloxacin	460	[144]
2009	China, Huangpu River	Sulfa methazine	313.44	[145]
2010	China, Haihe River	Sulfamethoxazole	137	[146]

compounds intercalation as mentioned in Sect. 1.3.3. Unlike phenolic compounds, antibiotic was soluble in water, which meant the sorption mechanism of antibiotics on clay minerals would be different from that of phenols. For example, bentonite, showing high surface area and pore volume, was reported to act as an superior adsorbent for antibiotic removal from waste water. According to Genç's [101] work, the adsorption of ciprofloxacin (CIP) from aqueous solution on bentonite was found to be excellent with the removal efficiency of 99% and the maximum capacity could reach 147.06 mg/g. Under acid pH condition, cation exchange was the dominant mechanism for CIP uptake. In neutral condition, electrostatic attraction and charge neutralization might dominate the adsorption process. However, in alkaline condition, proton transfer might happen. The results indicated that adsorption efficiency and behavior of CIP on bentonite was depended much on pH value. Similar to bentonite, the adsorption behavior of CIP on kaolin was also investigated by Li et al. [102] and the study found that cation exchange was the dominant mechanism for CIP adsorption on kaolinite. Batch experiment data and XRD results suggested that surface adsorption rather than intercalation occurred during the process because d spacing of kaolinite didn't changed after CIP adsorption. Aristilde et al. [103] investigated oxytetracycline (OTC) adsorption behavior on Na–montmorillonite. The results showed that in the presence of divalent metal cation ions, such as Ca^{2+}, Mg^{2+}, the adsorption capacity increased twice because the intercalation of OTC into montmorillonite layers was promoted with the present of cation ions and a ternary antibiotic–metal–clay complex was formed in which Mg^{2+} and Ca^{2+} served as bridges sites between oxytetracycline and montmorillonite. Sulfonamides were kinds of antibiotics widely used for decades and their extensively application as manure could result in soil contamination. The combination of sulfonic group with amino group in the molecular structure made sulfonamides properties special [104]. There were many minerals suitable for sulfonamides adsorption. For example, the adsorption behaviors of sulfonamides on phyllosilicate surfaces were studied by few researchers. Francisco-Márquez et al. [104] firstly investigated the adsorption of sulfonamides on (001) plane of pyrophyllite by using density functional theory (DFT) and Hartree-Fock methods. It was found that S1 and S2, which was sulfamethoxypyridazine and sulfamethoxydiazine, respectively, are likely to be adsorbed on the (001) plane of phylliosilicate due to lower energy through an exothermic process. The study also concluded that cations in the interlayer of phyllosilicate would affect adsorption process via cation exchange. Li et al. [105] found that the ofloxacin adsorption on kaolin was pH and ionic strength dependent. The main adsorption mechanism could be explained by cation exchange between OFL^{+} and kaolinite. When solution pH was below 7, the presence of Cu^{2+} could improve the adsorption efficiency due to electrostatic interaction and the formation of kaolinite–OFL–Cu ternary surface complexes. However, on the other hand, Cu^{2+} could also compete with the adsorption site. It should be noted that the divalent cation in the cay minerals had contradictory effects on antibiotics adsorption.

Besides clay minerals and their modifications, ferric minerals also showed superior sorption capabilities to antibiotic. Rakshit et al. [106] prepared nano-sized magnetite to adsorb ciprofloxacin. Results showed the sorption efficiency of CIP on nano–Fe_3O_4(s) increased from 45 to 80% when pH increased from 3.44 to 5.97, suggesting pH value would influence the adsorption process greatly. At acidic pH values both CIP and magnetite were positively charged, thereby the electrostatic repulsion would inhibit CIP to interact with the magnetite. When the pH increased to 6.0, the charge of CIP gradually turned to neutral, making the interaction between CIP and positively charged magnetite more favourable. When pH above 6.5, CIP was negatively charged due to deprotonation of the carboxyl group and repulsion between the adsorbate and adsorbent predominate again. Additionally, from the analysis of in situ ATR–FTIR, there was complexation between –COOH moiety of CIP and ferric oxide, suggesting inner sphere coordination also contribute to CIP immobilization during the adsorption process. Guo et al. [107] prepared goethite–biochar composite to remove tylosin and found that goethite played an important role in tylosin adsorption. From XRD and FTIR results, metal–ligand complex could form between goethite and tylosin. Moreover, the change of pH could affect present forms of tylosin and the surface properties of adsorbents, resulting sorption mechanism more complicated. Indeed, hydrophobic, electrostatic attraction, H–bonding, cation exchange and π–π interaction could affect the adsorption process over a wide range of pH. The maximum capacities of tylosin on the goethite-biochar composites were 5386.76 (μg/g)/(mg/L)^n described by Freundlich model.

Struvite can also be used as effective adsorbents for antibiotic adsorption. Struvite, namely magnesium ammonium phosphate hexahydrate ($MgNH_4PO_4{\cdot}6H_2O$), was a kind of phosphate minerals within crystal water. It was widely used as generous manure due to its rich contents of phosphate and nitrogen. The morphology of struvite is isoaxial or wedge–shaped, meanwhile the short columnar or thick plate like structure are also common type. Different from clay minerals, struvite had low solubility in water, which made it a controlled–released fertilizer. It has been reported that the precipitation of struvite was an effective method for phosphorus recovery and reuse [108, 109]. Many researchers were working on phosphate adsorption during the formation of struvite [110]. However, the adsorption of antibiotic on struvite were rarely studied. Sulfonamides, quinolones, and tetracyclines had been recently proved to tend to form stable complexes with Mg^{2+} and Ca^{2+} in the layers or on the surface of struvite [111]. Sibel et al. [112] investigated the interaction process of tetracyclines (TCs) with struvite and found that the adsorption capacity of tetracycline on struvite was depended on pH condition because the solubility and lipophilicity of tetracycline varied under different pH condition. In detail, when pH < 3.3, more than 50% of tetracycline exists as a cation. When pH between 3.3 and 7.68, tetracycline existed as a zwitterion, and as a monovalent anion when pH above 7.68. When pH further increased to 9.69, tetracycline present in the form of divalent anion. The various formation of tetracycline in solution directly affected its adsorption behavior. Moreover, it was proved that Ca^{2+} in the struvite could promote tetracycline adsorption because the complexion of Ca–tetracycline formed during struvite precipitation. The adsorption behavior could be better described by Freundlich isotherm and Redlich–Peterson models. Similarly, Ye et al. [112] investigated transportation process of tetracyclines during the crystallization of struvite. The findings revealed that the maximum adsorption capacities of TCs onto struvite crystals ranged from 1494.7 to 2160.0 μg/kg, which depended on the forms of TCs species. The main adsorption mechanism was ascribed to electrostatic adherence. The presence of Mg^{2+} interfered TCs adsorption by competing adsorption sites with TCs, which was also determined from pH change. However, Once Mg/P molar ratio was above 1.8, drastic increment of adsorption capacities was observed which was ascribed to complexation of Mg–TCs. Therefore, It should be noted that divalent cation existing in the struvite could affect the antibiotics adsorption behavior greatly.

In conclusion, the adsorption process for antibiotics on minerals were dominated by cation exchange and complexation, which was quite different from phenolic compounds that hydrophobic interaction was the main mechanism. Further works aiming at developing more novel and effective adsorbents should be conducted. In the meanwhile, the mechanism should be further investigated in order to provide theoretical basis for adsorbents designing.

5 Adsorption of Perfluorooctane Sulphonate (PFOS) on Minerals

5.1 The Introduction of Perfluorooctane Sulphonate (PFOS)

Perfluorooctane sulphonate, shorted as PFOS, was a typical persistent organic pollutant defined in 2009. PFOS were widely used in producing surface antifouling agent, detergent and electronic devices. From its chemical structure (Fig. 4), it can be seen that PFOS was composed of carbon chain without any aromatic rings, which was quite different from other organic compounds. The thermodynamic stability and environmental persistence deriving from strong bonding energy of C–F made it difficult to biodegrade. Moreover, the fluorinated alkyl chain exhibited special oleophobic and hydrophobic properties, leading to lower surface tension of PFOS in water system [113]. Hence, it was a big problem to dispose it effectively.

Up to now, many technologies have been developed to remove PFOS in aqueous system but respective drawbacks still existed. Taniyasu et al. [114] proved that perfluorocarboxylic acids (PFCAs) and perfluorosulfonic acids (PFSAs) could be photodegraded in the environment but the shorter chain which were more persistent would be produced. Hisao Hori et al. [115] found PFOS could be adsorbed on zero valent iron effectively and be decomposed at much higher temperature like 350 °C, indicating that more energy was cost during treatment process. Ruiz Beatriz Gomez et al. removed perfluoroalkyl substances (PFAs) by electrochemical treatment and high TOC removal efficiency (>90%) could be achieved. However, the shorter chain PFCAs occurred during electrochemical process and appropriate anode material was difficult to prepare [116]. Some investigations reported that PFOS could be removed by coagulation but the removal efficiency was less than 20%, thus limiting their applications in real wastewater treatment. Other methods, such as sonochemical treatment, oxidation under alkali condition or advanced oxidation process also suffered from the similar limitation like high cost or relatively low removing efficiency [113]. Therefore, adsorption, as one of the most conventional methods of decontamination has shown its advantages and often be used as commercial method for PFOS removal. The adsorbents were various, including carbon based materials, synthesized nanomaterials or their composites. Most of them have

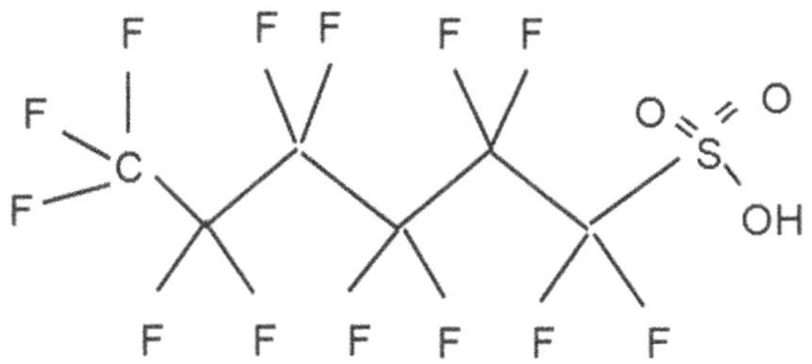

Fig. 4 The chemical structure of PFOS

been proved to have good ability for PFOS adsorption. In this chapter, we will focus on the minerals which were earth abundant with high ion exchanging capacity and large number of adsorption sites. Various kinds of minerals and their modifications as adsorbents for PFOs removal would be given and detailed discussion would also be shown as follows.

5.2 *Minerals for PFOS Adsorption*

Generally, many clay minerals were good candidates for PFOS adsorption. It was reported that electrostatic interaction and hydrophobic effect could contribute to PFOS adsorption on the surface of clay minerals. Zhang et al. [117] investigated the adsorption mechanism of PFOS on montmorillonite and Kaolinite at molecular level and found that PFOS molecules could intercalate into the layers of montmorillonite but could not into the kaolinite's, suggesting that minerals structure affected the adsorption performance towards PFOS. The sulfonate groups were Lewis hard base while the clay minerals were Lewis hard acid, therefore there were relatively strong interaction between PFOS and the minerals according to hard soft acid base (HSAB) theory [118]. Ligand exchange and surface complexion could also occur during chemisorption process. Additionally, humic (HA) substances in natural aqueous systems were found to have adverse effect on PFOS adsorption because HA could occupy the adsorption sites of montmorillonite and kaolinite. Ramona et al. [119] found that kaolinite was more appropriate for PFOS removal than goethite and electrostatic interaction played an important role in adsorption process. The adsorption behavior could be better described by Langmuir model, suggesting that monolayer adsorption dominated in the whole process. Xiao et al. [120] found kaolinite showed more uptake amount towards longer–chained perfluoroalkyl acids (PFAAs) than shorter–chained PFAAs due to stronger hydrophobic interaction present in longer–chained PFAAs system. From the results of zeta potential of kaolinite before and after adsorption, it could be concluded that PFAAs molecules were adsorbed on the electrical double layers of the kaolinite. He et al. studied the transformation process of PFOS in water and rutile interface by using molecular dynamics simulations. They found the adsorption behavior was quitely depended on initial concentrations of PFOS. All PFOS molecules directly interacted with the (110) plane of rutile by electrostatic attraction when the initial concentration of PFOS was low. Then the molecules began to aggregate in a complex multi layered structure when the initial concentration increasing. The outer layer of perfluoroalkyl chains interacted with adsorbents through van der Waals [121]. It should be noted that PFOS were kinds of anionic surfactants and the ionic strength in solution could affect the adsorption behavior greatly. The results in several studies showed $CaCl_2$ could change the multi–layered structure of minerals and increased the critical concentration of PFOS. Therefore, to sum up, the adsorption behavior of PFOS on minerals were depended on ion strength,pH condition as well as the length of the chain [121].

Besides common clay minerals, there were other minerals also showing good abilities to adsorb PFOS in aqueous system. Qian et al. [122] remove PFOS and phosphate simultaneously by boehmite and their adsorption behavior on boehmite were investigated in detail. Boehmite was composed of abundant of γ–AlOOH so that the surface of boehmite was highly activite for hydroxylation. The presence of large amount of surface hydroxyl made it easy to adsorb oxyanions and organic compounds. The adsorption batch experiment suggested that the adsorption behavior of PFOS on boehmite was described better by Langmuir model and the maximum adsorption capacity could reach 0.182 $\mu g/m^2$. Adsorption thermodynamic experiment showed that increased temperature didn't improve PFOS adsorption. The uptake amount of PFOS decreased with pH increasing when initial concentration of phosphate below 30 mg/L, indicating that electrostatic interaction played an important role in adsorption process and phosphate could occupy the adsorption site at low concentration. Du et al. [123] prepared a novel adsorbent via ball milling of Fe_3O_4 and fluorinated vermiculite. The magnetic adsorbent showed excellent selective adsorption character and fast adsorption behaviors towards PFOS in the complex water system. The maximum sorption capacity for PFOS could reach 1127 mg/g, which was better than any other adsorbents which has been previously reported. The related adsorption mechanism was also investigated. On the one hand, it was found that the nano sized Fe_3O_4 was embed homogeneously into the fluorinated adsorbent, making the surface of adsorbents more positive, thus facilitating the adsorption of PFOS. On the other hand, the fluorinated surfactant was intercalated into the layers of adsorbent and the C–F chain of the surfactant could selectively adsorb PFOS via fluorophilic interaction, which made the adsorbents composite exhibited excellent selectivity.

Minerals and their modifications, showed a good perspective in PFOS adsorption. Hydrophobic effect and electrostatic interaction predominated the adsorption process by most minerals. In the meanwhile, ligand exchange or complexion might also contribute PFOS removal on minerals when divalent metals exist in the aqueous system. Through surfactant modification or compounding with other nano materials, the minerals ability to adsorb PFOS could be improved further.

References

1. Belfroida, A.C., Van der Horst, A., Vethaak, A.D., Schaferc, A.J., Rijsc, G.B.J., Wegener, J., Cofino, W.P.: Analysis and occurrence of estrogenic hormones and their glucuronides in surface water and waste water in The Netherlands. Sci. Total Environ. **225**(1–2), 101–108 (1999)
2. Snyder, S.A., Keith, T.L., Verbrugge, D.A., Snyder, E.M., Gross, T.S., Kannan, K.: Analytical methods for detection of selected estrogenic compounds in aqueous mixtures. Environ. Sci. Technol. **33**(16)
3. Reif, J.S., Schaefer, A.M., Bossart, G.D., Fair, P.A.: Health and environmental risk assessment project for bottlenose dolphins tursiops truncatus from the southeastern USA. II. Environmental aspects. Dis. Aquat. Organ. **125**(2), 155–166 (2017)

4. Adams, C., Wang, Y., Loftin, K., Meyer, M.: Removal of antibiotics from surface and distilled water in conventional water treatment processes. J. Environ. Eng. **128**(3), 253–260 (2002)
5. Bolong, N., Ismail, A.F., Salim, M.R., Matsuura, T.: A review of the effects of emerging contaminants in wastewater and options for their removal. Desalination **239**(1–3), 229–246 (2009)
6. Mfrekysy, Jasim: Occurrences of endocrine disrupting compounds and pharmaceuticals in the aquatic environment and their removal from drinking water: challenges in the context desalination of the developing world. Sci. Total Environ. **408**(21), 5147–5157 (2010)
7. Rodil, R., Quintana, J.B., Concha-Grana, E., Lopez-Mahia, P., Muniategui-Lorenzo, S., Prada-Rodriguez, D.: Emerging pollutants in sewage, surface and drinking water in Galicia (NW Spain). Chemosphere **86**(10), 1040–1049 (2012)
8. HallingSørensen, B.: Algal toxicity of antibacterial agents used in intensive farming. Chemosphere **40**(7), 731–739 (2000)
9. Conder, J.M., Wolf, W.D., Russell, M.H., Buck, R.C.: Are PFCAs bioaccumulative? A critical review and comparison with regulatory criteria and persistent lipophilic compounds. Environ. Sci. Technol. **42**(4), 995–1003 (2007)
10. Zhang, K., Cao, Z., Huang, J., Deng, S., Wang, B., Yu, G.: Mechanochemical destruction of Chinese PFOS alternative F–53B. Chem. Eng. J. **286**, 387–393 (2016)
11. Deng, S., Nie, Y., Du, Z., Huang, Q., Meng, P., Wang, B., et al.: Enhanced adsorption of perfluorooctane sulfonate and perfluorooctanoate by bamboo-derived granular activated carbon. J. Hazard. Mater. **282**, 150–157 (2015)
12. Westerhoff, Paul, Yoon, Yeomin, Snyder, Shane, Wert, Eric: Fate of endocrine-disruptor, pharmaceutical, and personal care product chemicals during simulated drinking water treatment processes. Environ. Sci. Technol. **39**(17), 6649–6663 (2005)
13. Kiso, Y., Sugiura, Y., Kitao, T., Nishimura, K.: Effects of hydrophobicity and molecular size on rejection of aromatic pesticides with nanofiltration membranes. J. Membr. Sci. **192**(1–2), 1–10 (2001)
14. Nakada, N., Shinohara, H., Murata, A., Kiri, K., Managaki, S., Sato, N., et al.: Removal of selected pharmaceuticals and personal care products (PPCPs) and endocrine–disrupting chemicals (EDCs) during sand filtration and ozonation at a municipal sewage treatment plant. Water Res. **41**(19), 4373–4382 (2007)
15. Qiu, P., Xu, C., Chen, H., Jiang, F., Wang, X., Lu, R., et al.: One step synthesis of oxygen doped porous graphitic carbon nitride with remarkable improvement of photo–oxidation activity: role of oxygen on visible light photocatalytic activity. Appl. Catal. B-Environ. **206**, 319–327 (2017)
16. Pignatello, J.J., Mitch, W.A., Xu, W.: Activity and reactivity of pyrogenic carbonaceous matter toward organic compounds. Environ. Sci. Technol. **51**(16), 8893–8908 (2017)
17. Boardman, S.J., Lad, R., Green, D.C., Thornton, P.D.: Chitosan hydrogels for targeted dye and protein adsorption. J. Appl. Polym. Sci. **134**(21) (2017)
18. Kanmani, P., Aravind, J., Kamaraj, M., Sureshbabu, P., Karthikeyan, S.: Environmental applications of chitosan and cellulosic biopolymers: a comprehensive outlook. Bioresour. Technol. **242**, 295–303 (2017)
19. Das, N., Kandimalla, S.: Application of perovskites towards remediation of environmental pollutants: an overview. Int. J. Environ. Sci. Technol. **14**(7), 1559–1572 (2017)
20. Yang, L., Hu, J., He, L., Tang, J., Zhou, Y., Li, J., et al.: One-pot synthesis of multifunctional magnetic N-doped graphene composite for SERS detection, adsorption separation and photocatalytic degradation of Rhodamine 6G. Chem. Eng. J. **327**, 694–704 (2017)
21. Xu, J., Wang, L., Zhu, Y.: Decontamination of bisphenol A from aqueous solution by graphene adsorption. Langmuir: ACS J. Surf. Colloids **28**(22), 8418–8425 (2012)
22. Sadeghi-Kiakhani, M., Arami, M., Gharanjig, K.: Preparation of chitosan-ethyl acrylate as a biopolymer adsorbent for basic dyes removal from colored solutions. J. Environ. Chem. Eng. **1**(3), 406–415 (2013)

23. Liu, S., Wu, P., Chen, M., Yu, L., Kang, C., Zhu, N., et al.: Amphoteric modified vermiculites as adsorbents for enhancing removal of organic pollutants: bisphenol A and tetrabromobisphenol A. Environ. Pollut. **228**, 277–286 (2017)
24. Liu, C., Wu, P., Zhu, Y., Tran, L.: Simultaneous adsorption of Cd(2)(+) and BPA on amphoteric surfactant activated montmorillonite. Chemosphere **144**, 1026–1032 (2016)
25. Shi, C., Lv, C., Wu, L., Hou, X.: Porous chitosan/hydroxyapatite composite membrane for dyes static and dynamic removal from aqueous solution. J. Hazard. Mater. **338**, 241–249 (2017)
26. Li, Y., Bi, H.-Y., Jin, Y.-S.: Facile preparation of rhamnolipid-layered double hydroxide nanocomposite for simultaneous adsorption of p-cresol and copper ions from water. Chem. Eng. J. **308**, 78–88 (2017)
27. Munagapati, V.S., Kim, D.S.: Equilibrium isotherms, kinetics, and thermodynamics studies for congo red adsorption using calcium alginate beads impregnated with nano-goethite. Ecotoxicol. Environ. Saf. **141**, 226–234 (2017)
28. Gamba, M., Olivelli, M., Lázaro-Martínez, J.M., Gaddi, G., Curutchet, G., Sánchez, R.M.T.: Thiabendazole adsorption on montmorillonite, octadecyltrimethylammonium- and Acremonium sp.-loaded products and their copper complexes. Chem. Eng. J. **320**, 11–21 (2017)
29. Wang, F., Gao, Y., Sun, Q., Chen, Z., Megharaj, M., Naidu, R.: Degradation of microcystin-LR using functional clay supported bimetallic Fe/Pd nanoparticles based on adsorption and reduction. Chem. Eng. J. **255**, 55–62 (2014)
30. Yang, Y., Chen, T., Sumona, M., Gupta, B.S., Sun, Y., Hu, Z., et al.: Utilization of iron sulfides for wastewater treatment: a critical review. Rev. Environ. Sci. Biotechnol. **16**(2), 289–308 (2017)
31. Ezzatahmadi, N., Ayoko, G.A., Millar, G.J., Speight, R., Yan, C., Li, J., et al.: Clay-supported nanoscale zero-valent iron composite materials for the remediation of contaminated aqueous solutions: a review. Chem. Eng. J. **312**, 336–350 (2017)
32. Liu, Y., Jiang, X., Li, B., Zhang, X., Liu, T., Yan, X., et al.: Halloysite nanotubes@reduced graphene oxide composite for removal of dyes from water and as supercapacitors. J. Mater. Chem. A **2**(12), 4264 (2014)
33. Yu, L., Wang, H., Zhang, Y., Zhang, B., Liu, J.: Recent advances in halloysite nanotube derived composites for water treatment. Environ. Sci. Nano **3**(1), 28–44 (2016)
34. de Paiva, L.B., Morales, A.R., Valenzuela Díaz, F.R.: Organoclays: properties, preparation and applications. Appl. Clay Sci. **42**(1–2), 8–24 (2008)
35. Kumar, V., Kim, K.-H., Park, J.-W., Hong, J., Kumar, S.: Graphene and its nanocomposites as a platform for environmental applications. Chem. Eng. J. **315**, 210–232 (2017)
36. Apul, O.G., Wang, Q., Zhou, Y., Karanfil, T.: Adsorption of aromatic organic contaminants by graphene nanosheets: comparison with carbon nanotubes and activated carbon. Water Res. **47**(4), 1648–1654 (2013)
37. Yu, Y.-H., Yeh, J.-M., Liou, S.-J., Chang, Y.-P.: Organo-soluble polyimide (TBAPP–OPDA)/clay nanocomposite materials with advanced anticorrosive properties prepared from solution dispersion technique. Acta Mater. **52**(2), 475–486 (2004)
38. Bilgic, C.: Investigation of the factors affecting organic cation adsorption on some silicate minerals. J. Colloid Interface Sci. **281**(1), 33–38 (2005)
39. Yao, W., Yu, S., Wang, J., Zou, Y., Lu, S., Ai, Y., et al.: Enhanced removal of methyl orange on calcined glycerol-modified nanocrystallined Mg/Al layered double hydroxides. Chem. Eng. J. **307**, 476–486 (2017)
40. Philippe, A., Schaumann, G.E.: Interactions of dissolved organic matter with natural and engineered inorganic colloids: a review. Environ. Sci. Technol. **48**(16), 8946–8962 (2014)
41. Parkinson, G.S.: Iron oxide surfaces. Surf. Sci. Rep. **71**(1), 272–365 (2016)
42. Chen, H., Koopal, L.K., Xiong, J., Avena, M., Tan, W.: Mechanisms of soil humic acid adsorption onto montmorillonite and kaolinite. J. Colloid Interface Sci. **504**, 457–467 (2017)
43. Dillon, D., Combes, R., Zeiger, E.: Activation by caecal reduction of the azo dye D & C Red No. 9 to a bacterial mutagen. Mutagenesis **9**(4), 295–299 (1994)

44. Anupam Kaur, R.S.S., Grover, I.S.: Screening of azo dyes for mutagenicity with ames/salmonella assay. Environ. Mol. Mutagen. **22**, 188–190 (1993)
45. Chung, K.T., Cerniglia, C.E.: Mutagenicity of azo dyes: structure–activity relationships. Mutat. Res. **277** (1992)
46. Oliveria, G.A.R., Ferraz, E.R.A., Chequer, F.M.D., Grando, M.D., Angeli, J.P.F., Tsuboy, M.S., et al.: Chlorination treatment of aqueous samples reduces, but does not eliminate, the mutagenic effect of the azo dyes Disperse Red 1, Disperse Red 13, and Disperse Orange 1. Mutat. Res. **703**, 200–208 (2010)
47. Marrs, T.C., Colgrave, H.F., Rice, P., Edginton, J.A.G., Morris, B.: The repeated dose toxicity of a smoke containing Disperse Blue 180, an anthraquinone dye mixture. J. Hazard. Mater. **21**(1), 73–88 (1989)
48. Srivastava, S.J., Singh, N.D., Srivastava, A.K., Sinha, R.: Acute toxicity of malachite green and its effects on certain blood parameters of a catfish, Heteropneustes fossilis. Aquat. Toxicol. **31**(3), 241–247 (1995)
49. He, X., Chen, F.: Adsorption of rectorite to basic safranine in water. Ind. Saf. Environ. Prot. **35**(8), 11–13 (2009)
50. Chen Y.W.Y., Sun J., Luo, J., Duan, Y., Chen P.: Adsorption mechanism of modified rectorite adsorbing methylene blue simulation wastewater. J. Wuhan Inst. Technol. **33**(9), 68–71 (2011)
51. Zheng, P., Chang, P.R., Ma, X.: Preparation and characterization of rectorite gels. Ind. Eng. Chem. Res. **52**(14), 5066–5071 (2013)
52. Zhao Wang, G.Q.: Research on the adsorptive properties of bentnite to methylene blue. Ion Exch. Adsorpt. **8**(2), 156–160 (2002)
53. Ihssane, B., Benamar, M.: Adsorption of Bezathren dyes onto sodic bentonite from aqueous solutions. J. Taiwan Inst. Chem. E **75**, 105–111 (2017)
54. Chakraborty, U., singha, T., Chianelli, R.R., Hansda, C., Paul, P.K.: Organic–inorganic hybrid layer-by-layer electrostatic self-assembled film of cationic dye Methylene Blue and a clay mineral: spectroscopic and atomic force microscopic investigations. J. Lumin. **187**, 322–332 (2017)
55. Wang, M., Wang, L., Wang, A.: Characterization and Congo Red uptake capacity of a new lignocellulose/organic montmorillonite composite. Desalin Water Treat. **51**(37–39), 7120–7129 (2013)
56. Kang, S., Zhao, Y., Wang, W., Zhang, T., Chen, T., Yi, H., Rao, F., Song, S.: Removal of methylene blue from water with montmorillonite nanosheets/chitosan hydrogels as adsorbent. Appl. Surf. Sci. **448**, 203–211 (2018)
57. Wang, W., Yunliang, Z., Haoyu, B., Tingting, Z., Valentin, I.G., Shaoxian, S.: Methylene blue removal from water using the hydrogel beads of poly(Vinyl Alcohol)-sodium alginate-chitosan-montmorillonite. Carbohyd. Polym. **198**, 518–528 (2018)
58. Du, S., Wang, L., Xue, N., Pei, M., Sui, W., Guo, W.: Polyethyleneimine modified bentonite for the adsorption of amino black 10B. J. Solid State Chem. **252**, 152–157 (2017)
59. Elsherbiny, A.S., El-Hefnawy, ME., Gemeay, A.H.: Linker impact on the adsorption capacity of polyaspartate_montmorillonite composites towards methyl blue removal. Chem. Eng. J. **315**, 142–151 (2017)
60. Aguiar, J.E., Cecilia, J.A., Tavares, P.A.S., Azevedo, D.C.S., Rodríguez Castellón, E., Lucena, S.M.P., Junior, I.S.: Adsorption study of reactive dyes onto porous clay heterostructures. Appl. Clay Sci. **135**, 35–44 (2016)
61. Abidi, N., Duplay, J., Jada, A., Baltenweck, R., Errais, E., Semhi, K., et al.: Toward the understanding of the treatment of textile industries' effluents by clay: adsorption of anionic dye on kaolinite. Arab J. Geosci. **10**(16), 373 (2017)
62. Yagub, M.T., Sen, T.K., Afroze, S., Ang, H.M.: Dye and its removal from aqueous solution by adsorption: a review. Adv. Colloid Interface **209**, 172–184 (2014)
63. Song, H.J., You, S., Jia, X.: Synthesis of fungus-like MoS_2 nanosheets with ultrafast adsorption capacities toward organic dyes. Appl. Phys. A **121**(2), 541–548 (2015)

64. Geng, X., Zhang, Y., Han, Y., Li, J., Yang, L., Benamara, M., et al.: 2D water-coupled metallic MoS_2 with nanochannels for ultrafast supercapacitor. Nano Lett. **17**(3), 825–1832 (2017)
65. Wang, Z., Mi, B.: Environmental applications of 2D molybdenum disulfide (MoS_2) nanosheets. Environ. Sci. Technol. **51**(15), 8229–8244 (2017)
66. Li, Z.Z., Meng, X.C., Zhang, Z.S.: Equilibrium and kinetic modelling of adsorption of Rhodamine B on MoS_2. Mater. Res. Bull. **111**, 238–244 (2019)
67. Han, S., Liu, K., Hu, L., Teng, F., Yu, P., Zhu, Y.: Superior adsorption and regenerable dye adsorbent based on flower-like molybdenum disulfide nanostructure. Sci. Rep-Uk **7**, 43599 (2017)
68. Fang, Y., Huang, Q., Liu, P., Shi, J., Gang, X.: A facile dip-coating method for the preparation of separable MoS_2 sponges and their high-efficient adsorption behaviors of Rhodamine B. Inorg. Chem. Front. **5**, 827–834 (2018)
69. Aharon, G., Sadhanala, H.K., Subrata, S., Krishna, H., Kar, N.K.: Green synthesis of MoS_2 nanoflowers for efficient degradation of methylene blue and crystal violet dyes under natural sun light conditions. New J. Chem. **42**(17), 14318–14324 (2018)
70. Li, L., Yin, X., Sun, Y.: Facile synthesized low-cost MoS_2/CdS nanodots-on-nanorods heterostructures for highly efficient pollution degradation under visible-light irradiation. Sep. Purif. Technol. **212**, 135–141 (2018)
71. Wei Xiang, Z.K., Wang, H., Yang, J., Juan, Ma.: Study on decolirization of dyeing wastewater with modified zeoltes. Chem. Adhes. **33**(6), 55–59 (2003)
72. Muthoosamy, K., Manickam, S.: State of the art and recent advances in the ultrasound-assisted synthesis, exfoliation and functionalization of graphene derivatives. Ultrason. Sonochem. **39**, 478–493 (2017)
73. Cheng, Z.-L., Li, Y., Liu, Z.: Fabrication of graphene oxide/silicalite-1 composites with hierarchical porous structure and investigation on their adsorption performance for rhodamine B. J. Ind. Eng. Chem. **55**, 234–243 (2017)
74. Zhang, J., Ping, Q., Niu, M., Shi, H., Li, N.: Kinetics and equilibrium studies from the methylene blue adsorption on diatomite treated with sodium hydroxide. Appl. Clay Sci. **83–84**, 12–16 (2013)
75. Ge-shan, Z., Hong-hai, X., Xiao-jian, T., Fei, P., Chun-li, K.: Adsorption of anionic dyes onto chitosan-modified diatomite. Chem. Res. Chin. U **27**(6), 1035–1040 (2011)
76. Al-Ghouti, M.A., Al-Degs, Y.S., Khraisheh, M.A., Ahmad, M.N., Allen, S.J.: Mechanisms and chemistry of dye adsorption on manganese oxides-modified diatomite. J. Environ. Manage. **90**(11), 3520–3527 (2009)
77. Liu, L., Zhang, B., Zhang, Y., He, Y., Huang, L., Tan, S., et al.: Simultaneous removal of cationic and anionic dyes from environmental water using montmorillonite-pillared graphene oxide. J. Chem. Eng. Data **60**(5), 1270–1278 (2015)
78. Nafees, M., Waseem, A.: Organoclays as sorbent material for phenolic compounds: a review. Clean–Soil Air Water **42**(11), 1500–1508 (2014)
79. Nayak, P.S., Singh, B.K.: Removal of phenol from aqueous solutions by sorption on low cost clay. Desalination **207**(1–3), 71–79 (2007)
80. Park, Y., Ayoko, G.A., Kurdi, R., Horvath, E., Kristof, J., Frost, R.L.: Adsorption of phenolic compounds by organoclays: implications for the removal of organic pollutants from aqueous media. J. Colloid Interface Sci. **406**, 196–208 (2013)
81. Lee, S.M., Tiwari, D.: Organo and inorgano-organo-modified clays in the remediation of aqueous solutions: an overview. Appl. Clay Sci. **59–60**, 84–102 (2012)
82. Dutta, N.N., Borthakur, S., Patil, G.S.: Phase transfer catalyzed extraction of phenolic substances from aqueous alkaline stream. Sep. Sci. Technol. **27**(11), 1435–1448 (2006)
83. Liu, H.L., Wang, X.Y., Wang, X.R., Yu, H.X.: Toxicity of BPA and TBBPA to Daphnia magna and Zebrafish Brachydanio rerio. Environ. Sci. **28**(8), 1874–1878 (2007)
84. Fengru Jiao, Z.P.: Production and market analysis of bisphenol A. Chem. Ind. **26**(9), 21–23 (2004)

85. Bhatnagar, A., Anastopoulos, I.: Adsorptive removal of bisphenol A (BPA) from aqueous solution: a review. Chemosphere **168**, 885–902 (2017)
86. Hu, J.-Y., Aizawa, T., Ookubo, S.: Products of aqueous chlorination of bisphenol A and their estrogenic activity. Environ. Sci. Technol. **36**(9), 1980–1987 (2002)
87. Rivas, A., Lacroix, M., Olea-Serranoa, F., La¨ıos, I., Leclercqb, G., Olea, N.: Estrogenic effect of a series of bisphenol analogues on gene and protein expression in MCF-7 breast cancer cells. J. Steroid. Biochem. **82**, 45–53 (2002)
88. Takemura, H., Ma, J., Sayama, K., Terao, Y., Zhu, B.T., Shimoi, K.: In vitro and in vivo estrogenic activity of chlorinated derivatives of bisphenol A. Toxicology **207**(2), 215–221 (2005)
89. Ake, C.L., Wiles, M.C., Huebner, H.J., McDonald, T.J., Cosgriff, D., Richardson, M.B., et al.: Porous organoclay composite for the sorption of polycyclic aromatic hydrocarbons and pentachlorophenol from groundwater. Chemosphere **51**(9), 835–844 (2003)
90. Akcay, M., Akcay, G.: The removal of phenolic compounds from aqueous solutions by organophilic bentonite. J. Hazard. Mater. **113**(1–3), 189–193 (2004)
91. Zhu, L., Wang, L., Xu, Y.: Chitosan and surfactant co-modified montmorillonite: a multifunctional adsorbent for contaminant removal. Appl. Clay Sci. **146**, 35–42 (2017)
92. Wang, G., Zhang, S., Hua, Y., Su, X., Ma, S., Wang, J., et al.: Phenol and/or Zn^{2+} adsorption by single- or dual-cation organomontmorillonites. Appl. Clay Sci. **140**, 1–9 (2017)
93. Shen, H.-M., Zhu, G.-Y., Yu, W.-B., Wu, H.-K., Ji, H.-B., Shi, H.-X., et al.: Fast adsorption of p-nitrophenol from aqueous solution using β-cyclodextrin grafted silica gel. Appl. Surf. Sci. **356**, 1155–1167 (2015)
94. Al Bakain, R.Z., Abu-urayk, R.A., Hamadneh, I., Khalili, F.I., Al-Dujaili, A.H.: A study on removal characteristics of o-, m-, and p-nitrophenol from aqueous solutions by organically modified diatomaceous earth. Desalin Water Treat. **56**(3), 826–838 (2014)
95. Zhu, J., Qing, Y., Ma, L., Zhu, R., He, H.: The structure of montmorillonites modified with zwitterionic surfactants and their sorption ability. Miner. Petrol. **109**(3), 349–355 (2014)
96. Sarkar, B., Xi, Y., Megharaj, M., Krishnamurti, G.S.R., Naidu, R.: Synthesis and characterisation of novel organopalygorskites for removal of p-nitrophenol from aqueous solution: isothermal studies. J. Colloid Interface Sci. **350**(1), 295–304 (2010)
97. Shah, J., Jan, M.R., Zeeshan, M., Iqbal, M.: Removal and extraction of para-nitrophenol from aqueous samples using graphene magnetic nanocomposite. Arab. J. Sci. Eng. **42**(10), 4465–4472 (2017)
98. Wang, Z.M., Ooga, H., Hirotsu, T., Wang, W.-L., Wu, Q.Y., Hu, H.-Y.: Matrix-enhanced adsorption removal of trace BPA by controlling the interlayer hydrophobic environment of montmorillonite. Appl. Clay Sci. **104**, 81–87 (2015)
99. Xie, Q., Xie, J., Wang, Z., Wu, D., Zhang, Z., Kong, H.: Adsorption of organic pollutants by surfactant modified zeolite as controlled by surfactant chain length. Microporous Mesoporous Mater. **179**, 144–150 (2013)
100. Zhang, G.F., Zhang, S., Pan, B., Liu, X., Feng, L.S.: 4-Quinolone derivatives and their activities against Gram positive pathogens. Eur. J. Med. Chem. **143**, 710–723 (2018)
101. Genc, N., Can Dogan, E., Yurtsever, M.: Bentonite for ciprofloxacin removal from aqueous solution. Water Sci. Technol. **68**(4), 848–855 (2013)
102. Li, Z., Hong, H., Liao, L., Ackley, C.J., Schulz, L.A., MacDonald, R.A., et al.: A mechanistic study of ciprofloxacin removal by kaolinite. Colloid Surf. B **88**(1), 339–344 (2011)
103. Aristilde, L., Lanson, B., Miehe-Brendle, J., Marichal, C., Charlet, L.: Enhanced interlayer trapping of a tetracycline antibiotic within montmorillonite layers in the presence of Ca and Mg. J. Colloid Interface Sci. **464**, 153–159 (2016)
104. Francisco-Márquez, M., Soriano-Correa, C., Sainz-Díaz, C.I.: Adsorption of sulfonamides on phyllosilicate surfaces by molecular modeling calculations. J. Phys. Chem. C **121**(5), 2905–2914 (2017)
105. Li, Y., Bi, E., Chen, H.: Sorption behavior of ofloxacin to kaolinite: effects of pH, ionic strength, and Cu(II). Water Air Soil Pollut. **228**(1) (2017)

106. Rakshit, S., Sarkar, D., Elzinga, E.J., Punamiya, P., Datta, R.: Mechanisms of ciprofloxacin removal by nano-sized magnetite. J. Hazard. Mater. **246–247**, 221–226 (2013)
107. Guo, X., Dong, H., Yang, C., Zhang, Q., Liao, C., Zha, F., et al.: Application of goethite modified biochar for tylosin removal from aqueous solution. Colloid Surf. A **502**, 81–88 (2016)
108. Wang, H., Wang, X., Xia, P., Song, J., Ma, R., Jing, H., et al.: Eco-friendly synthesis of self-existed magnesium oxide supported nanorod-like palygorskite for enhanced and simultaneous recovery of nutrients from simulated wastewater through adsorption and in-situ struvite formation. Appl. Clay Sci. **135**, 418–426 (2017)
109. Shih, Y.J., Abarca, R.R., de Luna, M.D., Huang, Y.H., Lu, M.C.: Recovery of phosphorus from synthetic wastewaters by struvite crystallization in a fluidized-bed reactor: Effects of pH, phosphate concentration and coexisting ions. Chemosphere **173**, 466–473 (2017)
110. Wang, H., Wang, X., Ma, J., Xia, P., Zhao, J.: Removal of cadmium (II) from aqueous solution: a comparative study of raw attapulgite clay and a reusable waste-struvite/attapulgite obtained from nutrient-rich wastewater. J. Hazard. Mater. **329**, 66–76 (2017)
111. Başakçılardan-Kabakcı, S., Thompson, A., Cartmell, E., Corre, K.L.: Adsorption and precipitation of tetracycline with struvite. Water Environ. Res. **79**(13), 2551–2556 (2007)
112. Ye, Z.-L., Deng, Y., Lou, Y., Ye, X., Zhang, J., Chen, S.: Adsorption behavior of tetracyclines by struvite particles in the process of phosphorus recovery from synthetic swine wastewater. Chem. Eng. J. **313**, 1633–1638 (2017)
113. Li, Y.M., Zhang, F.S.: Characterization of a cetyltrimethyl ammonium bromide-modified sorbent for removal of perfluorooctane sulphonate from water. Environ. Sci. Technol. **35** (17–20), 2556–2568 (2014)
114. Taniyasu, S., Yamashita, N., Yamazaki, E., Petrick, G., Kannan, K.: The environmental photolysis of perfluorooctanesulfonate, perfluorooctanoate, and related fluorochemicals. Chemosphere **90**(5), 1686–1692 (2013)
115. Hori, H., Yamamoto, A., Sano, T., Yamashita, N., Taniyasu, S., Kutsuna, S.: Efficient decomposition of environmentally persistent perfluorooctanesulfonate and related fluorochemicals using zerovalent iron in subcritical water. Environ. Sci. Technol. **40**(3), 1049–1054 (2006)
116. Gomez-Ruiz, B., Gómez-Lavín, S., Diban, N., Boiteux, V., Colin, A., Dauchy, X., et al.: Efficient electrochemical degradation of poly- and perfluoroalkyl substances (PFASs) from the effluents of an industrial wastewater treatment plant. Chem. Eng. J. **322**, 196–204 (2017)
117. Zhang, R., Yan, W., Jing, C.: Mechanistic study of PFOS adsorption on kaolinite and montmorillonite. Colloids Surf. A **462**, 252–258 (2014)
118. Pearson, R.G.: Hard and soft acids and bases. J. Am. Chem. Soc. **85**, 3533–3539 (1963)
119. Ramona, L., Johnson, A.J.A., Smolen, Jean M., Simcik, Matt F., Lee Pen, R.: The adsorption of perfluorooctane sulfonate onto sand, clay, and iron oxide surfaces. J. Chem. Eng. Data **52**(4), 1165–1170 (2007)
120. Xiao, F., Zhang, X., Penn, L., Gulliver, J.S., Simcik, M.F.: Effects of monovalent cations on the competitive adsorption of perfluoroalkyl acids by kaolinite: experimental studies and modeling. Environ. Sci. Technol. **45**(23), 10028–10035 (2011)
121. He, G., Zhang, M., Zhou, Q., Pan, G.: Molecular dynamics simulations of structural transformation of perfluorooctane sulfonate (PFOS) at water/rutile interfaces. Chemosphere **134**, 272–278 (2015)
122. Qian, J., Shen, M., Wang, P., Wang, C., Hu, J., Hou, J., et al.: Co-adsorption of perfluorooctane sulfonate and phosphate on boehmite: influence of temperature, phosphate initial concentration and pH. Ecotox. Environ. Saf. **137**, 71–77 (2017)
123. Du, Z., Deng, S., Zhang, S., Wang, W., Wang, B., Huang, J., et al.: Selective and fast adsorption of perfluorooctanesulfonate from wastewater by magnetic fluorinated vermiculite. Environ. Sci. Technol. **51**(14), 8027–8035 (2017)
124. Auta, M., Hameed, B.H.: Modified mesoporous clay adsorbent for adsorption isotherm and kinetics of methylene blue. Chem. Eng. J. **198–199**, 219–227 (2012)

125. Wibulswas, R.: Batch and fixed bed sorption of methylene blue on precursor and QACs modified montmorillonite. Sep. Purif. Technol. **39**(1–2), 3–12 (2004)
126. Gurses, A., Dogar, C., Yalcin, M., Acikyildiz, M., Bayrak, R., Karaca, S.: The adsorption kinetics of the cationic dye, methylene blue, onto clay. J. Hazard. Mater. **131**(1–3), 217–228 (2006)
127. Vimonses, V., Lei, S., Jin, B., Chow, C.W.K., Saint, C.: Kinetic study and equilibrium isotherm analysis of Congo Red adsorption by clay materials. Chem. Eng. J. **148**(2–3), 354–364 (2009)
128. Hong, S., Wen, C., He, J., Gan, F., Ho, Y.S.: Adsorption thermodynamics of Methylene Blue onto bentonite. J. Hazard. Mater. **167**(1–3), 630–633 (2009)
129. Gücek, A., Şener, S., Bilgen, S., Mazmancı, M.A.: Adsorption and kinetic studies of cationic and anionic dyes on pyrophyllite from aqueous solutions. J Colloid Interface Sci. **286**(1), 53–60 (2005)
130. Weng, C.H., Pan, Y.F.: Adsorption of a cationic dye (methylene blue) onto spent activated clay. J. Hazard. Mater. **144**(1), 355–362 (2007)
131. Walker, G.M., Hansen, L., Hanna, J.A., Allen, S.J.: Kinetics of a reactive dye adsorption onto dolomitic sorbents. Water Res. **37**(9), 2081–2089 (2003)
132. Meshko, V., Markovska, L., Mincheva, M., Rodrigues, A.E.: Adsorption of basic dyes on granular activated carbon and natural zeolite. Water Res. **35**(14), 3357–3366 (2001)
133. Phan, T.N.T., Bacquet, M., Morcellet, M.: Synthesis and characterization of silica gels functionalized with monochlorotriazinyl β-cyclodextrin and their sorption capacities towards organic compounds. J. Incl. Phenom. Macrocycl. **38**(1), 345–359 (2000)
134. Ho, Y.S., Chiang, C.C., Hsu, Y.C.: Sorption kinetics for dye removal from aqueous solution using activated clay. Sep. Sci. Technol. **36**(11), 2473–2488 (2001)
135. Wiegel, S., Aulinger, A., Brockmeyer, R., Harms, H., Löffler, J., Reincke, H., et al.: Pharmaceuticals in the river Elbe and its tributaries. Chemosphere **57**(2), 107–126 (2004)
136. Kolpin, D.W., Furlong, E.T., Meyer, M.T., Thurman, E.M., Zaugg, S.D., Barber, L.B., et al.: Pharmaceuticals, hormones, and other organic wastewater contaminants in U.S. Streams, 1999–2000: a national reconnaissance. Environ. Sci. Technol. **36**(6), 1202–1211 (2002)
137. Calamari, D., Zuccato, E., Castiglioni, S., Bagnati, R., Fanelli, R.: Strategic survey of therapeutic drugs in the rivers Po and Lambro in Northern Italy. Environ. Sci. Technol. **37** (7), 1241–1248 (2003)
138. Kim, S.C., Carlson, K.: Temporal and spatial trends in the occurrence of human and veterinary antibiotics in aqueous and river sediment matrices. Environ. Sci. Technol. **41**(1), 50–57 (2004)
139. Managaki, S., Murata, A., Takada, H., Tuyen, B.C., Chiem, Nguyen H.: Distribution of macrolides, sulfonamides, and trimethoprim in tropical waters ubiquitous occurrence of veterinary antibiotics in the Mekong Delta. Environ. Sci. Technol. **41**(23), 8004–8810 (2007)
140. Xu, W.H., Zhang, G., Zou, S.C., Li, X.D., Liu, Y.C.: Determination of selected antibiotics in the Victoria Harbour and the Pearl River, South China using high-performance liquid chromatography-electrospray ionization tandem mass spectrometry. Environ. Pollut. **145**(3), 672–679 (2007)
141. Xu, W., Zhang, G., Zou, S., Ling, Z., Wang, G., Yan, W.A.: Preliminary investigation on the occurrence and distribution of antibiotics in the Yellow River and its Tributaries, China. Water Environ. Res. **81**(3), 248–254 (2009)
142. Gulkowska, A., He, Y., So, M.K., Yeung, L.W.Y., Leung, H.W., Giesy, J.P., et al.: The occurrence of selected antibiotics in Hong Kong coastal waters. Mar. Pollut. Bull. **54**(8), 1287–1293 (2007)
143. Tamtam, F., Mercier, F., Le Bot, B., Eurin, J., Tuc Dinh, Q., Clement, M., et al.: Occurrence and fate of antibiotics in the Seine River in various hydrological conditions. Sci. Total Environ. **393**(1), 84–95 (2008)

144. Zou, S., Xu, W., Zhang, R., Tang, J., Chen, Y., Zhang, G.: Occurrence and distribution of antibiotics in coastal water of the Bohai Bay, China: impacts of river discharge and aquaculture activities. Environ. Pollut. **159**(10), 2913–2920 (2011)
145. Jiang, L., Hu, X., Yin, D., Zhang, H., Yu, Z.: Occurrence, distribution and seasonal variation of antibiotics in the Huangpu River, Shanghai, China. Chemosphere **82**(6), 822–828 (2011)
146. Gao, L., Shi, Y., Li, W., Liu, J., Cai, Y.: Occurrence, distribution and bioaccumulation of antibiotics in the Haihe River in China. J. Environ. Monit. **14**(4), 1248–1255 (2012)

Adsorption of Microorganisms to Minerals

Ling Xia, Liyuan Ma, and Delong Meng

Abstract Adsorption of microorganisms (bacteria, fungi and microalgae, etc.) and biomacromolecules (humic substances, extracellular substances, proteins and nuclear acids, etc.) to minerals are universal in the nature environments. These processes also governed the inorganic and organic pollutants migration and conversion. This chapter discusses the interface reactions of microorganisms and minerals regarding the surface properties of both sides, leading to the microorganism's survival and growing, minerals' transformation and formation as well as polluted or harmful substances' retention and removal. The content broads the knowledge of bacteria, fungi and algae and biomacromolecules adsorption onto natural minerals and highlights the application of microorganism-mineral interaction on environmental engineering such as heavy metals and polluted organic matters removal.

Keywords Clay · Goethite · Bacteria · Fungi · Microalgae · Biomacromolecules

L. Xia (✉)
Hubei Key Laboratory of Mineral Resources Processing and Environment, Wuhan University of Technology, Luoshi Road 122, Wuhan, Hubei Province 430070, China
e-mail: xialing@whut.edu.cn

School of Resources and Environmental Engineering, Wuhan University of Technology, Luoshi Road 122, Wuhan, Hubei 430070, China

L. Ma
School of Environmental Studies, China University of Geosciences, Wuhan 430074, China

D. Meng
School of Minerals Processing and Bioengineering, Central South University, Changsha, China

Key Laboratory of Biometallurgy of Ministry of Education, Changsha, China

S. Song and B. Li (eds.), *Adsorption at Natural Minerals/Water Interfaces*,
Engineering Materials,
https://doi.org/10.1007/978-3-030-54451-5_7

1 Introduction

Microbes exist everywhere, and play essential roles in the global biogeochemical cycles such as generation and degradation of solid phases including biomineralization, bioweathering and bioleaching. It wasn't until the last few years of the twentieth century that scientists began to understand that the importance of microbes to the solid phases of the earth. Well, all kinds of microorganisms, including prokaryotes (bacteria), eukaryotes (fungi and algae), can contribute actively to these geological and mineral phenomena. During all these processes, the first step for the microbes is to contact with the solid and adsorb on the minerals. In addition, widespread humic acids and bioactive molecules secreted by the microorganisms can efficiently promote the adsorption process and pollutants retention. Thus, this chapter summarized the present state of knowledge in the adsorption of bacteria, fungi, algae as well as the adsorption promotion on minerals as well as environmental applications.

Microorganisms are widely involved in the biogeochemical cycle, and play an indispensable role. Microorganisms can directly or indirectly enhance mineral dissolution, disaggregation or secondary mineral formation and further affect the fate of the pollutants. Biomineralization, bioweathering and biohydrometallurgy are common phenomenon in nature. The interaction between microorganisms and minerals is a complex process, it is determined by mineral surface properties, microbial cell structure as well as the environmental conditions. Adsorption of microorganisms to the surface of mineral particles is the fundamental process. Research of the microbe-mineral interface represent one of the most fundamental and difficult areas of the microbiology, geology and mineralogy. The microbe-mineral interface studies include how do microbes attach, how do they chemically alter minerals once attached, how and when do they alter the minerals surface while attached and vice versa. The tools used for these studies include molecular biology, microbial physiology, geochemistry, electrochemistry and imaging of many different kinds of those interfaces. Microbes attach to mineral surfaces, form biofilms on them, and often alter the surface properties as a result of the interaction, leading to the destruction, transformation, or even formation of the minerals, and finally affecting the migration and transformation of the organic and inorganic pollutants, which in turn, as well, affecting the survival and reproduction of the microorganisms.

2 Bacteria Adsorption

2.1 Adsorption Mechanism

The surface interaction of microorganisms with solid particles involves four steps: (i) transport to the surface, (ii) contact and initial adhesion, (iii) firmer attachment, and then (iv) growth, to form adhering microcolonies or biofilms [1]. The initial adhesion is rapid and can be reversible or irreversible. It is a physicochemical process, which is described and reasonably well predicted by theories of colloid chemistry, such as the Derjaguin Landau Verwey Overbeek (DLVO) theory for electrostatic interactions [2, 3] and Lewis acid/base hydrophobic interactions [4]. Most of the data on microbial adhesion are obtained with bacteria and show that the adhesion depends on the surface properties of the cells and on their physiological state [5, 6]. Since microbial adhesion to mineral substrates and the formation of biofilms are important events in microbe-mineral interaction processes, any consideration of the mechanisms whereby microorganism adhere to solid surfaces must take account of the surface properties of the microorganism as well as those of the substrate concerned. Thus, it is essential to understand the basic microbiological characteristics of different microorganisms.

The structure and architecture of the microbial cell wall play a prominent role in adhesion to mineral surfaces. The size of bacteria ranges from about 0.2 μm to several micrometers in length, and the majority are about 1.0 μm in length or diameter. Bacteria are classified into two categories with reference to cell wall structure, namely, Gram-positive and Gram-negative. The chemical composition and structural features of cell walls are different in these two types. Gram-positive bacteria possess a well-defined, rigid outer cell wall, 15–30 nm thick, and an inner, closely held, cell-limiting plasma membrane. A major polymeric component is peptidoglycan and it may also contain one or more secondary polymers such as teichoic and teichuronic acids. On the contrary, the cell walls of Gram-negative bacteria contain a more general architecture having an outer membrane placed above a thin peptidoglycan layer. The bilayer membranes contain proteins, phospholipids, and lipopolysaccharides and also separate the external environment from the periplasm [7].

As revealed by electrophoretic measurements, bacteria possess a net negative surface charge at pH values found in most natural habitats due to the presence of peptidoglycan and teichoic acids. The cells acquire charge through the ionisation of surface polymeric groups, which is pH dependent. Stable suspensions of bacteria in dilute electrolytes result from a mutual electrostatic repulsion between like charges on the bacterial surfaces. A charge reversal at low pH values is indicative of the presence of some charged basic (amino) groups, which are revealed when dissociation of acidic (carboxyl, phosphate) groups is suppressed at the low pH values [8]. Competition between electrical and chemical forces may control surface charge neutralisation. An electrical double layer is established at the bacteria-solution interface. Similar to mineral particles, the surface chemical properties of microorganisms can be characterised by zeta potential and the isoelectric point (IEP). Increasing solution ionic strength may favour bacterial adhesion to a mineral substrate [9]. Bacterial cell dispersion and flocculation are also governed by electrical forces as in the case of mineral particles.

The bacteria also exhibit changes in the overall surface free energy of the cells, some of which have relatively hydrophilic surfaces and others have relatively hydrophobic surfaces. Bacterial surface hydrophobicity is also controlled by its cell membrane polymers. Hydrophobicity is conferred by hydrophobic molecules such as lipids present on cell surfaces. Hydrophobic bacterial cells are known to adhere to surfaces due to repulsion from the polar water molecule. Electrostatic repulsive forces decrease higher surface hydrophobicity. Since the bacterial surface charge enhances the possibilities for polar interactions with water molecules, the more charged the cell surfaces, the less hydrophobic it become. A decrease in the cell negative charge may then enhance cell surface hydrophobicity. Cells having higher hydrophobicity and lower electrophoretic mobility are more adherent. The distribution of hydrophobic sites on bacterial cells is not necessarily uniform and may result in a preferred orientation of certain bacteria at interfaces [10].

Since interfaces offer a better environment for bacterial nutrition and growth, bacterial adhesion at mineral–solution interfaces is necessitated under environmental conditions [9]. As chemical composition and architecture of the cell wall outer layers play a significant role in bacterial adhesion, the surface proteins and polysaccharides may be involved in bacterial adhesion to minerals. Bacteria are not inert colloidal particles, but are living organisms capable of metabolism, growth, and, in some instances, independent motion. An expression of the metabolism of bacteria is the production of variable amounts of polymers that are bound to the exterior of the cell envelope [11]. These polymers as well as flagellar, pili and fimbriae are important in the final process of bacterial adhesion to the solid surface. The formation of these extracellular polymers can be modified depending on the environmental conditions and therefore may have an effect on the efficiency and selectivity of adhesion between different species in natural habitats. It is relevant to consider the means of bacteria transporting from the bulk liquid to the vicinity of the solid-water interface before solving the problem of how the bacteria adheres to the solid surface.

2.1.1 Conditions in the Aqueous Phase

Adhesion of bacteria to solid surfaces occurs both in quiescent waters and turbulent flow. The transport of bacteria to substrate surfaces are certainly different under different conditions [12]. Bryers and Characklis [13] have shown that the rate of overall bacterial film development on surfaces in turbulent flow conditions increases with the biomass concentration dispersed in the bulk phase. This indicates that the particle flux to a surface is directly proportional to the bulk particle concentration as predicted by mass transport and particle deposition theories. Consequently, in very dilute suspensions of bacteria transport is probably the rate-limiting step in the process of bacterial deposition at the surface [14]. Under turbulent flow conditions a zone of relatively still water exists near a solid surface. This is termed the viscous sublayer and its thickness is dependent upon the magnitude of the fluid shear rate. Bacteria in the bulk liquid must penetrate this viscous sublayer in order to be deposited at the solid surface [7].

2.1.2 Transport Mechanisms

Transport of molecules and small particles may be described satisfactorily in terms of diffusion.

1. Sedimentation

This process is only of significance in low-shear systems and with relatively large particles, such as very large bacteria or aggregates (flocs) of normal bacteria. Most bacteria form stable suspensions, and only sediments when the system becomes destabilized, for example, in floc formation. Sedimentation of bacteria is unlikely to occur under turbulent flow conditions. Li et al. [15] analyze the contribution of sedimentation to bacterial mass transport in a parallel plate flow chamber, and conclude that the sedimentation is the major mass transport mechanism.

2. Chemotaxis

Many bacteria are motile in the bulk aqueous phase as a result of the propulsive action of flagella. Motile bacteria are capable of displaying a positive chemotactic response to certain nutrient sources and, as such, can respond to a nutrient concentration gradient that exists at a solid-water interface. Thus, bacteria may actively swim toward a surface and ultimately be held by the attractive forces operative near the surface. The role of chemotactic responses in the formation of primary surface films by bacteria has been considered by Young and Mitchell [16]. Bacteria may also exhibit a negative chemotactic response where inhibitory substances, such as hydrogen ions or antibiotics, accumulate at the solid-water interface [17]. Chemotactic responses are probably not of significance in turbulent flow conditions, but are probably important in transport through the viscous sublayer [14].

3. Brownian motion

Most bacteria have an effective radius of less than 1.0 μm and exhibit a significant degree of Brownian displacement when viewed under quiescent conditions under a microscope. Larger bacteria do not show this form of motion. Brownian motion probably contributes little to the transport of bacteria in turbulent flow, but should be a significant form of transport within the viscous sublayer [18].

4. Cell surface hydrophobicity

Mudd [19] demonstrated that bacteria can vary appreciably in their degree of cell surface hydrophobicity, and these observations have been confirmed and extended by more recent studies on the relative hydrophobicities of bacterial surfaces [20]. Hydrophobic bacteria could reasonably be considered as being rejected from the aqueous phase and attracted toward any nonaqueous phase, including solid surfaces. Consequently, hydrophobicity could provide a means whereby bacteria are attracted toward solid surfaces and would provide a means for a particular orientation of some bacteria at interfaces [10, 21].

5. Fluid Dynamic Forces

Bacteria often live in dynamic fluid environments and flow can affect fundamental microbial processes such as nutrient uptake and infection. Bacteria in turbulent flow systems are dispersed by eddy diffusion in the turbulent core region, thus maintaining a uniform concentration in the bulk of fluid. Such eddy diffusion may only transport the bacteria to the region of the viscous sublayer. If the bacteria are traveling faster than the fluid in the region of the wall, a lift force directs the bacteria toward the wall. In the viscous sublayer, the bacteria encounter significant frictional drag forces, which gradually slow a bacterium down as it approaches the surface. Microfluidic experiments show that fluid shear produces strong spatial heterogeneity in suspensions of motile bacteria, characterized by up to 70% cell depletion from low-shear regions due to 'trapping' in high-shear regions. Shear-induced trapping directly impacts widespread bacterial behaviours, by hampering chemotaxis and promoting surface attachment [22]. These spontaneous bursts of turbulence penetrate the viscous sublayer and provide a significant fluid mechanical force to direct the bacteria to the solid surface.

The above mechanisms provide the means for transporting bacteria from the bulk aqueous phase to the vicinity of the wall. The final adhesion of bacteria to solid surfaces involves a consideration of various aspects of colloid chemistry along with an understanding of some of the physiological responses of different bacteria to the conditions existing both in the bulk liquid and at the solid-water interface.

2.1.3 Adhesion Force

Interfacial interaction of microbes and minerals involves a series of physical and chemical process, that is to say, the adsorption depends not only on biological characteristics but also depends on the properties of various interface. The driving force and combination mode for the attachment of microorganisms to mineral surface have been paid a lot of attention [23]. Ghauri et al. [24] studied the adsorption of *Acidothiobacillus ferrooxidans* to pyrite surface, and put forward that the initial motivation for microorganisms attached to mineral is electrostatic force. Rodríguez et al. [25] determined the adhesive force between *A. ferrooxidans* and pyrite using atomic force microscope, and they speculated that the combination between them is a kind of irreversible chemical combination as the adhesive force reach up to 467 pN. The extracellular polymers is an important medium for the combined of microbes and mineral, as the secreted extracellular polymer of the attached cells are much higher than that of the free cells [26]. Generally, the adsorption force and the combination mode of microbes to minerals surface could be divided into long-range force, hydrophobic force, electrostatic force, hydrogen bonding force and chemical bonds force [27]. Because of the obvious variability in the extracellular polymers produced by different bacteria, different interactions will occur during firm attachment of these bacteria and it is possible that various types of interactions take place simultaneously with a certain bacterial type [7].

1. Long-range force: A long-range force interaction belongs to reversible adhesion, where an instantaneous attraction by long-range forces holding bacteria near a surface, so that the bacteria continue to exhibit Brownian motion and can

be readily removed from the surface by the shearing effects of a water jet or by the violent rotational movements of motile bacteria [28].

2. Hydrophobic interaction: Hydrophobic interactions are weak noncovalent effect between nonpolar molecules that tend to sink from each other in aqueous environments. In the process of microbial cells adsorbed to the surface of minerals, hydrophobic interaction means that hydrophobic cells can avoid water and tend to interact with hydrophobic minerals. Preston Devasia et al. [29] noted that *A. ferrooxidans* grown in sulfur, pyrite and chalcopyrite exhibit stronger hydrophobicity than that grown in ferrous iron, and the isoelectric point is higher than the latter. More than 50% of the bacteria grown on the mineral substrate can bind to octyl agarose, whereas only about 18% of the bacteria grown in the ferrous ionic solution bind to the octyl agarose, which indicated that a stronger hydrophobicity is favorable for adsorption. Devasia also studied the adsorption of *A. ferrooxidans* on the surface of the pyrite, which indicated that increasing the hydrophobicity of *A. ferrooxidans* was beneficial to the adsorption process and that the hydrophobic bacteria had greater adsorption than the hydrophilic bacteria. Naoya Ohmura et al. [30] studied the selective adsorption of *A. ferrooxidans* in bioleaching of pyrite, quartz, chalcopyrite and galena, using *Escherichia coli* as a control, with contact angle as a hydrophobicity index. The *E. coli* tends to adsorb to hydrophobic mineral surfaces through hydrophobic interactions, while *A. ferrooxidans* is selectively adsorbed onto iron-containing mineral surfaces such as pyrite and chalcopyrite.
3. Electrostatic interaction: Although hydrophobic interactions are dominant in the adsorption of hydrophobic bacteria, electrostatic interactions often exist and affect the process. The functional groups on the cell wall including carboxyl (-COOH), amino (-NH) and hydroxyl (-OH). These functional groups present different ionization constants in the solution, with the result that the cells are positively charged or negatively charged. Blake et al. [31] discovered this electrostatic effect when *A. ferrooxidans* adsorbs to pyrite and sulfur. Marshall [32] also found that the isoelectric point of pyrite and sulfur powder shifted significantly to the right after the action with *A. ferrooxidans* by electrophoresis, indicating that the surface electrical properties of the bacteria changed the surface properties of the mineral. Sulfur, pyrite and chalcopyrite show an apparent displacement after the interaction with *A. ferrooxidans* [33]. The results of these studies show that the isoelectric point of both bacteria and minerals migrates to a higher value, indicating that the bacteria is positively charged in a wider range of pH, and that the electronegativity of the mineral is also gradually increased in a wider pH Values, these changes will promote the adsorption of bacteria to minerals.
4. Hydrogen bonding: In addition to the hydrophobic interactions and electrostatic interactions discussed above, hydrogen bonding is generally considered to be a common adsorption mechanism. There are not only differences in the number of hydrogen bonds on different mineral surfaces, but also in bond energy, which can be attributed to the presence of different types of hydroxyl groups (M-OH). As the surface of polysaccharides and other substances have a number of

hydroxyl, carboxyl, and carbonyl groups, they can form hydrogen bonds with the mineral surface and the hydration layer. Santhiya et al. [34] studied the adsorption of *Acidithiobacillus thiooxidans* on mineral surface by FTIR, and the wave migration of the two sulfide deposits were detected, which provided the basis for the hydrogen bonding mechanism. Fu [33] studied the differences of the infrared spectra of sulfur culture and ferrous culture bacteria, and pointed out that the polyhydroxy peptide bond, C-O bond and hydrogen bond play an important role in the adsorption process. Subramanian et al. [35] examined the interaction between metabolites of *Bacillus subtilis* and galena/sphalerite by flotation and flocculation test, and found that the hydroxyl groups of polysaccharide compounds in cell metabolites interact with the metal hydroxide on mineral surface by hydrogen bonding and chemical forces. Rong et al. [36] reported that hydrogen bonding plays an important role in the process of *Pseudomonas putida* cells interacting with kaolinite and montmorillonite.

5. Chemical bonds interaction: According to the theory of the interaction between organic and mineral, there is chemical bonding between the organic groups on the cell surface and the metal elements on the mineral surface during the adsorption process. Most of the literature is based on the analysis of cell surface groups to infer the chemical action between cells and mineral surfaces. Namita et al. [37] studied the adhesion mechanism of *Paenibacillus polymyxa* on hematite, corundum and quartz surfaces, and concluded that interfacial energy is the main adsorption force of bacteria to hematite and corundum, and the chemical interaction is the main force that bacteria adsorb onto the quartz surface. Wang et al. [38] studied the influence of bacteria adsorption on the floatability of sulfide minerals, noting that one of the adhesion forces between bacteria and pyrite is the binding of the active groups distributed in the mucosal surface of the cell to the surface of the mineral, such as hydroxyl (-OH), Carboxyl (-COOH) and sulfydryl (-SH).

Despite a great deal of research currently under way on adhesion of bacteria to inert surfaces, there are many areas where much more information is required before we have a satisfactory understanding of the overall processes involved. The mechanisms whereby bacteria are transported to surfaces are not entirely clear [7].

2.2 Factors Affecting Adsorption of Bacteria on Mineral Surface

The adsorption of microorganisms to minerals surface is determined by various environmental factors including mineral type, pH value, ionic strength and conditioning time etc. Schalm et al. [39] studied the adsorption of *A. ferrooxidans* on the surface of pyrite and chalcopyrite. The results show that the adsorption of bacteria on the mineral surface is a rapid process, and the adsorption capacity of bacteria on pyrite surface is greater than that of chalcopyrite. Santhiya et al. [40] reported that the amount of bacteria adsorbed on galena is one order of magnitude higher than that on sphalerite. Farahat et al. [41] showed that the adsorption capacity of *E. coli* on quartz was the highest at pH less than 4.3, and the adsorption capacity on corundum was also higher between pH 4.5 and 8.5, while, the adsorption of *E. coli* on hematite surface is very low within the whole pH ranges. Yee et al. [42] reported that the adsorption of *B. subtilis* on the surface of quartz (SiO_2) and corundum (α-Al_2O_3) was a reversible process in which the reaction reached equilibrium within one hour. *B. subtilis* show a high affinity for corundum, and its adsorption capacity is significantly correlated with pH and ionic strength. However, its adsorption on quartz is weak, and independent of the pH value and ionic strength [43]. Increasing the ionic strength can change the zeta potential of hematite and *Bacillus polymyxa*, but the point of zero charge does not change. Adsorption experiments show that *B. polymyxa* adsorption on hematite and quartz surface is controlled by the conditioning time, pH and ionic strength. The adsorption of *B. polymyxa* on hematite surface is greater than that on quartz surface. This selective adsorption can be used for hematite flotation and flocculation [44]. Many bacteria and fungi have the ability to produce extracellular polymeric substance (EPS) in pure culture. In laboratory experiments, microbial polysaccharides of net zero, negative, or positive charge have been shown to adsorb rapidly and irreversibly to clay minerals [45, 46]. Electrostatic interactions are also involved in the case of charged EPS, and di- or trivalent cations enhance the adsorption of negatively charged EPS to negatively charged clay minerals [47]. Extracellular polysaccharides constitute the outermost surface of those microorganisms that produce them. The adsorption of EPS to mineral surfaces causes or strengthens the attachment of bacteria to solid particles.

2.3 *Influences on Minerals and Bacteria*

Interaction between microbes and minerals changed surface properties of the bacteria and the mineral particle. Consequences of microbial adhesion on minerals need to be understood. Electrokinetic properties of minerals are modified by microorganisms and their metabolic products. Significant surface chemical changes are thus brought about on minerals by bacterial interaction. The surface chemical properties such as hydrophobicity, flocculation and dispersion can be controlled through microbial interaction. The dispersion and flocculation of the mineral can be controlled by interaction with proteins, amino acids and polysaccharides, which are bacterial metabolic products. Besides, the formation of polymeric hydrous oxides of iron and aluminium due to microbial weathering result in changes in electrokinetic properties [7].

As for the bacteria, the metabolic activity will change after adsorption on mineral, including substrate utilization, nitrification, respiration, secretion and growth [48]. But whether the existence of the solid phase is beneficial or detrimental to the microbial survival has no definite conclusion. Zobell [49] reported that solid phase surface can absorb nutrients, so as to promote the growth of bacteria adsorbed on solid surface under the poor nutrition conditions. Martin et al. [50] found that the adsorption of metabolic waste by montmorillonite can strengthen metabolic activity of *Streptomyces* sp. and *Micromonospora* sp. Lavie and Stotzky [51] reported that the respiration of *Histoplasma capsulatum* significantly reduced with the presence of low concentrations of soil mineral (montmorillonite, kaolinite and attapulgite); with the increment of mineral concentration, the microbial respiration gradually decreased, as a result of the combination between clay minerals and hyphae, which interfering nutrients and metabolites transfer and the permeability of the cell wall. In addition, the adsorbed siderophore on clay mineral surface fixed a large number of iron nutrient, causing the iron nutritional deficiencies to fungus.

2.4 Bioweathering by Bacteria

Mineral nutrients are cycled between living and dead organic components of soil, and mineral weathering is a primary source of most of the essential elements for microorganisms and plants. Soil microorganisms influence weathering through physical and chemical processes [52] and they play an essential role in releasing the key nutrients from primary minerals that are required for their own and plants' nutrition [53]. Microbial attachment to mineral surfaces leads directly to the formation of a low pH microenvironment that contains extracellular polymers. This increases the water retention time on the mineral surface and the diffusion of ions away from the mineral; therefore it increases mineral dissolution and enhances secondary mineral production [52, 54]. In these microenvironments, mineral nutrients can be chelated directly from the soil minerals or shared amongst the surrounding microorganisms [55]. The surface attached microorganisms contribute to mineral dissolution by using different mechanisms such as lowering pH, complexing with surface ions, catalyzing redox reactions and by physical forcing [56].

Silicates and oxides are ubiquitous earth materials, and these two groups of minerals are among those first and best studied in terms of mineral-microbe interactions. In earth surface environments where oxygen is abundant, aerobic microorganisms interact with oxide and silicate minerals to obtain essential nutrients and to use them as protection from lethal habitats. The production of surface polymers and siderophores help them attach and dissolve minerals as a way to extract nutrients [57–59]. Many rocks and minerals contain nutrients essential to microbial growth, and microbes actively extract nutrients from solid materials. Phosphorus and potassium in soil are mostly in stable state, like aluminosilicate and apatite, which can not be directly absorbed and used for crops. Some of the soil microorganisms can produce organic acids and polysaccharides by metabolism, and then release soluble phosphorus, potassium, silicon and other elements by dissolution of soil minerals, which is conducive to the absorption and utilization of plants. The microorganisms mainly include bacteria, fungi and actinomycetes according to microbial species, among them, most research are focused on bacteria, including *Bacillus circulans, Bacillus sp., Pseudomonas and B. polymyxa,* etc. [60–62]. Iron is another essential nutrient that is often limited because of formation of insoluble iron oxides under oxic conditions. Microbes can recognize Fe-containing minerals and accelerate the release of Fe from silicate mineral, such as hornblende [63] and feldspars [61]. In order to attach to mineral surfaces and use Fe as an nutrient, certain bacteria (such as *Shewanella oneidensis* MR-1) seems to be able to modify the molecular arrangement at its outer cell membrane and react to molecular configurations present at the crystal surface of goethite [64].

2.5 Biobeneficiation by Bacteria

Microbe-mineral interactions can be used as an efficient removal of undesirable constituents in an ore through selective dissolution or modification of surface hydrophobicity or hydrophilicity to bring about selective flotation or flocculation. Consequences of microbe–mineral interactions relevant to microbially induced mineral beneficiation are the following [9]:

1. Adhesion to mineral substrates resulting in biofilm formation.
2. Bio-catalysed oxidation or reduction reactions and secretion of biopolymers.
3. Interaction of cells and bioreagents with ore matrix.
4. Surface chemical changes on minerals or dissolution.

Several types of autotrophic and heterotrophic bacteria, fungi, yeasts and algae as well as their metabolic products can be used in mineral beneficiation processes. In practice, mineral biobeneficiation can be devided into biodissolution, bioleaching, bioflocculation and bioflotation.

Biodissolution has been considered as an efficient way for the removal of undesirable constituents in an ore through selective microbial dissolution. Microorganisms bring about selective mineral dissolution through direct and indirect mechanisms. Direct mechanism involves microbial adhesion to the mineral surfaces leading to enzymatic attack. Cells can attach and colonise on minerals forming biofilms, which consist of various microbial consortia along with their metabolic and metal-reacted products. Subsequent to surface adhesion, attached organisms secrete various biopolymers that serve as adhesives facilitating irreversible attachment. Such biopolymers contain amino acids, exopolysaccharides and organic acids. Spore-forming microbial cells when exposed to mineral and metal ion containing environments form protective capsules around the cell walls.

Indirect mechanism of mineral dissolution involves the role of microbial metabolites acting as mineral solvents. Both oxidative and reductive biodissolution can take place. A typical example of oxidative dissolution is dissolution of sulfides such as pyrite in the presence of sulfur and iron oxidising bacteria such as *A. ferrooxidans* and *Acidithiobacillus thiooxidans*. Since the acidic metabolites of *A. thiooxidans* also contain reducing ions such as HS^-, thiosulphates and tetrathionates, the dissolution of minerals through reductive processes could also occur. The use of iron-reducing bacteria such as *Shewanella* and *Desulfuromonas*, on the other hand, bring about reduction of ferric to the bivalent state facilitating dissolution of ferric oxides such as hematite in an acid medium or in the presence of chelating agents.

Direct bioreduction of iron oxides by iron-reducing bacteria is a treatment option for removal of ferric oxides from kaolin clays and bauxite. For commercial quality kaolin, ferric-reduction of hematitic and goethitic iron by using indigenous ferric-reducing bacteria would enhance the quality. Development of a flexible biotechnology for kaolin beneficiation, either through microbial ferric reduction or ferrous oxidation, can be considered depending on the iron mineralogy. Microorganisms also play an important role in the dissolution of silicate structures. *Bacillus* spp. are active and can remove both free and bound iron occurring in kaolin. Biodegradation of iron oxyhydroxides and partial destruction of mica structure can be brought about by microbial interaction. Significant chemical and biological weathering of silicates can occur in bauxites deposits. Bacterial consortia occurring in bauxite deposits dissolve primary rock-forming minerals and serve as nucleation sites for the precipitation of secondary minerals. For example, the presence of various types of microorganisms in gray-colored bauxites has been well documented. Many metals can be enzymatically enriched and dispersed by organisms [65]. Microbial removal of pyrite from low-grade clay [66] and coal [67] using sulfur and iron-oxidising bacteria such as *A. ferrooxidans* and *A. thiooxidans* has been demonstrated.

Bioleaching is a simple and effective technology for valuable metal extraction from low-grade ores and mineral concentrates by selective dissolution with the help of bacteria. Metal recovery from sulfide minerals is based on the activity of chemolithotrophic bacteria, mainly *A. ferrooxidans* and *A. thiooxidans*, which convert insoluble metal sulfides into soluble metal sulfates. Non-sulfide ores and minerals can be treated by heterotrophic bacteria. In these case metal extraction is due to the production of organic acids and chelating and complexing compounds excreted into the environment. Bioleaching of minerals has opened up new opportunities for extractive metallurgy and biohydrometallurgy taken practice in the copper and uranium industries, especially for the treatment of low-grade ores. Besides the metals recovered in the leachate, there is an increasing interest in the insoluble metals left in the residues, e.g., lead. Inferior lead sulfide concentrates can be transformed into high-value concentrates by leaching of metals (e.g., zinc, cadmium, copper) that interfere with conventional processes for the recovery of the lead [68]. Similar procedures are being investigated for the extraction of silver and other precious metals that are finely disseminated in iron, arsenic, copper and zinc sulfides.

Other than selective biodissolution, interfacial changes brought by microorganisms due to mineral adsorption can induce surface hydrophobicity or hydrophilicity leading to selective flotation or flocculation. Such microbially induced surface chemical changes occur at a faster rate compared with biodissolution, facilitating rapid separation of different mineral constituents present in an ore matrix. All the bacteria used in biobeneficiation bring about significant surface chemical changes on interacted minerals such as kaolinite, silica, alumina, calcite, apatite, hematite, pyrite, chalcopyrite, sphalerite and galena. Significant shifts in mineral IEP values and measured zeta potentials can be observed after microbial interaction. Electrophoretic mobilities of pyrite, chalcopyrite and arsenopyrite are observed to be affected after interacting with *A. thiooxidans*. The highest surface adsorption of *A. thiooxidans* could be observed on pyrite, followed by chalcopyrite and arsenopyrite [9].

Bioflocculation is the use of bacteria and fungi or their metabolites as flocculants reagents in minerals processing. In some selective flocculation methods, kaolin containing titania (and other fine impurities) interacted with flocculants to promote settling of titania impurities, leaving the product kaolin to be recovered from the supernatant in a dispersed form. Biopolymer-producing microbes can flocculate phosphatic clay-slimes. Polymers isolated from *Leuconostoc mesenteroides* and *Xanthomonas* sp. were found to flocculate phosphatic clay-slimes. Microbially induced selective flocculation of hematite from kaolinite was also demonstrated with respect to interaction with *B. subtilis* [69]. Selective bioflocculation is another biological approach to remove calcium and iron from bauxites [70]. Microbial interactions enhanced surface hydrophobicity of silica and kaolinite while hematite and calcite were rendered more hydrophilic. Increased settling (flocculation) of hematite and calcite and dispersion of silica and kaolin clays could be observed after interacting with microbial cells and their metabolic products. Adsorption of bacterial proteins conferred enhanced surface hydrophobicity (as in the case of sphalerite, silica and kaolinite) and of exopolysaccharides rendered minerals such hematite, calcite and galena increasingly hydrophilic. During bioflocculation, mineral interface containing the biofilm is modified and interparticle bridging through surface biopolymers takes place [9].

Bioflotation is the utilization of bacteria and their secretions as flotation collectors or modifiers. It has been suggested that the bacteria and their secretions, mainly polysaccharides and lipids, adsorb on the pyrite surface and render it more hydrophilic, resulting in its depression [71]. Some microorganisms will also function under certain conditions as flotation activators, especially when the adsorbed microorganisms have active surface sites onto which a collector might adsorb [72]. Acidophiles such as *A. ferrooxidans* and *A. thiooxidans* were observed to alter the surface chemistry of sulfide minerals such as pyrite, chalcopyrite, arsenopyrite, sphalerite and galena. Different adsorption behaviours of *Acidithiobacillus* cells led to selective depression of pyrite and galena among the above sulfides. Microbially induced flotation of sulfide minerals using *A. thiooxidans* has also been reported [34]. The differential flotation of pyrite–chalcopyrite mixtures after conditioning with xanthate followed by cells of *A. thiooxidans* resulted in 85% flotation recovery for chalcopyrite while only about 15% of pyrite could be recovered. In the presence of copper activation, chalcopyrite–arsenopyrite recoveries could be substantially increased to more than 90% with significant pyrite depression from a ternary mixture of the three sulfides. Flotation behaviour of galena and sphalerite was also studied before and after interaction with *A.*

thiooxidans. Flotation recovery of sphalerite was observed to be significantly enhanced while galena was depressed. Significant differences in surface adsorption of cells onto galena and sphalerite along with the nature of the surface reaction products are responsible for the observed flotation behaviour of galena and sphalerite. Bacterial interaction also promoted the dispersion of sphalerite particles while flocculating galena [9]. The selective separation of pyrite from quartz and calcite can be achieved through microbially induced flotation and flocculation in the presence of *P. polymyxa* [73]. Similarly, chalcopyrite could be selectively separated from quartz and calcite by treatment with *P. polymyxa* and extracellular protein [74].

2.6 Application of Bacteria-Minerals Interaction in Environmental Engineering

Interactions between soil minerals and microbes leading to a series of profound impact on ecological processes. Bacteria and soil mineral have strong adsorption ability to heavy metals and organic pollutants [75], and the pollutants transportation in soil is closely related to bacteria migration. The microorganisms always interact with minerals in the environment forming microbe-mineral complexes. Generally, the adsorption of heavy metals and the degradation of organic pollutants by the complexes is different from that of the individual microbes or minerals. Both positive and negative influences of mineral-microorganism interactions on the biodegradation of hydrocarbons and biosorption of heavy metal ions are reported.

Huang et al. [76] found that the bacteria can promote the adsorption of heavy metal on kaolinite and soil, and the surface area of the soil colloid and kaolinite increased by 3.0–8.0% after mixes with bacteria, meantime its electronegativity increases. In clay minerals and bacteria coexistence system, the adsorption of heavy metals by microbes is more easily and stable. Flemming et al. [77] pointed out that, when Ag^+ adsorbed on the cell walls-kaolinite complexes, the EDTA desorption rate of Ag^+ is similar to that of the cell walls, but less than the kaolinite, which suggests that the cell walls play the key role in the adsorption process. Similar phenomenon is also found by Ohnuki et al. [78] in the process of uranium ions adsorption by kaolinite and *B. subtilis*. Zou et al. [79] studied the adsorption and desorption of heavy metal ions (Cu^{2+}, Zn^{2+}, Cd^{2+}, Cr^{3+} and Pb^{2+}) by *Bacillus mucilaginosus*-contained mineral (zeolite and diatomite) composite adsorbents, and the results show that the adsorption of the bacteria and mineral composites were better than the simple bacteria adsorption.

Some success stories of the clay supported microbial degradation of organic pollutions include PAHs (mainly LMW compounds) and VOCs (e.g., organophosphorus (fenamiphos) and organocarbamate (carbaryl) [80]. Microbial degradation can be enhanced by adding nutrients and minerals or inoculating additional microbial agents [81]. In this regard, clay minerals are efficient support materials and provide a protective habitat for microorganisms by forming biofilm. As a result, microorganisms could degrade the adsorbed contaminants on clay minerals more effectively. The diatomite or modified diatomite has been used as microbe carrier in wastewater biotreatment. Nan et al. [82] reported that the using of mixed white-rot fungi F1 and denitrifying bacterial H1 loaded by modified diatomite to treat dye wastewater can overcome the drawback that F1 strain could reduce the chromaticity best single and H1 strain could reduce the CODcr best single. It needs to be emphasized that the attachment of microbial cells on clay minerals is the dominant mechanism that enhances the rate of biodegradation and it is inconclusive which microorganism performs the best on which clay mineral. Chen et al. [83] studied the adsorption and biodegradation of carbaryl by *Pseudomonas putida* on montmorillonite, kaolinite and goethite. The presence of montmorillonite enhanced the activity of *P. putida* and ultimately stimulated the bioavailability of carbaryl. Goethite displayed an inhibitory effect on bacterial activity and reduced carbaryl degradation. The biodegradation of mineral-adsorbed carbaryl was mainly controlled by the activity of the degrading microorganisms. Bioremoval of diethylketone by the synergistic combination of microorganisms and clays were carried out by Quintelas et al. [84] and they demonstrated that *Arthrobacter viscosus* and *Streptococcus equisimilis*, by themselves or interacting with bentonite, sepiolite, kaolinite and vermiculite clays, are able to remove efficiently diethylketone from aqueous solutions. Almost complete removal of diethylketone was achieved when using the bacteria mixed with clays, as a result of that diethylketone is degraded by the bacteria and then the intermediates/sub-products may be adsorbed by the functional groups present on the cell wall and/or clay surfaces.

Distribution of heavy metal ions in the complex of mineral-bacteria is affected by pH and ionic strength conditions. When the pH is less than 5.5, at least 50% of the Pb^{2+} is combined with the bacteria surface in the microbe-goethite complex, however, when pH is higher than 6, about 70% of Pb^{2+} adsorb on the surface of goethite [85]. Ionic strength significantly influences strontium adsorption by bacteria or hydrated ferric oxide, while its influence on the adsorption by bacteria-ferric oxide complex is small. When $NaNO_3$ concentration increased from 0 to 0.1 mol/L, the adsorption quantity of strontium by bacteria-ferric oxide complex decreased from 34 to 24 μmol/g.

3 Fungi Adsorption

Fungi are a class of single or multi-cell eukaryotic microorganisms that contain no chloroplast and no plastid, which are typical heterotrophic microorganism. In the Earth's lithosphere, fungi are of fundamental importance as decomposer organisms, animal and plant pathogens and symbionts (e.g. lichens and mycorrhizas), being ubiquitous in subaerial and subsoil environments [86]. The ability of fungi to interact with minerals, metals, metalloids and organic compounds through biomechanical and biochemical processes, makes them ideally suited as biological weathering agents of rock and building stone. They also play a fundamental role in biogeochemical cycling of nutrients, (e.g. C, N, P and S) and metals (e.g. Na, Mg, Ca, Mn, Fe, Cu, Zn, Co and Ni) essential for the growth of living organisms in the biosphere. In addition, they play an integral role in the mobilization and immobilization of non-essential metals (e.g. Cs, Al, Cd, Hg and Pb). Most studies on mineral-microbe interactions and microbial involvement in geological processes have concentrated on bacteria and archaea, fungi have to a certain extent, been neglected. The role of fungi in geomicrobiological processes is an important aspect of the geomicrobiology. The roles include their deteriorative potential on rock, building stone and mineral surfaces and involvement in the formation of secondary mycogenic minerals, and the migration and fixation of nutrients or metal ions during above processes.

3.1 Bioweathering by Fungi

Fungi usually rely on two kinds of synergy function to promote rock or mineral weathering, namely biomechanical weathering and biochemical weathering [87, 88].

1. **Biomechanical weathering**

The rocks or minerals could be weathered by fungus directly or indirectly. The direct biomechanical weathering is mainly completed through the hyphae. For example, the hyphae could penetrate to the rocks from rock surface pores and fractures, forming a loose rock structure, which lead to the increase in specific surface area and the rate of erosion. This invasive effect of fungal hyphae is mainly based on its nutritional requirements, and achieved because of the expansion stress caused by expansion of cells. Compared with the hyphae with a higher expansion, those small expansion of hyphae tends to penetrate the surface of the tough matrix easily. The growth and distribution of fungi hyphae in rocks or minerals are related to the chemical composition and surface micro-morphology of the matrix. The presence of hyphae was more likely to be observed at the crack or grain junction of the rock surface. For dense mineral materials, hyphae can cut out a "invasive tunnel" along the surface of the mineral facies. These phenomena indicate that the hyphae can detect the environmental space in order to find the best territory for intrusion [89]. This detection and selection of the invasion direction can be achieved through a series of stress reactions, in which the touch is an important stress mechanism. It can guide the fungal growth along the weak direction of the matrix. The chemical orientation make the fungus avoid areas that are not conducive to growth (such as dry, heavy metal-rich areas) and tend to grow in nutrient-rich areas [86].

Indirect biomechanical weathering is usually associated with the extracellular mucus produced by the fungus, which facilitates immobilization of the fungal membrane so that it can stably adhere to the surface of the substrate and produce a mechanical pressure by the shrinkage and expansion of the bacterial membrane, causing mineral wear. Both direct and indirect biomechanical weathering will be synergistic with chemical weathering, which is thought to play a more important role in biological weathering [87].

2. **Biochemical weathering**

Fungi can promote chemical weathering of minerals by two mechanisms, namely proton promotion and ligand progression. Proton-promoting weathering substances can be metabolized to inorganic or organic acids, or carbon dioxide by the excretion of CO_2. The protons in these acidic substances will be effectively replaced with the metal ions in the mineral, and ligand-induced weathering mainly rely on organic ligands to complex metal ions, such substances include organic acids, iron carriers and extracellular polymers and other metabolites with complex sites. Proton-promoting and ligand-promoting weathering are affected by fungal metabolic activity [88].

Balogh-Brunstad et al. [90] studied the attachment of ectomycorrhizal fungi on the surface of biotite, and found that the rate of biotite decomposition by fungus is 2 to 3 times higher than chemical reaction with the same temperature and pH, but the immobilization of fungi on the mineral surface has a certain selectivity. Gleeson [91] studied the microecological characteristics of granite outcrops and found that the distribution of fungi is significantly selective for the chemical composition of minerals. Scanning electron microscopy of field-abandoned ores showed that the fungus attached on the surface of the pyrite, and fungi closely relate to the mineral surface of secondary sulfite minerals [92].

3.2 *Biobeneficiation by Fungi*

Fungi such as *Aspergillus niger* secrete several organic acids such as citric, oxalic and gluconic acids promoting the dissolution of the mineral in the medium. Fungi have been proved to have better leaching effects than bacteria in bioleaching experiments. Castro et al. [93] found that the extraction efficiency of Zn and Ni from the calamine and silicon-magnesium-nickel ore by fungi (aspergillus and penicillium) is much higher than by bacteria, and the strong leaching ability of fungi is related to its metabolite which containing a large number of low-molecular-weight organic acids. On the other hand, the valance variation of metal ions caused by the process of fungi metabolism also leading to a better leaching efficiency, such as Fe(III) and Mn(IV) can be reduced to Fe(II) and Mn(II) which has a higher solubility. In addition, extracellular ligands secreted by *Aspergillus* or *Penicillium* are also widely used to extract metals such as zinc, copper, nickel and cobalt from various solid materials, including from low-quality ores [94]. The leaching ability of heterotrophic fungi, through secretion of organic acids, of various laterite ores, including saprolite, weathered saprolite, limonite and nontronite was studied by Valix et al. [95]. Strains of *Aspergillus* and *Penicillium* were found to be the most efficient organisms. Biotechnological methods for leaching kaolin using fungi such as *A. niger* can also be considered.

3.3 Application of Mineral-Fungi Interaction in Environmental Engineering

Both clays and fungal biomass have been investigated due to their properties to sorb toxic metals and pollutants. Clay minerals with their excellent adsorption, ion exchange and co-precipitating capacities are widely used for cleaning up polluted soil, sediment and water. Fungi have been used as adsorbent in water treatment, which is mainly focused on two aspects, decolorizing and elimination of heavy metals. Attention has also been paid to the sorption of metals to mixtures of soil fungi with clays. Removal of divalent metal ions Cu^{2+}, Cd^{2+} and Zn^{2+} from pH buffered solutions by the soil fungi *Rhizopus arrhizus* and *Trichoderma viride* has been examined in the presence and absence of the clay minerals montmorillonite and kaolinite [96]. Accumulation of the metals by all of the sorbents was dependent on the external pH and decreased with pH. *A. niger* hyphae and potassium mineral powder form *A. niger*-mineral aggregates with the help of the polysaccharide and other metabolic products by intertwining, adsorption, adhesion and other role. The polysaccharide concentration and structure were significantly changed after the formation of aggregates, which was helpful for chelating metal ions and adsorbing water molecules, thus providing beneficial microenvironment for fungi to utilize mineral nutrition [97]. Huang et al. [98] studied the adsorption of copper and cadmium on the complexes of rhizobia and kaolin, goethite, amorphous iron and other minerals. The results showed that the addition of bacteria significantly increased the adsorption affinity of kaolin and goethite to cadmium. Biomass of melanin-producing microfungi of the genus *Cladosporium* and *Aureobasidium pullulans*, clay minerals and fungal mycelia grown in clay-containing medium were compared for their equilibrium Cu and Cd uptake from pH buffered solutions by Fomina and Gadd [99]. The presence of kaolinite and palygorskite in the medium generally reduced both Cu and Cd sorption capacity and the metal-binding ability of the fungal–clay mixtures. In contrast, addition of bentonite into the medium did not appreciably alter the Cd sorption ability but increased the sorption of Cu. And they propose that the mechanisms underlying the changed sorption abilities of the complex biomineral sorbents are blocking and masking of binding sites in the case of decreased sorption ability, and modification of binding sites and emergence of new ones in the case of increased sorption ability. The ability of microorganisms to degrade the adsorbed contaminants on clay minerals, lies in the specificity of both the microorganism(s) and the contaminant(s) and the bio-physico-chemical characteristics of the environment. These include the associated flora and fauna, soil type, temperature, pH, salinity, soil organic matter (SOM), moisture content and other predominating factors, such as sorption, desorption and diffusion of contaminants and dissolution of minerals in the contaminated sites [100].

4 Algae Adsorption

4.1 Algal Surface Property

Algae are ubiquitous single to multi cellular chlorophyll containing organisms existing everywhere from water to land that can convert solar energy to chemical energy. They are a diverse group of photosynthetic organisms that can produce a number of economically valuable bioproducts, which are widely used as food, drugs and cosmetics. Algae play important roles in mineral colonisation and weathering, and the algae-mineral interactions have many applications in the environmental engineering. The surface properties of algae and minerals are important in the mineral-algal interaction. Given that mineral surface property is discussed above, the surface property of algal cells will be emphasized in this chapter.

Algal and bacterial surfaces consist of a complex, heterogeneous mixture of possible binding sites including proton-active carboxylic, phosphoric, phosphodiester, hydroxyl and amine functional groups, which play a major role in surface binding capacity, biomineralisation, biofilm formation and toxic adsorption. Because of the deprotonation of the surface functional groups, algal cells generally present an electron donor parameter and a negatively charged surface. A review reported by Henderson et al. [101] demonstrated that characteristics of algal cells in terms of morphology, motility, surface charge are phyla specific. Hadjoudja et al. [102] compared the surface characterisation of *Microcystis aeruginosa* and *Chlorella vulgaris*, found that *M. aeruginosa* showed lower hydrophobic character, while higher carboxyl functional groups concentration and total functional groups concentration. Ozkan and Berberoglu [103] further evaluated 12 different microalgae including fresh and seawater species of green algae, diatoms and cyanobacteria for their physico-chemical surface properties, the results showed that the electron donor parameter correlated well with the free energy of cohesion in all groups of microalgae and species known to form colonies and exhibit benthic culture shad distinctly hydrophobic surfaces compared to microalgae preferring planktonic growth. Xia et al. [104] investigated six oleaginous green microalgae for their surface physicochemical properties and showed the similar results.

The surface properties of algae are not only species specific, the growth stage will also alter the surface character of the algal cells. Zhang et al. [105] found that net zeta potential of *Chlorella zofingiensis* decreased while the total functional groups increased with the growth phase from exponential, stationary to declining phases. Li et al. [106] reported that the phosphoryl groups decreased with the aged culture and disappeared in the declining phase, which are important for the metals binding. Xia et al. [107] investigated three green microalgae for their surface characteristics changes during the growth periods and demonstrated that the surface properties like particle size, the degree of hydrophobicity, and the total concentration of functional groups increased while net surface zeta potential decreased. Thus, with the culture aged and the nutrient depleted, the cellular biochemical component as well as the surface properties changed. It is not difficult to speculate that the growth conditions can also influence on the surface properties of algal cells and further affect the interaction with minerals. Further work is needed to understand the environmental manipulation of the algal surface property for the mineral-algal/bacterial interactions.

4.2 Algal Bioweathering and Biomineralization

All organisms including oxygenic phototrophs (algae and cyanobacteria) living in a certain environment adapt to the given conditions and interact with their surroudings to some extent. Like bacteria and fungi, cyanobacteria have also been clearly associated with biomineralization and bioweathering. However, unlike bacteria and fungi, the growth and biofilm formation of algae and cyanobacteria need solar energy as their energy source. Many species of microalgae are also able to fix nitrogen gas from the atmosphere.

It is widely accepted that biofilm formation on clean surfaces usually starts with phototrophic organisms (algae, cyanobacteria) which use CO_2 from the atmosphere and sunlight as their carbon and energy sources. They can grow on the stone surface or may penetrate some millimeters into the pore system. To inhabit terrestrial rocky environments, including very extreme locations, algae got many evolved mechanisms to adapt to the unfavorable environment. Among different genus of alga, cyanobacteria are known for their ability to withstand many stressful environments such as freezing, high salinity and low water activity. For example, water-stressed cyanobacteria can produce osmoprotectants inside of the cells to protect them from the effects of dehydration [108]. They also excrete extracellular polymeric substances (EPS) to the outside environment to resist the radiation and drought. Besides physiological responses, the peculiar water-induced motility is another strategy for the algal cells to deal with the drought or desert soils.

In general, the mechanisms of mineral weathering as those mentioned above for bacterial weathering, i.e. biophysical and biochemical, can also apply for cyanobacteria-induced process. Biophysical weathering includes physical alteration of the rock substrate through EPS produced by the cyanobacterial cells and formation of dark crusts at the surface of rocks. Herrera et al. [109] reported that cyanobacteria contribute to deterioration of peridotite due to the formation of dark crust. Biochemical activity such as changes in pH are considered as specific to cyanobacteria. In contrast with heterotrophic bacteria as mentioned above, cyanobacterial metabolism usually results in pH increase in the vicinity of the cells via their photosynthetic activity (e.g. due to transfer of OH^- out of the cell in exchange for HCO_3^- as a carbon source for photosynthesis). Some researchers believed that the increased pH is the most likely cause of the silicate mineral deterioration [110, 111]. However, others supposed that decreased pH conditions due to the EPS secreted by cyanobacteria may be the reason for enhanced rock dissolution [111]. Garcia-Pichel [112] also suggested that pH decrease due to the activity of carbon dioxide from respiration at night might contribute to cyanobacterial mineral dissolution. Besides, production of organic acids is another proposed mechanism for cyanobacteria mineral deterioration. These compounds affect the mineral weathering through proton-based and ligand-based mechanisms, which is pH dependent.

Biomineralization is a process by which living organisms produce minerals, which could be silicates in algae and diatoms, carbonates in invertebrates and calcium.

4.3 *Application of Algal-Minerals Interaction in Environmental Engineering*

4.3.1 Algae-Mineral Interaction for Heavy Metals Retention

Cyanobacteria are considered as the primary colonizers in the mine tailings due to their capability of carbon and nitrogen fixation and the capacity of resistance to the heavy metals toxicity [113]. When attached to the toxic minerals, algal cells usually develop a biofilm to resistant the harsh environment. García Meza et al. (2005) [114] proved that algae survive on mine tailings with a variety of heavy metals trough the mechanisms of forming biofilms. It is believed that EPS production was a critical feature for the survival of the biofilm. Other strategy of algal cells for dealing with metal toxicity are exclusion from the cells, binding onto the cell wall, sequestration within the cells or by complexation with organic compounds and storage in the vacuole [115]. Besides live organisms, algal biomass from died bodies can also be perspective adsorbents due to the large capacity for extensive heavy metals retention. It has been found that algal biomass is rich in organic ligands or functional groups, such as carboxyl, hydroxyl, phosphate, amine groups, and so on, which play dominant roles in the heavy metals removal. The mechanisms for heavy metals adsorbed onto algal biomass can be explained by ion-exchange and surface complexation. This ability to accumulate metals, and other pollutants can be harnessed in creating remediation solutions using actively growing photosynthetic biofilms.

4.3.2 Harmful Algal Removal

Algae have high photosynthetic efficiency and growth rate as well as strong vitality. Once there is solar radiation and nutrients, the algae occur. Just because of this, the algal bloom or red tide erupted frequently and caused a great economic loss in recent years [116]. These harmful algal blooms (HABs) pose a serious threat to commercial fisheries and aquaculture, human health, and also coastal aesthetics. Several approaches, such as chemical (using toxic regents), mechanical (harvesting algal cells by floating, fishing out, or centrifugation) and biological techniques, have been studied in removal of HABs in water systems. But they've got a lot of problems. For example, chemical methods may have adverse effects on other organisms and the expedited release of microcystins [117]. The high cost of mechanical methods inhibits its application to large natural waters. Whereas biological and ecological restoration methods [118] need huge engineering backup and the relatively long time which make it actually impractical for abrupt HABs over natural large areas.

The application of natural clays over the surface of a bloom to remove the algae from the water column through co-flocculation and sedimentation is regarded the most promising control strategy, with respect to maximizing effectiveness and minimizing costs and environmental impacts [119–122]. They can not only adsorb the algal cells but also sediment the dissolved toxins [123]. Many nature minerals can be natural, nontoxic and inexpensive clays to flocculate and remove algal cells. This method has been used successfully in the field in Japan [124], South Korea [125], China [126] and the United States [127], demonstrating the effectiveness of clay to remove a number of algal species from the water column. The flocculation process can be divided into two stages: the adsorption of algal cells onto the surface of clay minerals and the subsequent formation of flocs. Thus, the adsorption of algal cells on clay minerals plays crucial roles in the removal of algal cells.

There are many factors affecting the adsorption process and further affect the removal efficiency of harmful algal cells, including characteristics of algal species and the clay minerals selected. Pan et al. [128] using 26 commercially available clays and minerals to evaluate their removal efficiency of cyanobacterial blooms from Taihu Lake, China as listed in. According to the performances in terms of algal removal efficiency and removal rate, the 26 clays/minerals were classified into three categories: Type I clays (sepiolite, talc, ferric oxide, and kaolinite) showed the highest removal efficiency and removal rate; followed by Type II clays ang Type III.

Besides nature clays or local soils, the chemical modified clay minerals could further enhance the removal ability of harmful algae. The surfactants, polyalummum chloride (PAC) [119, 129], chitosan [130–132], hexadecyl trimethyl ammonium bromide (CTAB) [133], xanthan [134], gemini surfactant [135], have been widely used in clay modification for harmful algae removal. When the clay minerals were modified by these surfactants, the surface charge of surfactant-modified clay could be switched from negative to positive in natural water environment with pH less than 10.5. This is essential in creating flocculation potential and hence the destabilization of the algae suspension since algal cells are normally negatively charged in natural waters [136].

4.3.3 Cyanobacterisation to Counteract Desertification

Cyanobacteria are regarded as useful biological agents in remediation and amelioration of soil environments. The practice of using cyanobacteria as inoculants has been extensively studied for the past decades. Using cyanobacteria as soil conditioners has been a resurgence recently due to the increasing awareness of the rolc they play in the formation of biocrusts. Biocrusts are composed of different proportions of cyanobacteria, microalgae, fungi, lichens and bryophytes, in which cyanobacteria are commonly major parts of biocrusts and represent a pioneer of arid sandy soils. The formation of biocrusts in arid or semiarid environments can potentially determine a decrease in abiotic levels, shifting the state of the ecosystem, modification of soil characteristics, soil-water content and hydrological cycles [137]. They can fix carbon and nitrogen and profound effects on plat germination. Cyanobacterisation was successfully applied over large hyper arid areas in Qubqi Desert, China, seeding large sand dune areas (up to 30 km^2) and significantly improved the local soil conditions [138]. Thus, inoculating cyanobacteria to build artificial soil crusts is considered as a perspective strategy to desertification reversal [139].

To build biocrusts, the first step is to screen cyanobacterial isolates. The capability to grow on soil is typical of filamentous cyanobacteria of genera such as *Cylindrospermum*, *Nostoc*, *Anabaena*, *Oscillatoria*, *Phormidium*, *Microcoleus* and *Schizothrix*. EPS productivity and soil stabilizing capability is the two key screening criteria. EPS can help the strain withstand stressful environments such as drought, UV radiation and protozoan attack, promote cell cohesion and adhesion to the sands, ease the formation of stable biofilms. Soil stability is an index of soil degradation, which can impede wind in the desert. Besides, the capability to tolerate UV radiation, low temperature, high salinity and drought is also highlighted. Among EPS producing cyanobacteria, *Microcoleus vaginatus*, *Scytonema javanicum*, *Phormidium tenue* and the genus *Leptolyngbya* have been tested as inoculants in laboratory or open field trials and showed different characters. Xie et al. [140] reported that the co-inoculation of *M. vaginatus* and *P. tenue* produced a firmer soil structure than soly using of *M. vaginatus*. Li et al. [141] investigated a co-inoculation of *M. vaginatus* and *S. javanicum* with different ratios and found that a ratio of 10/1 was the optimal condition for the large-scale inoculation.

After inoculating the selected cyanobacteria, the sand particles were bound and cemented with the microalgal filaments and secreted EPS, thereby forming a stable biofilm crust structure on the sand surface. Besides the selection of inoculants for the tolerance of harsh environment, the soil characteristics such as soil texture, sand grain size, microstructures, and the potential to capture dust change with developments in biocrust composition and morphology throughout the biocrust succession [142–144]. Ram and Aaron [145] reported that topsoil texture changes with biocrust succession; as the biocrusts become more mature, they contain higher proportions of fine grained particles. Others further suggested a positive feedback between the biocrusts and the amount of fine particles [144, 146]. Hu et al. [147] demonstrated that dust incorporation into the biocrusts increases the growth of cyanobacteria and strengthens the cohesion of biocrusts. Rozenstein et al. [148] reported the biocrusts developed more rapidly on the fine fraction (<125 lm) than on the coarser fractions and suggested that the difference in the pore size between the different fractions is the reason for these discrepancies in biocrust development.

5 Biomacromolecules Adsorption

5.1 Humic Substances

Humic substances (HS) are found in all soils and water that contain organic matter. They are the result of biological and chemical processes, and the amounts generated in soil are several times greater than those in waters. Humic substances are the major components of the mixture of materials that comprise soil organic matter, and these substances, which are by far the most abundant organic materials in the environment, are themselves complicated mixtures of biologically transformed organic debris [149]. Humic substances are described as a mixture of polydisperse, heterogeneous, polyelectrolytes [150] that can be divided into three fractions: fluvic acid, humic acid and humin. These fractions are differentiated on the basis of their solubility at different pH [151]. Humic acids remain insolution at pH2 and higher, fulvics even at lower pH and humin is not soluble under alkaline or acid conditions. Humic substances are predominantly negatively charged due to the abundance of carboxylic and phenolictype of groups [151]. Hydration of the charged groups and the electrostatic repulsion between the charges results in an extended conformation of the humic acid molecules [152]. Due to the addition of indifferent electrolyte the charges are shielded, resulting in a reduction of the electrostatic repulsion and a more tightly coiled configuration.

Clays and metal oxides are the most abundant, finely divided inorganic minerals in natural systems. Humic substances-clay complexes are well known for their contribution to soil structure and environmental processes. The research of the adsorption of HSs by clays can be dated by 1950th [153]. Due to the negatively charged surface characterization of both humic substance and clay minerals in the nature environment, some adsorption can occur if polyvalent exchangeable cations are present [154]. When iron and aluminum hydroxides are polymerized at clay surfaces, opportunity arises for strong bonding between clay and organic materials to develop. This may involve "ligand exchange" reactions as well as simple anion exchange. Secondary, non-coulombic interactions may be more important for interactions between silanol surface and humic and fluvic acids [154].

It was also found that differences in structural and surface properties of clay minerals result in their different affinity for FA/HA. Feng et al. (2005) proposed that the adsorption of peat humic acid was higher on kaolinite than montmorillonite under the same solution conditions. However, Zhang et al. [155] investigated the adsorption phenomena and mechanisms of fulvic acid (FA) and humic acid (HA) on three clay minerals (kaolinite, smectite and vermiculite) and found that kaolinite showed the lowest adsorption for both FA and HA. It was concluded that differences in characteristics of FA and HA such as molecular weight, functional group compositions and hydrophobicity lead the different adsorption of FA and HA on kaolinite and smectite, while the electron transfer mechanism is much more important in determining the different adsorption behaviors of FA and HA on vermiculite.

Besides clay, metal oxides, mainly iron oxide particles play the major roles in the formation, structure and strength of aggregates in soils. HSs and metal oxides surfaces can both be regarded as geo-colloids. For the adsorption of humic substances on positively charged minerals, Gu et al. (1994) stated that six major mechanisms of organic matters adsorbing on iron (or aluminum) oxides might be involved: (1) electrostatic interaction, (2) ligand exchange-surface complexation, (3) hydrophobic interaction, (4) entropic effect, (5) hydrogen bonding and (6) cation binding [156]. Vermeer [152] proposed that the adsorption of HA molecules onto mineral surfaces is determined by two opposite processes, charge compensation and specific adsorption energy enhance the adsorption and the adsorption is inhibited by lateral interactions and loss of entropy in the adsorbed state. However, compared to HA, the adsorption of FA is less strong to goethite [157]. Fu and Quan [158] explored that the binding affinities of the iron oxides with the FA was in the order of hematite > goethite > akageneite, and concluded the interaction mechanism was to the ligand-exchange involving carboxylic functional groups of the FA and the surfaces sites of both hematite and goethite, while no complexation can be evidenced in the case of akaganeite, only surface adsorption.

It is generally accepted that the importance of the different mechanisms for a given system depends on the mineral particles under investigation and the solution conditions during the experiments [152]. Adsorption of humic substances to minerals is strongly influenced by pH and ion strength [157, 159]. It is generally found that the adsorption decreases at increasing pH (1–14) [154, 160]. Sometimes, an adsorption maximum was found around acid conditions. It has been shown that the pH dependency of the adsorption (of both small and big molecules) is directly linked to the mass balance of protons in the adsorption process. When protons are consumed (co-adsorbed) in the process, adsorption will decrease at increasing pH. On the contrary, when there is a net proton release, adsorption will increase with the increase of pH. It has been demonstrated that the proton coadsorption or release depends on the charge of the ions (particles) and its location in the electrostatic field near the surface. The difference in size and charging behavior of FA and HA may therefore lead to different electrostatic interactions with the oxide surface and therefore to a different proton co-adsorption and pH dependency of the adsorption.

HA sorption increased with increasing ionic strength and decreasing pH, and the presence of the background electrolyte Ca^{2+} largely enhanced sorption in comparison to Na^{+}. An attempt was made to separate the contributions of various modes of interactions to HA sorption, where ligand exchange was estimated to account for approximately 32% of PHA sorption on clay surfaces, van der Waals 22% and cation bridging 41% when Ca^{2+} was the background electrolyte [160]. Similar results were also observed by Kloster et al. [161]. This indicates that the surface of soil minerals acts as a nucleation centre for HA aggregation, adsorption and surface aggregation induced by Ca^{2+} [161].

The sorption of humic substances to mineral surfaces also play important roles in the contaminant binding. Many studies have proved that the mineral-humic acid interaction affects the metal sorption processes. Spark et al. [162] reported that sorption of trace metal is enhanced in the combined systems of humic acid and the minerals goethite and silica due to secondary reactions in which metal-humic acid complexes are adsorbed by the minerals, whereas not enhanced for α-alumina and kaolinite, possibly due to competing reactions associated with the sorption of the humic acid by these minerals. Liu and Gonzalez (1999) stated that [163] montmorillonite with preadsorbed humic acid does not show a significant change in the capacity of adsorbed metal ions. An increase in the ionic strength at a pH of 6.5 results in an increase in the adsorption of lead on montmotillonite in the presence of humic acid, while at a lower pH, the increase in ionic strength results in a decrease in metal adsorption. The bridging of bivalent metal ions between montmorillonite and humic acid is proposed as the dominant adsorption mechanism. However, Arias et al. (2002) documented that HAs were found to enhance the copper and cadmium adsorption capacity onto kaolin surface. Thus, the different source of HAs and the different minerals significantly affect the metals binding. Further investigation of the complex binary system is needed to elucidate the metals retention mechanisms.

5.2 Extracellular Polymeric Substances

EPS (extracellular polymeric substances) were defined as "organic polymers of microbial origin which in biofilm systems are frequently responsible for binding cells and other particulate materials together (cohesion) and to the substratum (adhesion)" [164]. EPS are either bound to the cell surface ("capsular"), or released into solution ("free") or associated with the hydrated matrix of biofilms [164]. EPS are composed dominantly of polysaccharides and proteins, with nucleic acids and lipids as minor constituents. They contain various weakly acidic functional groups, such as carboxyl, phosphoryl, amide, amino, and hydroxyl, which can react with the mineral surfaces through the mechanisms of hydrophobic, electrostatic, covalent, and polymer–polymer interactions [165]. Adsorption of EPS to minerals can alter the substrata physico-chemistry and influence initial bacterial adhesion processes via conditioning film formation, and also heavy metals accumulation in soil or aquatic environments [166, 167].

Omoike and Chorover (2006) [168] stated that adsorption of EPS from *B. subtilis* on goethite increased with the decrease of pH from 9.0 to 3.0 and the decrease of NaCl concentrations from 100 to 1 mM. Cao et al. [169] claimed that an increase in the concentration of cations and/or a decrease in the pH favored EPS adsorption on to minerals. They also found that EPS-P moieties predominantly from nucleic acids are adsorbed preferentially on goethite and EPS-N moieties mainly from proteins are adsorbed preferentially on clay minerals. Fang et al. [170] concluded to the same results that the adsorption energy constant (K) of EPS on goethite was in the sequence of EPS phosphate-containing moieties > nitrogen-containing moieties > carbon-containing moieties, indicating those containing phosphate were the most strongly adsorbed. Mikutta et al. [171] showed that bentonite sorbed much more EPS-C than ferrihydrite. During sorption, EPS were chemically and size fractionated with bentonite favoring the uptake of low-molecular weight components and EPS-N, and ferrihydrite selectively retaining high-molecular weight and P-rich components.

Macromolecules may bind to the mineral surface by inner-sphere (directly bonded to the surface) or outer-sphere (aqua-ion surrounded by water molecules and thus held to the surface of the sorbent by electrostatic attraction) complexation [172]. Kwon et al. [173] using the atomic force microscope (AFM) and electronic structure calculations suggested the phosphate-bearing polymers are the major components in EPS responsible for the adhesion on silica surfaces at low pH with H-bonds and electrostatic interactions as the dominant forces. Zhu et al. [174] using Quartz crystal microbalance with dissipation (QCM-D) revealed that deposition efficiencies of EPS from four bacterial strains on bare silica surfaces increased with increasing ionic strength in both monovalent and divalent solutions. Omoike et al. [175] using attenuated total reflectance fourier transform infrared (ATR-FTIR) spectroscopy found that EPS from *Pseudomonas aeruginosa* and *B. subtilis* bind to Fe centers on goethite via inner-sphere complexation of phosphate-containing macromolecules. Quantum chemical calculations results further revealed that the reaction is monodentate rather than bidentate complexation at pH 6.0 and 10 mM NaCl. Cao et al. [169] also found hydrogen bonding and the electrostatic interaction are the main forces governing the adsorption of EPS on clay minerals, while chemical bonding interactions (ligand exchange) also contribute to selective fractionation of EPS adsorption on goethite. Fang et al. [170] using Extended X-ray absorption fine structure (EXAFS) and X-ray absorption near edge structure (XANES) spectroscopy directly demonstrated P–O–Fe bonds were formed between EPS phosphoryl groups and goethite, and phosphate groups of EPS can form monodentate inner-sphere complexes at lower pH 3.0, while form bidentate inner-sphere complexes at higher pH 9.0 when adsorbing to goethite. Liu et al. [176] using scanning transmission X-ray microscopy (STXM) and nanoscale secondary ion mass spectrometry (NanoSIMS) showed the reaction of EPS with goethite led to a preferential adsorption of lipids and proteins and a strong S enrichment in geothite-EPS aggregates. They concluded that the reactivity of the EPS-covered mineral surfaces will change according to the sub μm spaced pattern of adsorbed lipid or protein-rich domains.

Mineral-EPS association may trigger a differential retention of heavy metals. Mineral-attached EPS might modify metal uptake to minerals by providing additional complexation sites [177], or by reducing metal immobilization by blockage of nanometer-sized pores or by influencing the aggregation status of the respective EPS-mineral association [178]. Fang et al. [177] reported that EPS either bound to goethite or montmorillonite increased the retention of Cu(II). They using potentiometric and microcalorimetric titration analysis concluded that the type of dominating EPS-mineral interaction, that is, innersphere complexation in the case of goethite and hydrophobic/van der Waals forces for montmorillonite, control the quantity of available surface sites in the respective composite and subsequently the sorption of copper. Mikutta et al. (2012) documented that mineral-bound EPS increased the extent and rate of Pb^{2+}, Cu^{2+}, and Zn^{2+} sorption for bentonite (following the order $Pb^{2+} > Cu^{2+} > Zn^{2+}$), whereas either no effect or a decrease in metal uptake was observed for ferrihydrite. EPS-mineral association modifies the physical properties of the mineral sorbents in terms of pore structure and causes a fractionation of EPS with respect to major chemical components and particle size, therefore, offering a variable site density available for metal complexation.

Another important question with regard to the migration of heavy metals in the soil and sediments is the relative ration of mineral to EPS. EPS, with different ratio, when sorbed to mineral can modify the properties of the mineral sorbents differently. Compared the component additivity principle, Yan et al. [179] elucidated the enhancement of Cd(II) adsorption by the EPS-montmorillonite composites under low weight ratios EPS (EPS/montmorillonite = 1/50 and 0.5/50), in contrast, it failed at high weight ratio (EPS/montmorillonite = 5/50). Further explorations showed the interested fact that, under low mass ratios, more surface reactive sorption sites on EPS released, resulting the increase of negative surface charges due to EPS promoting the stripping of montmorillonite. In addition, the Cd(II) bridging occurred between EPS and montmorillonite in the whole process. Hence, with the efforts of these three reactions, a low mass EPS/mineral proportion of the composites achieved the strengthening of cadmium adsorption.

Despite all this, these are many possibilities in this field remain to be explored.

5.3 *Proteins*

Proteins are composed of amino acids possessing flexible polymeric chains with lateral groups, and thus large affinities for water-solid interfaces. They have contrasting physicochemical properties: hydrophilic or hydrophobic, negatively, neutrally, or positively charged lateral chains [180], and therefore, show strong and complex interactions with the extensive mineral surfaces in nature environment. Protein-mineral complexes, especially protein-clay complexes, provide protection against degradation and inactivation, and retain much of their biological activity; e.g. bound insecticidal proteins from *Bacillus thuringiensis* kill insect larvae [181], and bound enzymes convert substrates [182]. Besides, the adsorption of proteins by clay minerals play active roles in the soil ecosystem safety, e.g. carbon and nitrogen availability [183]. They have also been recently emphasized on the development of novel hybrid structural and functional biomaterials applied to biosensing, biocatalysis, and drug delivery [184, 185].

Adsorption of proteins on solid phase clay may involve in a variety of physical and chemical interactions [186]. In general, small protein molecules with positive charge can be intercalated into the interlayer of montmorillonite through cation exchange [187]. In addition, electrostatic interactions occur between charged protein molecules and clays minerals since they have permanent negative and variable surface charges [188]. Besides, because clay minerals possess 'hydrophobic region' and 'hydrophilic region' on the surface, they are capable of binding polar protein molecules since their octahedral surface is hydrophilic while the tetrahedral surface with hydroxyl groups is hydrophobic. The hydrophilic character of the clay surface interacts with the positively charged side chains of the protein, while the hydrophobic portion of clay surface interacts with the positively charged side chains of the protein [189]. The hydrophobic interaction between the protein and the siloxane surface of clay mineral is a spontaneous thermodynamic process with saving energy so that it can compensate the electrostatic repulsion of negative charge between protein and clay mineral surface [190]. In addition, hydrogen bonding and van der Waal forces have also been proved to be important in adsorption of proteins onto clay minerals [186].

Different clay minerals have different adsorption capacities for protein molecules. Generally, the adsorption capacity of protein on montmotillonite was literally higher than that on kaolinite and illite [191, 192]. The different properties of clays such as surface area, cation exchange capacity, charge density and degree of swelling lead to different adsorption sites for proteins. For smectites (montmorillonite and saponite), adsorption usually occurs at both interlayer and external surface [193, 194], while for kaolinite and illite, only occurs at the external surface [183]. Besides, the protein properties in terms of weight and size also decide the different binding sites on clay minerals. Ralla et al. [195] demonstrated that proteins of smaller molecular weights revealed a relatively higher adsorption capacity. Zhou et al. [191] reported that smaller insecticidal protein molecules were intercalated into the gallery region of montmorillonite whereas larger insecticidal protein molecules were adsorbed on the surface of the montmorillonite. Barral et al. [196] also concluded that larger protein molecules had lower adsorption capacity than the smaller ones.

Besides, the external conditions play important roles on binding of proteins onto clay minerals. Solution pH, the most important factor, affects both the surface charge of clay minerals and the degree of ionization of protein moleculs. Helassa et al. [197] reported that adsorption capacity of toxin on montmorillonite and kaolinite decreased with increasing pH from 6.5 to 9, peaked at pH 6.5, the isoelectric point of the protein. Ionic strength is another important factor influencing protein adsorption. Ralla et al. [195] reported a decrease in adsorption of ovalbumin protein by the smectitic clay mineral with increasing ionic strength. However, Simonson and Brooks [198] claimed that the adsorbed amount of hemoglobin by bentonite increased with an increasing concentration of sodium salts. In addition, the adsorption temperature affects the rate and extent of adsorption of protein molecules onto clay minerals. Generally, higher temperatures were propitious for adsorbing more proteins because the protein adsorption is normally a diffusion-controlled process. However, Zhou et al. [191] claimed that the adsorption of was not significantly affected by the temperature between 10 and 50 °C. It was presumably explained by the fact that that the difference of agglomeration of adsorbed protein molecule and entanglement of unfolded protein molecule chains at lower or higher temperature. The dielectric constant of medium influences the conformation and the stability of protein molecules, and further affect the protein adsorption. It was believed that adsorption decreases with decreasing dielectric constant of medium. By changing dielectric constant of proteins, solvent may influence the protein adsorption. Sun et al. [199] reported that the amount of adsorbed protein on bentonite was enhanced by increasing ethanol concentration, then decreased sharply when further increasing ethanol concentration. The increase of ethanol content helped bentonite swell and its layers separate, making it easier for protein molecule to enter into its structure in ethanol solution than only in water. However, high alcohol concentration can weaken the hydrophobic force within protein molecule and lead to a decrease of hydrophobic interaction and low amount of adsorption.

5.4 Nuclear Acids

Nucleic acids, the storage molecules of hereditary information, are composed of repeating polymers of deoxyribonucleotides (in DNA) or ribonucleotides (in RNA), which are themselves composed of a base (a purine or a pyrimidine), a sugar (which is either ribose or deoxyribose) and a molecule of phosphoric acid. The interaction between nucleic acids and mineral surfaces is crucial for the preservation of genetic information by protecting them against enzymatic digestion, UV radiation and X-rays [200]. Thus, the adsorption of mono-, oligo-, and polynucleotides and their components on mineral surfaces may have been important for the origin of life. Many minerals were reported on the adsorption of DNAs and RNAs, while clay mineral is the most studied and discussed in the section.

The adsorption of nucleric acids like DNAs and RNAs on clay mineral are both involved in electrostatic forces, hydrogen bonding, ligand exchange and cation bridge, etc. as reviewed by Yu et al. [186]. DNA adsorption on well-characterized adsorbents and soil was affected by the ionic strength, pH of the medium, type and content of clay as well as characteristics and configurations of DNA molecules. The protonation of the amino groups of nucleic acid bases causes the adsorption of charged DNA or RNA by the charged clay surface. Ligand exchange is a reaction process which the phosphate groups at the two ends of the nucleic acid molecules are directly bound to the hydroxyl group on the surface of clay minerals. It was suggested that the nucleic acids are associated with several adsorption mechanisms onto one type of mineral [201]. Cai et al. [202] reported more than 50% of salmon sperm DNA was directly adsorbed on the positively charged edges of montmorillonite by electrostatic forces, while only 21.9% were adsorbed onto kaolinite by electrostatic forces and ligand exchange. Hashizume et al. (2010) claimed that the A, C and cytosine (U) bases of RNA components were bound to Mg^{2+}-montmorillonite by van der Waals. Sciascia et al. [203] revealed that the adsorption of protonated adenine (A) and cytosine (C) bases onto montmorillonite in an acidic solution occurred through electrostatic forces at the faces of the lamella of clay mineral and ionic exchange at the interlayer region. In addition, the existing multivalent cations such as Ca^{2+} and Mg^{2+} in the medium may be responsible for the cation bridge process for adsorption of nucleic aicds by clays [204]. Some polar DNA biomolecules may be strongly adsorbed through hydrogen bonding on the corner and broken edges which are amphoteric and able to act as a proton donor or acceptor of clay mineral particles such as montmorillonite and kaolinite [205].

Nucleic acids molecules are bound primarily to the edges or the planar surfaces of the clay minerals and does not significantly intercalate into the clay layers [186]. Cai et al. [206] found that DNA was adsorbed onto the planar surface of montmorillonite or mainly on the edges for kaolinite. Mignon et al. [207] analyzes the adsorption of RNA/DNA nucleobases on the Na^+-montmorillonite surfaces and showed that adsorption has been considered either on the side comprising the Na^+ counterion or on the opposite side, where only siloxane bonds are present.

The type of the minerals and the nature of DNA or RNA nucleic acids bases significantly affect the adsorption of nucleic acids on the clays. Generally, the A base was bound more strongly than C and U bases to montmorillonite. Benetoli et al. [208] found that montmorillonite gave a higher amount of adsorption of A base than kaolinite in artificial seawater at pH 2.0. Under similar conditions, the amount of the adsorbed nucleic acid bases of A, C and U by kaolinite were much lower than those by montmorillonite and bentonite. Besides, the adsorption of DNA or nucleic acid bases of RNA can be affected significantly by the pH of solution. The adsorption of DNA is related to the DNA isoelectric point (around pH 5) and the amount of adsorbed DNA considerably decreases with increasing solution pH from 3.0 to 9.0 [182]. Adsorption of the RNA bases of A, C and U by montmorillonite and kaolinite at pH 2 was greater than that at pH 7.2 in artificial seawater [208]. Ionic strength is another factor influencing the adsorption. Cai et al. [209] reported that the adsorption of DNA on montmorillonite, kaolinite and goethite increased with increasing concentration of $MgCl_2$. However, the adsorption of salmon sperm dsDNA molecules by the clay mineral was not affected by the solution NaCl concentrations in the range of 0.1 to 0.5 mol/L [210]. As for the RAN adsorption, Franchi et al. [211] found that divalent cations were more efficient than monovalent ones, and indicated that cations directly took part in the formation of bridges between the negative charges on the mineral surface and the phosphate groups of the genetic polymers.

Minerals can adsorb biomacromolecules through cation exchange, electrostatic interactions, hydrophobic/hydrophilic interaction, cation bridge, ligand exchange and van der Waals forces, and so forth. The adsorption varies with the surface, structure of both minerals and biomolecules. Thus, it is still necessary to get deep understanding of these reaction processes.

References

1. Huang, P.M., Bollag, J.-M., Senesi, N.: Interactions between soil particles and microorganisms: impact on the terrestrial ecosystem. Wiley (2002)
2. Loosdrecht, M.C.V., Lyklema, J., Norde, W., Zehnder, A.J.: Influence of interfaces on microbial activity. Microbiol. Rev. **54**, 75 (1990)
3. Hermansson, M.: The DLVO theory in microbial adhesion. Colloid Surface B. **14**, 105–119 (1999)
4. Bellon-Fontaine, M.N., Mozes, N., Hc, V.D.M., Sjollema, J., Cerf, O., Rouxhet, P.G., et al.: A comparison of thermodynamic approaches to predict the adhesion of dairy microorganisms to solid substrata. Cell Biophys. **17**, 93–106 (1990)
5. Dufrene, Y.F., Rouxhet, P.G.: Surface composition, surface properties, and adhesiveness of Azospirillum brasilense-variation during growth. Can. J. Microbiol. **42**, 548–556 (1996)
6. Hong, Z., Rong, X., Cai, P., Dai, K., Liang, W., Chen, W., et al.: Initial adhesion of Bacillus subtilis on soil minerals as related to their surface properties. Eur. J. Soil. Sci. **63**, 457–466 (2012)
7. Savage, D.C., Fletcher, M.: Bacterial adhesion: mechanisms and physiological significance. Q. Rev. Bio. **37** (1985)
8. Plummer, D.T., James, A.M.: Some physical investigations of the behaviour of bacterial surfaces III. The variation of the electrophoretic mobility and capsule size of Aerobacter aerogenes with age. Biochim. Biophys. Acta. **53**, 453–460 (1961)
9. Pandey, B.D., Natarajan, K.A.: Microbiology for minerals, metals. In: Materials and the Environment. Crc Press (2017)
10. Marshall, K., Cruickshank, R.: Cell surface hydrophobicity and the orientation of certain bacteria at interfaces. Arch. Microbiol. **91**, 29–40 (1973)
11. Costerton, J.W., Irvin, R.T., Cheng, K.J.: The bacterial glycocalyx in nature and disease. Annu. Rev. Microbiol. **35**, 299 (1981)
12. Garrett, T.R., Bhakoo, M., Zhang, Z.: Bacterial adhesion and biofilms on surfaces. Prog. Nat. Sci. Mater. **18**, 1049–1056 (2008)
13. Bryers, J.D., Characklis, W.G.: Kinetics of initial biofilm formation within a turbulent flow system. **15**(4), 483-491 (1981)
14. Characklis, W.G., Bryers, I.B.J.D.: Bioengineering report: Fouling biofilm development: A process analysis. Biotechnol. Bioeng. **102**, 310–347 (2009)
15. Li, J., Busscher, H.J., Mei, H.C.V.D., Norde, W., Krom, B.P., Sjollema, J.: Analysis of the contribution of sedimentation to bacterial mass transport in a parallel plate flow chamber: Part II: Use of fluorescence imaging. Colloid Surface B. **87**, 427–432 (2011)
16. Young, L.Y., Mitchell, R.: The role of chemotactic responses in primary film formation (1973)
17. Young, L.Y., Mitchell, R.: Negative chemotaxis of marine bacteria to toxic chemicals. Appl. Microbiol. **25**, 972–975 (1973)
18. Harvey, R.W., Garabedian, S.P.: Use of colloid filtration theory in modeling movement of bacteria through a contaminated sandy aquifer. Environ. Sci. Technol. **25**, 178–185 (1997)
19. Mudd, S., Mudd, E.B.: Certain interfacial tension relations and the behavior of bacteria in films. J. Exp. Med. 40:647–660 (1924)

20. Dahlbäck, B., Hermansson, M., Kjelleberg, S., Norkrans, B.: The hydrophobicity of bacteria -an important factor in their initial adhesion at the air-water interface. Arch. Microbiol. **128**, 267–270 (1981)
21. Rosenberg, M., Kjelleberg, S.: Hydrophobic interactions: role in bacterial adhesion. Adv. Microb. Ecol. 353–393 (1986)
22. Rusconi, R., Guasto, J.S., Stocker, R.: Bacterial transport suppressed by fluid shear. Nat. Phys. **10**, 212–217 (2014)
23. Xiancai, T., Bowen, Z., Tingting, L.: Attachment of microbes on mineral surface in microbial weathering and its significance (in Chinese). Geol. J. China Univ. **17**, 21–28 (2011)
24. Ghauri, M.A., Okibe, N., Johnson, D.B.: Attachment of acidophilic bacteria to solid surfaces: The significance of species and strain variations. Hydrometallurgy **85**, 72–80 (2007)
25. Rodríguez, A.G., Navarro, H.R.: Interfacial insights of pyrite colonized by Acidithiobacillus thiooxidans cells under acidic conditions. Hydrometallurgy **103**, 35–44 (2010)
26. Sand, W., Gehrke, T.: Extracellular polymeric substances mediate bioleaching/biocorrosion via interfacial processes involving iron(III) ions and acidophilic bacteria. Res. Microbiol. **157**, 49–56 (2006)
27. Hori, K.: Adhesion of bacteria. In: Biofilm and Materials Science. Springer, pp. 23–34 (2015)
28. Marshall, K., Stout, R., Mitchell, R.: Mechanism of the initial events in the sorption of marine bacteria to surfaces. Microbiology **68**, 337–348 (1971)
29. Devasia, P., Natarajan, K.A., Sathyanarayana, D.N., Rao, G.R.: Surface chemistry of thiobacillus ferrooxidans relevant to adhesion on mineral surfaces. Appl. Eeviron. Microb. **59**, 4051–4055 (1993)
30. Ohmura, N., Kitamura, K., Saiki, H.: Selective adhesion of thiobacillus ferrooxidans to pyrite. Appl. Eeviron. Microb. **59**, 4044–4050 (1993)
31. Blake, R.C., Shute, E.A., Howard, G.T.: Solubilization of minerals by bacteria: electrophoretic mobility of thiobacillus ferrooxidans in the presence of iron, pyrite, and sulfur. Appl. Eeviron. Microb. **60**, 3349–3357 (1994)
32. Marshall, K.C.: Interfaces in microbial ecology: Harvard University Press (1976)
33. Jianhua, F.: Bioleaching of copper sulfide minerals, the bacterial ultrastructure and the mechanisms of attachment, purifing SFORase of Thiobacillus ferrooxidans (in Chinese). PhD, Central South University (2004)
34. Santhiya, D., Subramanian, S., Natarajan, K.A.: Surface chemical studies on galena and sphalerite in the presence of Thiobacillus thiooxidans with reference to mineral beneficiation. Miner. Eng. **13**, 747–763 (2000)
35. Subramanian, S., Santhiya, D., Natarajan, K.A.: Surface modification studies on sulphide minerals using bioreagents. Int. J. Miner. Process. **72**, 175–188 (2003)
36. Rong, X., Huang, Q., He, X., Chen, H., Peng, C., Liang, W.: Interaction of Pseudomonas putida with kaolinite and montmorillonite: A combination study by equilibrium adsorption, ITC, SEM and FTIR. Colloid Surface B. **64**, 49–55 (2008)
37. Deo, N., Natarajan, K.A., Somasundaran, P.: Mechanisms of adhesion of Paenibacillus polymyxa onto hematite, corundum and quartz. Int. J. Miner. Process. **62**, 27–39 (2001)
38. Jun, W., Kangnian, Z. (1996) Influence of bacteria on flotation of sulfide mineral. Foreign mineral processing in metal ore pp. 4–10 (in Chinese)
39. Sharma, P.K., Rao, K.H., Forssberg, K.S.E., Natarajan, K.A.: Surface chemical characterisation of Paenibacillus polymyxa before and after adaptation to sulfide minerals. Int. J. Miner. Process. **62**, 3–25 (2001)
40. Santhiya, D., Subramanian, S., Natarajan, K.A., Rao, K.H., Forssberg, K.S.E.: Bio-modulation of galena and sphalerite surfaces using Thiobacillus thiooxidans. Int. J. Miner. Process. **62**, 121–141 (2001)

41. Farahat, M., Hirajima, T., Sasaki, K., Doi, K.: Adhesion of Escherichia coli onto quartz, hematite and corundum: Extended DLVO theory and flotation behavior. Colloid Surface B **74**, 140–149 (2009)
42. Yee, N., Fein, J.B., Daughney, C.J.: Experimental study of the pH, ionic strength, and reversibility behavior of bacteria–mineral adsorption. Geochim Cosmochim Ac **64**, 609–617 (2000)
43. Ams, D.A., Fein, J.B., Dong, H., Maurice, P.A.: Experimental Measurements of the Adsorption of Bacillus subtilis and Pseudomonas mendocina Onto Fe-Oxyhydroxide-Coated and Uncoated Quartz Grains. Geomicrobiol J. **21**, 511–519 (2004)
44. Shashikala, A.R., Raichur, A.M.: Role of interfacial phenomena in determining adsorption of Bacillus polymyxa onto hematite and quartz. Colloid Surface B **24**, 11–20 (2002)
45. Chenu, C., Pons, C.H., Robert, M.: Interaction of kaolinite and montmorillonite with neutral polysaccharides. In: International Clay Conference, pp. 375–381 (1985)
46. Malik, M., Letey, J.: Adsorption of Polyacrylamide and Polysaccharide Polymers on Soil Materials. Soil Sci. Soc. Am. J. **55**, 380–383 (1991)
47. Santoro, T., Stotzky, G.: Effect of electrolyte composition and pH on the particle size distribution of microorganisms and clay minerals as determined by the electrical sensing zone method*1. Arch. Biochem. Biophys. **122**, 664–669 (1967)
48. Jones, D.L., Edwards, A.C.: Influence of sorption on the biological utilization of two simple carbon substrates. Soil Biol. Biochem. **30**, 1895–1902 (1998)
49. Zobell, C.E.: The Effect of Solid Surfaces upon Bacterial Activity. J. Bacteriol. **46**, 39 (1943)
50. Martin, J.P., Filip, Z., Haider, K.: Effect of montmorillonite and humate on growth and metabolic activity of some actinomycetes. Soil Biol. Biochem. **8**, 409–413 (1976)
51. Lavie, S., Stotzky, G.: Interactions between clay minerals and siderophores affect the respiration of Histoplasma capsulatum. Appl. Environ. Microb. **51**, 74–79 (1986)
52. Banfield, J.F., Barker, W.W., Welch, S.A., Taunton, A.: Biological impact on mineral dissolution: Application of the lichen model to understanding mineral weathering in the rhizosphere. P Natl Acad Sci Usa **96**, 3404–3411 (1999)
53. Uroz, S., Calvaruso, C., Turpault, M.P., Sarniguet, A., Boer, W.D., Leveau, J.H.J., et al.: Efficient mineral weathering is a distinctive functional trait of the bacterial genus Collimonas. Soil Biol. Biochem. **41**, 2178–2186 (2009)
54. Ojeda, L., Keller, G., Muhlenhoff, U., Rutherford, J.C., Lill, R., Winge, D.R.: Role of glutaredoxin-3 and glutaredoxin-4 in the iron regulation of the Aft1 transcriptional activator in Saccharomyces cerevisiae. J. Biol. Chem. **281**, 17661–17669 (2006)
55. Rogers, J.R., Bennett, P.C.: Mineral stimulation of subsurface microorganisms: release of limiting nutrients from silicates. Chem. Geol. **203**, 91–108 (2004)
56. Ahmed, E.: Microbe-mineral interactions in soil: Investigation of biogenic chelators, microenvironments and weathering processes. Department of Geological Sciences, Stockholm University, Sweden, Doctor of philosophy (2015)
57. Liermann, L.J., Kalinowski, B.E., Brantley, S.L., Ferry, S.G.: Role of bacterial siderophores in dissolution of hornblende. Geochim. Cosmochim. Ac. **64**, 587–602 (2009)
58. Maurice, P.A., Haack, E.A., Mishra, B.: Siderophore sorption to clays. Biometals **22**, 649 (2009)
59. Roberts, J.A., Fowle, D.A., Hughes, B.T., Kulczycki, E.: Attachment Behavior of Shewanella putrefaciens onto Magnetite under Aerobic and Anaerobic Conditions. Geomicrobiol J. **23**, 631–640 (2006)
60. Bennett, P., Rogers, J., Choi, W., Hiebert, F.: Silicates, silicate weathering, and microbial ecology. Geomicrobiol J. **18**, 3–19 (2001)
61. Mauck, B.S., Roberts, J.A.: Mineralogic control on abundance and diversity of surface-adherent microbial communities. Geomicrobiol J. **24**, 167–177 (2007)
62. Mailloux, B.J., Alexandrova, E., Keimowitz, A.R., Wovkulich, K., Freyer, G.A., Herron, M., et al.: Microbial Mineral Weathering for Nutrient Acquisition Releases Arsenic. Appl Environ Microbl **75**, 2558–2565 (2009)

63. Kalinowski, B.E., Liermann, L.J., Brantley, S.L., Barnes, A., Pantano, C.G.: X-ray photoelectron evidence for bacteria-enhanced dissolution of hornblende. Geochim Cosmochim Ac **64**, 1331–1343 (2000)
64. Dong, H.: Mineral-microbe interactions: a review. Front Earth Sci. China **4**, 127–147 (2011)
65. Papassiopi, N., Vaxevanidou, K., Paspaliaris, I.: Effectiveness of iron reducing bacteria for the removal of iron from bauxite ores. Miner. Eng. **23**, 25–31 (2010)
66. Ryu, H.W., Cho, K.S., Yong, K.C., Sang, D.K., Mori, T.: Refinement of low-grade clay by microbial removal of sulfur and iron compounds using Thiobacillus ferrooxidans. J. Ferment. Bioeng. **80**, 46–52 (1995)
67. Juszczak, A., Domka, F., Kozłowski, M., Wachowska, H.: Microbial desulfurization of coal with Thiobacillus ferrooxidans bacteria. Fuel **74**, 725–728 (1995)
68. Bosecker, K.: Bioleaching: metal solubilization by microorganisms. FEMS Microbiol. Rev. **20**, 591–604 (1997)
69. Poorni, S., Natarajan, K.A.: Microbially induced selective flocculation of hematite from kaolinite. Int. J. Miner. Process. **125**, 92–100 (2013)
70. Deo, N., Natarajan, K.A.: Studies on interaction of Paenibacillus polymyxa with iron ore minerals in relation to beneficiation. Int. J. Miner. Process. **55**, 41–60 (1998)
71. Bagdigian, R.M., Myerson, A.S.: The adsorption of Thiobacillus ferrooxidans on coal surfaces. Biotechnol. Bioeng. **28**, 467–479 (1986)
72. Donati, E.R., Sand, W.: Microbial Processing of Metal Sulfides. Springer Ebooks (2007)
73. Patra, P., Natarajan, K.A.: Microbially-induced flocculation and flotation for pyrite separation from oxide gangue minerals. Miner. Eng. **16**, 965–973 (2003)
74. Patra, P., Natarajan, K.A.: Microbially induced flocculation and flotation for separation of chalcopyrite from quartz and calcite. Int. J. Miner. Process. **74**, 143–155 (2004)
75. Yee, N., Fein, J.: Cd adsorption onto bacterial surfaces: A universal adsorption edge? Geochim. Cosmochim. Ac. **65**, 2037–2042 (2001)
76. Huang, Q., Chen, W., Xu, L.: Adsorption of Copper and Cadmium by Cu- and Cd-Resistant Bacteria and Their Composites with Soil Colloids and Kaolinite. Geomicrobiol J. **22**, 227–236 (2005)
77. Flemming, C.A., Ferris, F.G., Beveridge, T.J., Bailey, G.W.: Remobilization of toxic heavy metals adsorbed to bacterial wall-clay composites. Appl. Environ. Microb. **56**, 3191–3203 (1990)
78. Ohnuki, T., Yoshida, T., Ozaki, T., Samadfam, M., Kozai, N., Yubuta, K., et al.: Interactions of uranium with bacteria and kaolinite clay. Chem. Geol. **220**, 237–243 (2005)
79. Zou, C.Y., Lian, B., Zang, S.Y., Wu, F.C.: Adsorption and desorption of heavy metal ions by bacteria-contained mineral composite adsorbents (in Chinese). J. Saf. Environ. **11**, 43–46 (2011)
80. Biswas, B., Sarkar, B., Rusmin, R., Naidu, R.: Bioremediation of PAHs and VOCs: Advances in clay mineral-microbial interaction. Environ. intl. **85**, 168 (2015)
81. Dong, H., Lu, A.: Mineral-Microbe Interactions and Implications for Remediation. Elements **8**, 95–100 (2012)
82. Nan, Q., Mingxing, G., Yanfu, Z., Gang, N., Ruijia, C.: Treatment characteristics of mixed microbial loaded on modified diatomite for dye wastewater. Bulletin of the chinese ceramic society **34**, 3451–3455 (2015)
83. Chen, H., He, X., Rong, X., Chen, W., Cai, P., Liang, W., et al.: Adsorption and biodegradation of carbaryl on montmorillonite, kaolinite and goethite. Appl. Clay Sci. **46**, 102–108 (2009)
84. Quintelas, C., Costa, F., Tavares, T.: Bioremoval of diethylketone by the synergistic combination of microorganisms and clays: uptake, removal and kinetic studies. Environ. Sci. Pollut. R. **20**, 1374–1383 (2013)
85. Templeton, A.S., Spormann, A.: Speciation of Pb(II) sorbed by Burkholderia cepacia/goethite composites. Environ. Sci. Technol. **37**, 2166–2172 (2003)
86. Gadd, G.M.: Geomycology: biogeochemical transformations of rocks, minerals, metals and radionuclides by fungi, bioweathering and bioremediation. Mycol. Res. **111**, 3–49 (2007)

87. Burford, Fomina E.P., Gadd, M., G, M.: Fungal involvement in bioweathering and biotransformation of rocks and minerals. Mineral. Mag. **67**, 1127–1155 (2003)
88. Li, H., Cai, L., Yao, Q.Z., Zhou, G.T.: Fungal involvement in biogeological processes and application to environmental bioremediation (in Chinese). Geol. J. China Univ. **3**, 382–394 (2015)
89. Jacobs, H., Boswell, G.P., Harper, F.A., Ritz, K., Davidson, F.A., Gadd, G.M.: Solubilization of metal phosphates by Rhizoctonia solani. Mycol. Res. **106**, 1468–1479 (2002)
90. Balogh-Brunstad, Z., Keller, C.K., Dickinson, J.T., Stevens, F., Li, C.Y., Bormann, B.T.: Biotite weathering and nutrient uptake by ectomycorrhizal fungus, Suillus tomentosus, in liquid-culture experiments. Geochim. Cosmochim. Ac. **72**, 2601–2618 (2008)
91. Gleeson, D.: Meter-Scale diversity of microbial communities on a weathered pegmatite Granite outcrop in the Wicklow Mountains, Ireland; Evidence for mineral induced selection? Geomicrobiol J. **27**, 1–14 (2010)
92. Gadd, G.M., Raven, J.A.: Geomicrobiology of eukaryotic microorganisms. Geomicrobiol J. **27**, 491–519 (2010)
93. CastroIM, Fietto J.L.R., Vieira, R.X., Trópia, M.J.M., Campos, L.M.M., Paniago, E.B., et al.: Bioleaching of zinc and nickel from silicates using Aspergillus niger cultures. Hydrometallurgy **57**, 39–49 (2000)
94. Mulligan, C.N., Galvez-Cloutier, R.: Bioremediation of metal contamination. Envion. Monit. Assess. **84**, 45–60 (2003)
95. Valix, M., Usai, F., Malik, R.: Fungal bio-leaching of low grade laterite ores. Miner. Eng. **14**, 197–203 (2001)
96. Morley, G.F., Gadd, G.M.: Sorption of toxic metals by fungi and clay minerals. Mycol. Res. **99**, 1429–1438 (1995)
97. Hu, J., Lian, B., Yu, J.P., Hu, X.: Formation of aspergillus niger-mineral aggregation and characterization of polysaccharide from aggregation (in Chinese). Acta Microbiol. Sin. **6**, 756–763 (2011)
98. Huang, Q., Jianmei, W.U., Chen, W., Xueyuan, L.I.: Adsorption of cadmium by soil colloids and minerals in presence of Rhizobia. Pedosphere **10**, 299–307 (2000)
99. Fomina, M., Gadd, G.M.: Metal sorption by biomass of melanin-producing fungi grown in clay-containing medium. J. Chem. Technol. Biot. **78**, 23–34 (2010)
100. Loy, A., Mandl, M., Barton, L.L.: Geomicrobiology: Molecular and environmental perspective. Springer Science & Business Media (2010)
101. Henderson, R., Parsons, S.A., Jefferson, B.: The impact of algal properties and pre-oxidation on solid-liquid separation of algae. Water Res. **42**, 1827–1845 (2008)
102. Hadjoudja, S., Vignoles, C., Deluchat, V., Lenain, J.F., Jeune, A.H.L., Baudu, M.: Short term copper toxicity on Microcystis aeruginosa and Chlorella vulgaris using flow cytometry. Aquat. Toxicol. **94**, 255–264 (2009)
103. Ozkan, A., Berberoglu, H.: Cell to substratum and cell to cell interactions of microalgae. Colloid Surface B **112**, 302–309 (2013)
104. Xia, L., Li, H., Song, S.: Cell surface characterization of some oleaginous green algae. J. Appl. Phycol. **28**, 2323–2332 (2016)
105. Zhang, X., Amendola, P., Hewson, J.C., Sommerfeld, M., Hu, Q.: Influence of growth phase on harvesting of Chlorella zofingiensis by dissolved air flotation. Bioresource Technol **116**, 477–484 (2012)
106. Li, Y., Song, S., Xia, L., Yin, H., Meza, J.V.G., Ju, W.: Enhanced Pb (II) removal by algal-based biosorbent cultivated in high-phosphorus cultures. Chem. Eng. J. **361**, 167–179 (2019)
107. Xia, L., Huang, R., Li, Y., Song, S.: The effect of growth phase on the surface properties of three oleaginous microalgae (*Botryococcus* sp. FACGB-762, *Chlorella* sp. XJ-445 and *Desmodesmus bijugatus* XJ-231). PLOS ONE 12 (2017) e0186434 (2017)
108. Wynn-Williams, D.D.: Cyanobacteria in Deserts-Life at the Limit?. Springer, Netherlands (2000)

109. Herrera, L.K., Arroyave, C., Guiamet, P., Saravia, S.G.D., Videla, H.: Biodeterioration of peridotite and other constructional materials in a building of the Colombian cultural heritage. Int Biodeter Biodegr **54**, 135–141 (2004)
110. Budel, B., Kuhl, B.W., Pfanz, H., Sultemeyer, D., Wessels, D.: Reshaping of sandstone surfaces by cryptoendolithic cyanobacteria: bioalkalization causes chemical weathering in arid landscapes. Geobiology **2**, 261–268 (2004)
111. De, L.R.A., Wierzchos, J., Sancho, L.G., Ascaso, C.: Acid microenvironments in microbial biofilms of antarctic endolithic microecosystems. Environ. Microbiol. **5**, 231–237 (2003)
112. Garcia-Pichel, F.: Plausible mechanisms for the boring on carbonates by microbial phototrophs. Sediment Geol **185**, 205–213 (2006)
113. Huang, L.N., Tang, F.Z., Song, Y.S., Wan, C.Y., Wang, S.L., Liu, W.Q., Shu, W.S.: Biodiversity, abundance, and activity of nitrogen-fixing bacteria during primary succession on a copper mine tailings. Fems. Microbiol. Ecol. **78**, 439–450 (2011)
114. García-Meza, J.V., Barrangue, C., Admiraal, W.: Biofilm formation by algae as a mechanism for surviving on mine tailings. Environ. Toxicol. Chem. **24**, 573–581 (2005)
115. Varshney, P., Mikulic, P., Vonshak, A., Beardall, J., Wangikar, P.P.: Extremophilic micro-algae and their potential contribution in biotechnology. Bioresour. Technol. **184**, 363 (2015)
116. Magaña, H.A., Contreras, C., Villareal, T.A.: A historical assessment of Karenia brevis in the western Gulf of Mexico. Harmful Algae **2**, 163–171 (2003)
117. Chorus, I., Bartram, J., Chorus, I., Bartram, J.: Toxic cyanobacteria in water: A guide to their public health consequences, monitoring and management. Limnol. Oceanoge. **45**, 255–258 (1999)
118. Sengco, M.R., Li, A., Tugend, K., Kulis, D., Anderson, D.M.: Removal of red-and brown-tide cells using clay flocculation. I. Laboratory culture experiments with Gymnodinium breve and Aureococcus anophagefferens. Mar. Eco. Progress Series **210**, 41–53 (2001)
119. Beaulieu, S.E., Sengco, M.R., Anderson, D.W.: Using clay to control harmful algal blooms: deposition and resuspension of clay/algal flocs. Harmful Algae **4**, 123–138 (2005)
120. Pierce, R.H., Henry, M.S., Higham, C.J., Blum, P., Sengco, M.R., Anderson, D.M.: Removal of harmful algal cells (Karenia brevis) and toxins from seawater culture by clay flocculation. Harmful Algae **3**, 141–148 (2004)
121. Lee, Y., Choi, J., Kim, E., Youn, S., Yang, E.: Field experiments on mitigation of harmful algal blooms using a Sophorolipid-Yellow clay mixture and effects on marine plankton. Harmful Algae **7**, 154–162 (2008)
122. Anderson, D.M., Garrison, D.J.: The ecology and oceanography of harmful algal blooms. Limnol. Oceanogr. **42**, 1009–1305 (1997)
123. Pierce, R.H., Blum, P., Sengco, M.R., Anderson, D.M.: Removal of harmful algal cells (Karenia brevis) and toxins from seawater culture by clay flocculation. Harmful Algae **3**, 141–148 (2004)
124. Shirota, A.: Red tide problem and countermeasures 2. Int. J. Aquat. Fish. Technol. **1**, 195–223 (1989)
125. Davidovich, N.A., Bates, S.S.: Sexual Reproduction in the Pennate Diatoms Pesudo-NitzschiaI Multlseries and P. Pseudodelicatissima (Bacillariophyceae). J. Phycol. **34**, 126–137 (2010)
126. Yu, Z.M., Zou, J.Z., Ma, X.N.: Application of clays to removal of red tide organisms I. Coagulation of red tide organisms with clays. Chin. J. Oceanol. Limn. **12**, 316–324 (1994)
127. Gobler, C.J., Lonsdale, D.J., Boyer, G.L.: A review of the causes, effects, and potential management of harmful brown tide blooms caused by Aureococcus anophagefferens (Hargraves et sieburth). Estuaries **28**, 726–749 (2005)
128. Pan, G., Zhang, M., Chen, H., Zou, H., Yan, H.: Removal of cyanobacterial blooms in Taihu Lake using local soils. I. Equilibrium and kinetic screening on the flocculation of Microcystis aeruginosa using commercially available clays and minerals. Environ. Pollut. **141**, 195–200 (2006)

129. Sengco, M.R., Anderson, D.W.: Controlling harmful algal blooms through clay Flocculation. J. Eukaryot. Microbiol. **51**, 169–172 (2004)
130. Pan, G., Zou, H., Chen, H., Yuan, X.: Removal of harmful cyanobacterial blooms in Taihu Lake using local soils III. Factors affecting the removal efficiency and an in situ field experiment using chitosan-modified local soils. Environ. Pollut. **141**, 206–212 (2006)
131. Zou, H., Pan, G., Chen, H., Yuan, X.: Removal of cyanobacterial blooms in Taihu Lake using local soils II. Effective removal of Microcystis aeruginosa using local soils and sediments modified by chitosan. Environ. Pollut. **141**, 201–205 (2006)
132. Li, H., Pan, G.: Simultaneous removal of harmful algal blooms and microcystins using Microorganism- and Chitosan-Modified Local Soil. Environ. Sci. Technol. **49**, 6249–6256 (2015)
133. Liu, G., Fan, C., Zhong, J., Zhang, L., Ding, S., Yan, S., Han, S.: Using hexadecyl trimethyl ammonium bromide (CTAB) modified clays to clean the Microcystis aeruginosa blooms in Lake Taihu, China. Harmful Algae **9**, 413–418 (2010)
134. Chen, J., Pan, G.: Harmful algal blooms mitigation using clay/soil/sand modified with xanthan and calcium hydroxide. J. Appl. Phycol. **24**, 1183–1189 (2012)
135. Wu, T., Yan, X., Cai, X., Tan, S., Li, H., Liu, J., Yang, W.: Removal of Chattonella marina with clay minerals modified with a gemini surfactant. Appl. Clay Sci. **50**, 604–607 (2010)
136. Li, L., Pan, G.: A universal method for flocculating harmful algal blooms in marine and fresh waters using modified sand. Environ. Sci. Technol. **47**, 4555–4562 (2013)
137. Chamizo, S., Cantón, Y., Miralles, I., Domingo, F.: Biological soil crust development affects physicochemical characteristics of soil surface in semiarid ecosystems. Soil Biol. Biochem. **49**, 96–105 (2012)
138. Lan, S., Zhang, Q., Wu, L., Liu, Y., Zhang, D., Hu, C.: Artificially accelerating the reversal of desertification: cyanobacterial inoculation facilitates the succession of vegetation communities. Environ. Sci. Technol. **48**, 307 (2014)
139. Rossi, F., Li, H., Liu, Y., Philippis, R.D.: Cyanobacterial inoculation (cyanobacterisation): Perspectives for the development of a standardized multifunctional technology for soil fertilization and desertification reversal. Earth-Sci. Rev. **171**, 28–43 (2017)
140. Xie, Z., Liu, Y., Hu, C., Chen, L., Li, D.: Relationships between the biomass of algal crusts in fields and their compressive strength. Soil Biol. Biochem. **39**, 567–572 (2007)
141. Li, H., Rao, B., Wang, G., Shen, S., Li, D., Hu, C., Liu, Y.: Spatial heterogeneity of cyanobacteria-inoculated sand dunes significantly influences artificial biological soil crusts in the Hopq Desert (China). Environ. Earth Sci. **71**, 245–253 (2014)
142. Felde, A.V.M.N., Peth, S., Uteau-Puschmann, D., Drahorad, S., Felix-Henningsen, P.: Soil microstructure as an under-explored feature of biological soil crust hydrological properties: case study from the NW Negev Desert. Biodivers. Conserv. **23**, 1687–1708 (2014)
143. Williams, A.J., Buck, B.J., Soukup, D.A., Merkler, D.J.: Geomorphic controls on biological soil crust distribution: A conceptual model from the Mojave Desert (USA). Geomorphology **195**, 99–109 (2013)
144. Williams, A.J.: Biological soil crusts in the mojave desert, USA: Micromorphology and Pedogenesis. Soil Sci. Soc. Am. J. **76**, 1685–1695 (2012)
145. Ram, A., Aaron, Y.: Negative and positive effects of topsoil biological crusts on water availability along a rainfall gradient in a sandy arid area. CATENA **70**, 437–442 (2007)
146. Pietrasiak, N., Drenovsky, R.E., Santiago, L.S., Graham, R.C.: Biogeomorphology of a Mojave Desert landscape-Configurations and feedbacks of abiotic and biotic land surfaces during landform evolution. Geomorphology **206**, 23–36 (2014)
147. Hu, C., Liu, Y., Song, L., Zhang, D.: Effect of desert soil algae on the stabilization of fine sands. J. Appl. Phycol. **14**, 281–292 (2002)
148. Rozenstein, O., Zaady, E., Katra, I., Karnieli, A., Adamowski, J., Yizhaq, H.: The effect of sand grain size on the development of cyanobacterial biocrusts. Aeolian Res. **15**, 217–226 (2014)
149. Hayes, M.H., Clapp, C.E.: Humic substances: Considerations of compositions, aspects of structure, and environmental Influences. Soil Sci. **166**, 723–737 (2001)

150. Fontes, M.R., Weed, S.B., Bowen, L. H.: Association of microcrystalline goethite and humic acid in some Oxisols from Brazil. Soil Sci. Soc. Am. J. **56**, 982–990 (1992)
151. Aiken, G.R., Mcknight, D.M., Wershaw, R.L., Maccarthy, P.M.: Humic substances in soil, sediment, and water: geochemistry, isolation and characterization. Q. Rev. Biol. **142**, (1985)
152. Vermeer, A.W.P.: Interactions between humic acid and hematite and their effects on metal ion speciation. Ph.D. Thesis, Wageningen Agricultural University (1996)
153. Evans, L.T., Russell, E.W.: The Adsorption of humic and fulvic acids by clays. J. Soil Sci. **10**, 119–132 (1959)
154. Greenland, D.J.: Interactions between humic and fulvic acids and clays. Soil Sci. **11**, 34–41 (1971)
155. Zhang, L., Luo, L., Zhang, S.: Integrated investigations on the adsorption mechanisms of fulvic and humic acids on three clay minerals. Colloid Surfaces A **406**, 84–90 (2012)
156. Gu, B., Schmitt, J., Chen, Z., Liang, L., Mccarthy, J.F.: Adsorption and desorption of natural organic matter on iron oxide: mechanisms and models. Environ. Sci. Technol. **28**, 38 (1994)
157. Weng, L., van Riemsdijk, W.H., Koopal, L.K., Hiemstra, T.: Adsorption of humic substances on goethite: comparison between humic acids and fulvic acids. Environ. Sci. Technol. **40**, 7494 (2006)
158. Fu, X., Quan, X.: Complexes of fulvic acid on the surface of hematite, goethite, and akaganeite: FTIR observation. Chemosphere **63**, 403–410 (2006)
159. Qin, X., Liu, F., Wang, G., Huang, G.: Adsorption of humic acid from aqueous solution by hematite: effects of pH and ionic strength. Environ. Earth Sci. **73**, 4011–4017 (2015)
160. Vermeer, A.W.P., Riemsdijk, W.H.V., Koopal, L.K.: Adsorption of humic acid to mineral particles. 1. Specific and electrostatic interacions. **14**, 2810–2819 (1998)
161. Kloster, N., Avena, M.: Interaction of humic acids with soil minerals: adsorption and surface aggregation induced by Ca^{2+}. Environ. Chem. **12**, 731–738 (2015)
162. Spark, K.M., Wells, J.D., Johnson, B.B.: Sorption of heavy metals by mineral-humic acid substrates. Soil Res **35**, 113–122 (1997)
163. Liu, A., Gonzalez, R.D.: Adsorption/Desorption in a System Consisting of Humic Acid, Heavy Metals, and Clay Minerals. J. Colloid Interface Sci. **218**, 225–232 (1999)
164. Wingender, J., Neu, T.R., Flemming, H.: What are Bacterial Extracellular Polymeric Substances? In: Wingender, J., Neu, T.R., Flemming, H. (eds.) Microbial Extracellular Polymeric Substances: Characterization, pp. 1–19. Structure and Function, Springer, Berlin Heidelberg, Berlin, Heidelberg (1999)
165. Tsuneda, S., Aikawa, H., Hayashi, H., Yuasa, A., Hirata, A.: Extracellular polymeric substances responsible for bacterial adhesion onto solid surface. FEMS Microbiol. Lett. **223**, 287–292 (2003)
166. Banfield, J.F., Barker, W.W., Welch, S.A., Taunton, A.: Biological impact on mineral dissolution: application of the lichen model to understanding mineral weathering in the rhizosphere. P Natl Acad Sci **96**, 3404–3411 (1999)
167. Sheng, G., Yu, H., Li, X.: Extracellular polymeric substances (EPS) of microbial aggregates in biological wastewater treatment systems: A review. Biotechnol. Adv. **28**, 882–894 (2010)
168. Omoike, A., Chorover, J.: Adsorption to goethite of extracellular polymeric substances from Bacillus subtilis. Geochim Cosmochim Ac **70**, 827–838 (2006)
169. Cao. Y., Wei. X., Cai, P., Huang, Q., Rong, X., Liang, W.: Preferential adsorption of extracellular polymeric substances from bacteria on clay minerals and iron oxide.Colloid Surf. B **83**,122–127 (2011)
170. Fang, L., Cao, Y., Huang, Q., Walker, S.L., Cai, P.: Reactions between bacterial exopolymers and goethite: A combined macroscopic and spectroscopic investigation. Water Res. **46**, 5613–5620 (2012)
171. Mikutta, R., Baumgärtner, A., Schippers, A., Haumaier, L., Guggenberger, G.: Extracellular Polymeric Substances from Bacillus subtilis Associated with Minerals Modify the Extent and Rate of Heavy Metal Sorption. Environ. Sci. Technol. **46**, 3866–3873 (2012)
172. Sparks, D.L.: Environmental Soil Chemistry (Second Edition) (2003)

173. Kwon, K.D., Vadillo-Rodriguez, V., Logan, B.E., Kubicki, J.D.: Interactions of biopolymers with silica surfaces: Force measurements and electronic structure calculation studies. Geochim. Cosmochim. Ac. **70**, 3803–3819 (2006)
174. Zhu, P., Long, G., Ni, J.R., Tong, M.: Deposition Kinetics of Extracellular Polymeric Substances (EPS) on Silica in Monovalent and Divalent Salts. Environ. Sci. Technol. **43**, 5699 (2009)
175. Omoike, A., Chorover, J., Kwon, K.D., Kubicki, J.D.: Adhesion of Bacterial Exopolymers to α-FeOOH: Inner-Sphere Complexation of Phosphodiester Groups. Langmuir **20**, 11108–11114 (2004)
176. Liu, X., Eusterhues, K., Thieme, J., Ciobota, V., Höschen, C., Mueller, C.W., Küsel, K., Kögel-Knabner, Rösch I.P., Popp, J.: STXM and NanoSIMS investigations on EPS fractions before and after adsorption to goethite. Environ. Sci. Technol. **47**, 3158–3166 (2013)
177. Fang, L., Huang, Q., Wei, X., Liang, W., Rong, X., Chen, W., Cai, P.: Microcalorimetric and potentiometric titration studies on the adsorption of copper by extracellular polymeric substances (EPS), minerals and their composites. Bioresource Technol **101**, 5774 (2010)
178. Jaisi, D.P., Dong, H., Kim, J., He, Z., Morton, J.P.: Nontronite particle aggregation induced by microbial Fe(III) reduction and exopolysaccharide production. Clay Clay Miner **55**, 96–107 (2007)
179. Yan, S., Cai, Y., Li, H., Song, S., Xia, L.: Enhancement of cadmium adsorption by EPS-montmorillonite composites. Environ. Pollut. **252**, 1509–1518 (2019)
180. Rigou, P., Rezaei, H., Grosclaude, J., Staunton, S., Quiquampoix, H.: Fate of prions in soil: adsorption and extraction by electroelution of recombinant ovine prion protein from montmorillonite and natural soils. Environ. Sci. Technol. **40**, 1497–1503 (2006)
181. Koskella, J., Stotzky, G.: Microbial utilization of free and clay-bound insecticidal toxins from bacillus thuringiensis and their retention of insecticidal activity after incubation with microbes. Appl. Environ. Microb. **63**, 3561 (1997)
182. Stotzky, G.: Persistence and biological activity in soil of insecticidal proteins from Bacillus thuringiensis and of bacterial DNA bound on clays and humic acids. J. Environ. Qual. **29**, 691–705 (2000)
183. Fiorito, T.M., Icoz, I., Stotzky, G.: Adsorption and binding of the transgenic plant proteins, human serum albumin, β-glucuronidase, and Cry3Bb1, on montmorillonite and kaolinite: Microbial utilization and enzymatic activity of free and clay-bound proteins. Appl. Clay Sci. **39**, 142–150 (2008)
184. Vaiana, C.A., Leonard, M.K., Drummy, L.F., Singh, K.M., Bubulya, A., Vaia, R.A., Naik, R.R., Kadakia, M.P.: Epidermal Growth Factor: Layered Silicate Nanocomposites for Tissue Regeneration. Biomacromol **12**, 3139–3146 (2011)
185. Ruizhitzky, E., Aranda, P., Darder, M., Ogawa, M.: Hybrid and biohybrid silicate based materials: molecular vs. block-assembling bottom-up processes. Cheminform **42**, 801–828 (2011)
186. Yu, W.H., Li, N., Tong, D.S., Zhou, C.H., Lin, C.X.C., Xu, C.Y.: Adsorption of proteins and nucleic acids on clay minerals and their interactions: A review. Appl. Clay Sci. **80**, 443–452 (2013)
187. Johnston, C.T., Premachandra, G.S., Szabo, T., Lok, J., Schoonheydt, R.A.: Interaction of Biological Molecules with Clay Minerals: A Combined Spectroscopic and Sorption Study of Lysozyme on Saponite. Langmuir **28**, 611–619 (2012)
188. Wang, X.C., Lee, C.: Adsorption and desorption of aliphatic amines, amino acids and acetate by clay minerals and marine sediments. Mar. Chem. **44**, 1–23 (1993)
189. Dumat, Camille, Hervé Quiquampoix, A., Staunton, S.: Adsorption of Cesium by Synthetic Clay − Organic Matter Complexes: Effect of the Nature of Organic Polymers. Environ. Sci. Technol. **34**, 2985–2989 (2000)
190. Aquino, A.J., Tunega, D., Haberhauer, G., Gerzabek, M.H., Lischka, H.: Adsorption of organic substances on broken clay surfaces: A quantum chemical study. J. Comput. Chem. **24**, 1853–1863 (2003)

191. Zhou, X., Huang, Q., Chen, S., Yu, Z.: Adsorption of the insecticidal protein of Bacillus thuringiensis on montmorillonite, kaolinite, silica, goethite and Red soil. Appl. Clay Sci. **30**, 87–93 (2005)
192. Awad, A.M., Shaikh, S.M., Jalab, R., Gulied, M.H., Nasser, M.S., Benamor, A., Adham, S.: Adsorption of organic pollutants by natural and modified clays: A comprehensive review. Sep. Purif. Technol. **228**, 115719 (2019)
193. Szabó, T., Mitea, R., Leeman, H., Premachandra, G.S., Johnston, C.T., Szekeres, M., Dékány, I., Schoonheydt, R.A.: Adsorption of protamine and papain proteins on saponite. Clay Miner. **56**, 494–504 (2008)
194. Fusi, P., Ristori, G.G., Calamai, L., Stotzky, G.: Adsorption and binding of protein on "clean" (homoionic) and "dirty" (coated with Fe oxyhydroxides) montmorillonite, illite and kaolinite. Soil Biol. Biochem. **21**, 911–920 (1989)
195. Ralla, K., Sohling, U., Riechers, D., Kasper, C., Ruf, F., Scheper, T.: Adsorption and separation of proteins by a smectitic clay mineral. Bioproc Biosyst Eng **33**, 847–861 (2010)
196. Barral, S., Villa-García, M.A., Rendueles, M., Díaz, M.: Interactions between whey proteins and kaolinite surfaces. Acta Mater. **56**, 2784–2790 (2008)
197. Helassa, N., Quiquampoix, H., Noinville, S., Szponarski, W., Staunton, S.: Adsorption and desorption of monomeric Bt (Bacillus thuringiensis) Cry1Aa toxin on montmorillonite and kaolinite. Soil Biol. Biochem. **41**, 498–504 (2009)
198. Simonson, T., Brooks, C.L.: Charge screening and the dielectric constant of proteins: Insights from molecular dynamics. J. Am. Chem. Soc. **118**, 8452–8458 (1996)
199. Sun, X.F., Li, C., Wu, Z.S., Xu, X.L, Ren, L.N., Zhao, H.S.: Adsorption of protein from model wine solution by different bentonites. Chinese J. Chem. Eng. **15**, 632–638 (2007)
200. Nd, C.H., Jonsson, C.M., Jonsson, C.L., Sverjensky, D.A., Hazen, R.M.: Adsorption of nucleic acid components on rutile (TiO_2) surfaces. Astrobiolog **10**, 311 (2010)
201. Saeki, K., Kunito, T., Sakai, M.: Effects of pH, ionic strength, and solutes on DNA adsorption by andosols. Biol Fert of Soils **46**, 531–535 (2010)
202. Peng, C., Huang, Q., Zhu, J., Jiang, D., Zhou, X., Rong, X., Liang, W.: Effects of low-molecular-weight organic ligands and phosphate on DNA adsorption by soil colloids and minerals. Colloid Surfaces B **54**, 53 (2007)
203. Sciascia, L., Liveri, M.L.T., Merli, M.: Kinetic and equilibrium studies for the adsorption of acid nucleic bases onto K10 montmorillonite. Appl. Clay Sci. **53**, 657–668 (2011)
204. Beall, G.W., Sowersby, D.S., Roberts, R.D., Robson, M.H., Lewis, L.K.: Analysis of oligonucleotide DNA binding and sedimentation properties of montmorillonite clay using ultraviolet light spectroscopy. Biomacromol **10**, 105–112 (2009)
205. Aquino, Adélia J.A., Tunega, Daniel, Georg Haberhauer, A., Gerzabek, Martin H., Lischka, Hans: Solvent Effects on Hydrogen Bonds A Theoretical Study. J. Phys. Chem. A **106**, 1862–1871 (2002)
206. Cai, P., Huang, Q., Zhang, X.: Microcalorimetric studies of the effects of $MgCl_2$ concentrations and pH on the adsorption of DNA on montmorillonite, kaolinite and goethite. Appl. Clay Sci. **32**, 147–152 (2006)
207. Mignon, P., Ugliengo, P., Sodupe, M.: Theoretical Study of the Adsorption of RNA/DNA Bases on the External Surfaces of Na^+-Montmorillonite. J. Phys. Chem. C **113**, 13741–13749 (2009)
208. Benetoli, L.O.D.B., Santana, H.D., Zaia, C.T.B.V., Zaia, D.A.: MnAdsorption of nucleic acid bases on clays: an investigation using Langmuir and Freundlich isotherms and FT-IR spectroscopy. Monatsh Chem - Chemical Monthly **139**, 753–761 (2008)
209. Cai, P., Huang, Q., Zhang, X.: Microcalorimetric studies of the effects of MgCl2 concentrations and pH on the adsorption of DNA on montmorillonite, kaolinite and goethite. Appli Clay Sci **32**, 147–152 (2006)
210. Saeki, K., Sakai, M., Ada, S.I.: DNA adsorption on synthetic and natural allophanes. Appl. Clay Sci. **50**, 493–497 (2010)
211. Franchi, M., Ferris, J.P., Gallori, E.: Cations as Mediators of the Adsorption of Nucleic Acids on Clay Surfaces in Prebiotic Environments. Origins Life Evol B **33**, 1–16 (2003)

Dewatering of Mineral Adsorbents

Yubiao Li

Abstract This chapter will comprehensively introduce the most common water types (free water, vicinal water, bound water, interstitial water, etc.) adsorbed by various mineral adsorbents such as montmorillonite. In addition, the most usually used dewatering strategies and the processes including thickening, filtration, centrifugation, incineration that can be applied to efficiently separate the liquid from the solid minerals will be demonstrated, achieving a low moisture content to meet the disposal or reuse purposes. Moreover, the dehydration mechanisms for each water type will also be illustrated. Finally, the most recently developed dewatering methods for other solid-liquid separation will also be introduced and explained as the promising potential means in dewatering mineral adsorbents.

Keywords Thermal treatment · Thickening · Filtering · Ultrasonic treatments

1 Water States and Types in Selected Mineral Adsorbents

Most adsorption processes occur by utilising as less as adsorbents to treat a large amount of contaminated solution, especially by using mineral adsorbents. The dewatering process, or called solid–liquid separation process, requires a high water removal efficiency to reduce the mineral solid volume for storage or reuse.

Mineral adsorbents can normally be used to remove both organic and inorganic or even radioactive materials from the aqueous. After adsorption, the solid mineral materials should be separated from the solution. Therefore, the water status with the solid minerals should be understood prior to the separation conducted. As most of

---

Y. Li (✉)
Hubei Key Laboratory of Mineral Resources Processing and Environment, Wuhan University of Technology, Luoshi Road 122, Wuhan 430070, Hubei, China
e-mail: Yubiao.Li@whut.edu.cn

Y. Li
School of Resources and Environmental Engineering, Wuhan University of Technology, Luoshi Road 122, Wuhan 430070, Hubei, China

S. Song and B. Li (eds.), *Adsorption at Natural Minerals/Water Interfaces*, Engineering Materials,
https://doi.org/10.1007/978-3-030-54451-5_8

the mineral adsorbents have big specific surface area that beneficial to the removal of contaminants, these surfaces would retain some water easily. In addition, the pores existing within the mineral adsorbents have high capacity in remaining water molecules, resulting in increased difficulty in separating the solid minerals from aqueous solution.

Among all the minerals, clay minerals are a big family that can be used to remove contaminants [1], due to its wide source, cheap price and high adsorption capacity. These properties generate a wide application of clay minerals in purifying pollutants. Clay minerals primarily consisting of illite, kaolinite and montmorilonite are normally present in a very fine status, presenting high ion-exchange and adsorption capabilities [2]. The easy adsorption of water molecules on the surface or interlayer space of clay minerals results in a hydration process, presenting a challenging in the dehydration of these water molecules.

Water associated with mineral solids can be generally divided into two types, i.e. free water and bound water. In a more detailed classification way, water can be classified into four types, i.e. free water, interstitial water, vicinal (or surface) water and hydration water (chemically bound) [3, 4]. The free water is not combined with the mineral solids and can be removed by a simple gravitational method such as sedimentation while the removal of interstitial water can be achieved by mechanical dewatering device including filters, centrifuges. The vicinal water is normally connecting with the mineral surface in terms of adsorption and adhesion by virtue of water molecular structure, which cannot be moved away via mechanical methods. The water of hydration on the solid mineral surface is chemically connected, therefore, can only be thermo-treated.

Peng et al. [5] applied a DFT study on the adsorption of water on Na^+-montmorillonite (001) basal and (010) edge surfaces. The results indicated that water molecular adsorbed on MMT-(001) surface was predominantly through electrostatic interaction between water and Na^+ cation while the adsorption on (010) edge was mainly via hydrogen bonding between water and –OH or $-OH_2$ groups. In addition, the adsorption energy of water on (010) was greater than that on (001) surface, indicating that water adsorption on (001) basal surface was favorable. Moreover, the partial density of states (PDOS) suggested that a portion of electron density was transferred from the MMT (001) and (010) surfaces to water molecules via the bonding between Na 3s–O_w 2p and H 1s–O 2p orbitals, respectively. This study demonstrate a new tool to investigate the water adsorption on mineral adsorbents, which will be beneficial to improving the efficiency of following dewatering process after using as adsorbent. Therefore, the removal of hydration bound water from the mineral surface limits the final water removal efficiency.

The water adsorption was also investigated onto the natural mineral dust particles such as aluminumsilicates including feldspars and clays (illite, kaolinite, montmorillonite) [6]. The surfaces of these dust minerals can be the seedbed for the adsorption of gas and water molecules as adsorbents. Goodman et al. [7] investigated the water adsorption on individual metal oxide surface, i.e. CaO, MgO, SiO_2 and established the relationship between water layers adsorbed and the relative humidity (RH). Other studies also revealed quantitative determination of water monolayer formation, e.g. first water monolayer was formed at approximately 12% RH while up to four water monolayer were formed when RH was around 80% [8].

2 Dewatering Methods and Processes

Removal water from mineral adsorbents with high water content is essential for their further processing due to decreased volume, the possibility to be reused, etc. A various methods can be applied to dewater the mineral adsorbents, including traditional sedimentation, filtering, thickening, centrifuging, vacuuming, thermal treatments and some other techniques that not used so widely but very efficient, such as eletrokinetic and ultrasonic treatments. The high efficient dewatering process applied is normally a combination of the above methods.

Approximately two decades ago, most of mineral processing plants applied a combination of vacuum filters and thermal dryers to dewater the mineral pulp. In developing to date, pressure filter has been widely used in separating water from the solid minerals [9]. Flocculant-assisted, gravity thickening and filtration techniques that currently applied in the industry are normally with low efficiency and require further improvements in processing rate and reducing costs. Although not many published works indicate the dewatering methods for mineral adsorbents, the understanding of the methods applied in the traditional mineral processing plants can be referred as the potential techniques used for dewatering mineral adsorbents, since both of them are the slurry with mineral solid particles and solution water. However, the selection of a suitable dewatering method used for one specific mineral adsorbent depends on various conditions, e.g. particle size, the water status with minerals. For instance, the sand bed and filter press technologies require large space, limiting their application in dewatering the mineral adsorbents at small scale. In contrast, centrifuging technology requires much less space and gives rise to a faster process [10], although the maintaining and operation costs might be higher.

To date, effective dewatering of mineral adsorbents challenges both the mineral processing and environmental related industries, as well as the manufacturers for dewatering equipment [11]. As noted above, water removal not only depends on the dewatering process but also on the water status associated with adsorbents [12]. For example, the dewatering of mineral adsorbent can refer to the water removal process in the mineral processing industry as they are both separating solid mineral from the liquid solution. Therefore, the dewatering methods applied in the mineral processing and water treatment plants are discussed below.

The predominantly cost-effective method applied in the dewatering system is sedimentation involving the setting of suspended particles under gravity or external forces including centrifugal force [13], due to its simple design and operation, high capacity and low operating and maintaining costs.

Sedimentation is efficient when difference is available between the liquid and solid. As the adsorption normally occurs to the contaminated water solution, the carrier liquid is therefore normally water. The bulk removal of water from the mineral solid is generally achieved by sedimentation, or thickening, producing a 55–65% solid by weight, followed by a filtration or thermal drying to give a much higher solid by weight, e.g. 95% [13].

The sedimentation rate of adsorbents is determined by Stokes' or Newton's laws, depending on the mineral solid particle size. If the mineral adsorbent particle is too fine, the settle would be very slow by gravity alone. In this case, centrifugal sedimentation can be applied. Alternatively, the adsorbents can be agglomerated or called flocculated into big size to increase the settling rates following coagulation and flocculation processes.

Coagulation and the addition of polymer flocculation with high molecular weight can be applied to increase the sedimentation rate of the mineral adsorbents, especially those of colloidally stable fine mineral particles. Coagulation screens the repulsive electrical double layer interactions and improve the approaching of mineral particles, leaving the van der Waals forces as being the predominant force between adsorbent particles. On the other hand, the addition of high molecular weight polymer flocculation increased the gravity settling due to the particle aggregation via the bridging mechanisms [14].

It has also been reported that the divalent ions of Ca^{2+} and Mg^{2+} improved the dewatering process of some minerals [15] while Nielsen and Keiding [16] found that Fe^{3+} played a more significant role in enhancing the dewatering process due to the apparently more greater settling rates. In addition to cations added, polyelectrolytes are another category for efficient flocculating due to the formed network to adsorb mineral particles, resulting in increased sedimentation rates [17].

Compared to the traditional thickeners, a pelletizing/thickening system where inorganic coagulant was applied to mix and neutralize the surface charge of solid particles, reduces the solid moisture and retention period significantly when using a belt press filter afterwards [18]. Moreover, thickeners can be combined with other dewatering techniques, e.g. Gálvez et al. [19] applied both hydrocyclones and thickeners to dewater in a copper concentration plant.

Different from sedimentation, filtering method involves the separation of solid from water through a porous medium keeping the solid but passing the treated water. Filtration can be applied in a small or big scale, depending on the processing scale of the mineral adsorbents. For instance, if the mineral adsorbents are applied in the dewatering of high concentration salts of flotation solution, the amount of used mineral adsorbents would be very high. Then filtration would follow thickening to optimise the whole dewatering process.

However, the presence of slime normally plays a negative role in filtration due to the block effects to the filter media. The addition of flocculants diminishes this effect due to the increased space between particles, resulting in an easier filtration. During filtering process, both pressure and vacuum can be used as the driving force, including drums, discs and belt filters. Li et al. [20] reported that vacuum dehydration can be used to remove water from the metakonlin based materials and inhibit micro cracks formation and sample shrinkage. Pressure is normally beneficial to dewatering process, but high pressure is not always a good choice to dewater mineral adsorbents. Some results demonstrated that a threshold pressure existed. In other words, the dewatering process can be improved by increasing pressure until a point after which no significant improvement could be obtained [21, 22]. In addition, the filter with magnetic pre-treatment would be a choice for dehydration to the mineral adsorbents with magnetic properties [23].

In addition to the sedimentation and filtering methods, centrifugal technique has also been applied to separate water from solid matters including activated sludge [24]. The centrifugation rotational speed is normally crucial for efficient dewatering while the centrifuging time apparently affects the dewatering results [10]. Moreover, eletrokinetic dewatering method has been successfully applied to increase the water removal content from very fine solid at micrometer scale or even smaller via electro-osmosis. For instance, Raats et al. [25] applied an electric potential of 30 V to increase the solid content by up to 24%, as compared to the conventional dewatering process. Although the above two methods have not been widely applied in the industry, they present a potential in dewatering of mineral adsorbents. Therefore, further improvements in manufacturing suitable equipment should be paid more attention.

Ultrasonication has been widely applied in treating the fine sludge [26]. The moisture of sludge is normally greater than 90%, which is significantly greater than the mineral solids. The dewatering process for bio-sludge containing a huge amount of water attracts much attention from researcher doing work in various research areas. Therefore, the applied technology for dewatering sludge would be also possible to remove water from mineral adsorbents. Bien et al. [27] investigated a positive polyelectrolyte and ultrasound to treat mineral sludge and found that the CST was reduced from 390 to 20 s, significantly increased the settling rates. In addition, the final water content was decreased from 87 to 78%. In addition, Bien and Wolny [28] found that the ultrasonic field increased the removal of water significantly, but a critical ultrasonic power level exists. Ruiz-Hernando et al. [12] treated sewage sludge using ultrasonic successfully. As ultrasonic technology can be used to pre-treat fine biosludge to reduce the water percentage [29], it is expected to be attempted in the dewatering process of mineral adsorbents.

In addition, thermal treatment can eventually enhance dewaterability effectively of mineral adsorbents especially for those with high percentage of chemically bound water. Thermal dewatering at low temperature requires longer processing time than that at higher temperature. This technique reduces the water moisture to a content as low as 5%. Rotary thermal dryers composing of a relatively long cylindrical shell are often applied. Heating within the equipment is to increase the temperature to remove water from the mineral adsorbents, leaving a relatively dry solid. However, this method requires high energy consumption, limiting its wide application in industry.

3 Dewatering Mechanisms

The normal dewatering technologies to separate water from the solid mineral adsorbents include thickening, filtering, centrifuging, vacuuming and pressuring, thermal treatments [19], similar to the dewatering process for the pulp in the mineral processing department. To a common sense, effective separation of the mineral adsorbents from the treated solution requires high sedimentation rate, clear supernatant and compact consolidation of the mineral adsorbents if gravity is applied, including gravity-assisted thickening. Although few studies focus on the dewatering mechanisms of mineral adsorbents, the previous research on the water removal from the solid minerals can be referred. The understanding of dewatering mechanisms of the mineral slurry would be beneficial to enhancing the dewatering process of the mineral adsorbents.

The dewaterability of mineral adsorbents depends on various attractive and repulsive forces between mineral particles. The control of the interaction between minerals may be achieved by adjusting the interfacial chemistry, mineral dispersion, particle interaction, including solution pH and ionic strength [11, 30]. Minerals with high adsorption capacity should be with high specific surface area, porous properties, active sites on the surface, etc. Specifically, the removal of pollutants by clay minerals are highly likely due to their large cation exchange capacity and high surface area, for instance, Li et al. [1] successfully applied montmorillonite in treating an important antihistamine drug diphenhydramine (DPH) from the contaminated wastewater. The specific structure and swelling properties of some clay minerals give rise to the present intractable dewatering issues due to a large amount of water retained within the mineral adsorbents.

As many different minerals belong to clay mineral family, their different properties would result in various dewatering effects, e.g., the dewaterability of kaolinite ($Al_2Si_2O_5(OH)_4$) as a non-swelling clay while smectite, a swelling clay would not be the same [11]. The kaolinite basal face compromises an inert tetrahedral siloxane and hydrated octahedral alumina structures having permanent negative charges due to the isomorphous substitution. While, smectites are layered aluminosilicates with a 2:1 dioctahedral structure, readily taking up both the water and organic molecules [31]. Their two thirds of the octahedral sites are occupied by trivalent cations and normally Na^+ or Ca^{2+} are present within the interlayer. Within this family montmorillonite is one predominant compound having TOT layers, i.e. two silica tetrahedral sheets sandwiched one alumina octahedral sheet (O). Si^{4+} in the tetrahedral siloxane sheets may be isomorphously substituted by Al^{3+} while the substitution of Al^{3+} by other cations such as Mg^{2+} normally happens to the octahedral sheet, leading to a negative charge compensated by the interlayer exchangeable cations. The interlayer cations normally hydrated with the interlayer water molecular, expanding the crystalline layers. According to these mechanisms, montmorillonite related adsorbents have been used to remove pollutants from the contaminated solution.

Actually, there are two swelling mechanisms for the smectites, depending on the basal spacing between two sheets. For instance, the crystalline swelling is normally due to the adsorption of water molecules on the basal crystal surfaces. When polyvalent cations are within the interlayer, the expansion is normally smaller than 10 Å, since the repulsive force of the hydration of the adsorbed cations is compromised by the electrostatic attraction between cations and silicate layers. When monovalent cations are located within the interlayer, osmotic swelling can occur, resulting in an expansion up to 30–40 Å or higher [11, 30].

In contrast, the edge surfaces of the clay minerals having reactive silanol and aluminol species are reported to adsorb or interact with H^+ or OH^- in aqueous solution, depending on the solution pH. The pH dependent properties of the edge surfaces contributes to the variation of electrokinetic zeta potential, flocculant adsorption, dispersion yield stress and colloid stability, thereby influencing the dewatering process [25].

Therefore, the dewatering of clay minerals would be challenging due to their high gelation or swelling and space-filling structures, due to low settling rates and poor supernatant clarity [11]. Therefore, better understanding of the hydration and dehydration of the interlayer cations in the layered aluminosilicates is crucial to the downstream dewatering process after using as an adsorbent [32, 33], especially in the nuclear management due to their good plasticity, low permeability and high capacity in adsorption [34].

McFarlane et al. [30] discussed the influence of interfacial chemistry and shear on the dewatering and interaction between clay mineral particles. The results show that the modification of pulp chemistry through adding Ca^{2+} and Mn^{2+} resulted in a dramatically reduced zeta potential of mineral particles, complete suppression of swelling, collapse of honeycomb network structure and an accompanying decrease in shear yield stress. Although the addition of coagulants and flocculants improved the settling rates, the final sediment solid content did not enhance significantly. Further shear application to the pre-sedimented pulp increased the consolidation by 5–7%. Furthermore, Addai-Mensah [11] comprehensively investigated the slurry pH, hydrolysable metal ion, temperature, polymeric flocculant structure type and shear on the interfacial chemistry, flocculation, particle interaction and the dewatering effects of both kaolinite and smectite. Therefore, the swelling degree in the smectite clay can be reduced by increasing the ionic strength of the supernate by multivalent cations.

Studies show that several factors influence the water amount contained in the interlayer of montmorillonite, e.g. the nature of the intercalated cations and their hydration energies, the layer charge and the chemical potential [35]. Although many published work have reported the application of empirical force field methods to measure the swelling behavior [36–39], few studies reveal the fundamental mechanisms for water removal from the interlayer.

When filtering is applied for dewatering adsorbents, the flowing of the liquid can be regarded through the capillary tubes existed in the solid cakes. The qualification of liquid flow in porous media follows the Darcy's equation:

$$\frac{dV}{dt} = K\frac{A\Delta P}{\eta L} \tag{1}$$

where V is the filtrate volume at time t while ΔP is the difference in pressure across the media, L is the thickness of the filter cake, A is the area of the cross-section of the porous material, η is viscosity of the filtrate while K is the permeability constant [40, 41]. This indicates that both high pressure and dewatering areas are beneficial to water removal. In addition, low cake thickness and filtrate viscosity can improve filtration efficiency.

The separation of water from the solid mineral adsorbents can be calculated based on Eq. 2:

$$\frac{dV}{dt} = \frac{\pi r^4 \Delta P}{8\eta L} \tag{2}$$

where r is the radius of the capillaries [42].

The above equations indicate that the filtration is linear to processing time, which is however not always correct in practice. A parabolic performance is normally observed during the filtration experiment, highly likely due to the modified pore size within the filter cake. The changed pore size will therefore result in changed pressure. As pressure given in Eqs. 1 and 2 is regarded as an integer, but actually it should be a distribution and a range of integers. The differential pressure needed for dewatering process can be calculated using Eq. 3 which stands for the capillary pressure across the media.

$$\Delta P = \frac{2\gamma \cos\theta}{r} \tag{3}$$

where γ is the surface tension of the filtrate, $\cos\theta$ is the contact angle of the mineral sample, r is the capillary radius among particles. As the moisture within the capillary is pretty hard to be removed, the applied pressure should be greater than the pressure calculated according to Eq. 3. Once applying pressure, the filtrate will flow fast until an irreducible saturation is achieved when the moisture is kept in a constant value. The value of this saturation depends on various parameters such as the pressure, filter media, cake thickness, particle size and shape, viscosity.

It should be noted that water molecules tend to be easily adsorbed in the interlayer of clay minerals, which is a hydration process. The removal of water from the interlayer of clay minerals therefore should overcome some forces. In order to better understand how to remove the adsorbed water from adsorbents including clay minerals, the status of water molecular should be investigated firstly.

Normally, thermal dehydration mechanisms of minerals can be investigated by thermogravimetric analysis (TGA) and differential thermal analysis (DTA) or differential scanning calorimetry (DSC). From a more theoretical point of view, Fonseca et al. [35] applied DFT to investigate the dehydration process of interlayer water molecules from Na^+-montmorillonite with different hydration degrees.

The results indicated that dewatering one water layer to fully dehydration was easier than that from two to one water layer, with a temperature of 47 °C rather than 125 °C, consistent with the experimental results, demonstrating a more fundamental strategy to predict the dehydration process under desired conditions.

The thermal dehydration process of the swelling clays involves the removal of interlayer water which depends on the temperature. The dehydration process provides useful information regarding the interlayer configuration as the dehydration is related to the composition of the counterbalancing cations within the interlayer and the tetrahedral sheet. Koster and Guggenheim [43–45] reported that the dehydration of montmorillonite was divided into two stages, i.e. the dehydration of voluminous but weakly bounded water from the outer hydration shell of the interlayer cations and the water strongly bonded to the inner hydration shell, with the dehydration being determined by the interlayer cation chemistry. In other words, the dehydration temperatures may be controlled by the radius, charge and hydration energy of the cations, although other factors may exist. Bray and Redfern [31] investigated the dehydration of Ca-montmorillonite and found that the rate-determining step of the first stage is due to the hindered migration of water molecules from the Ca^{2+} outer hydration spheres. In addition, the second dehydration stage was diffusion controlled, determining the whole dehydration process.

The ultrasound is a relatively new available approach to dewater the mineral adsorbents. Some published work indicated that low frequency ultrasound can put the fine particles together, thereby easing the dewatering process within a short time. Ultrasonic pretreatment of the slurry with both solution and solid minerals can migrate water via the natural channels within the solid or the channels created by ultrasonic wave propagation. In addition, the acoustic streaming, local heating, interface instabilities, agitation, etc., because ultrasonic may also be in favour of separating solid and liquid [46].

In addition, ultrasonication can normally disrupt the microbial cells by the ultrasonic waves, therefore, the entrapped water within the pores of the mineral adsorbents can also be extracted by ultrasonication due to its cavitation effects [47]. The cavitation induces gas-bubble in the liquid phase, with the bubble being growing to a critical size and violently collapse. This process gives rise to a high hydroshear strength, high temperature (5000 K) and pressure (500 atm) at a local sites on the mineral surface. In addition, free radicals with high activities can also be generated during the ultrasonication process. These formed free radicals can accelerate chemical reactions significantly. Therefore, the application of ultrasonication in dewatering mineral adsorbents is highly due to the cavitation effects.

Combination of ultrasound and other methods can agglomerate the mineral adsorbent particles together and enhance the dewatering process, giving a more dense solid products with less water contents.

References

1. Li, Z., Chang, P.-H., Jiang, W.-T., Jean, J.-S., Hong, H., Liao, L.: Removal of diphenhydramine from water by swelling clay minerals. J. Colloid Interf. Sci. **360**, 227–232 (2011)
2. Clay Minerals, S.: Clay Clay Minerals. Pergamon Press, Long Island City (1968)
3. Vesilind, P.A., Hsu, C.-C.: Limits of sludge dewaterability. Water Sci. Technol. **36**, 87–91 (1997)
4. Kopp, J., Dichtl, N.: Influence of the free water content on the dewaterability of sewage sludges. Water Sci. Technol. **44**, 177–183 (2001)
5. Peng, C., Min, F., Liu, L., Chen, J.: A periodic dft study of adsorption of water on sodium-montmorillonite (001) basal and (010) edge surface. Appl. Surf. Sci. **387**, 308–316 (2016)
6. Joshi, N., Romanias, M.N., Riffault, V., Thevenet, F.: Investigating water adsorption onto natural mineral dust particles: linking drifts experiments and bet theory. Aeol. Res. **27**, 35–45 (2017)
7. Goodman, A., Bernard, E., Grassian, V.: Spectroscopic study of nitric acid and water adsorption on oxide particles: enhanced nitric acid uptake kinetics in the presence of adsorbed water. J. Phys. Chem. A **105**, 6443–6457 (2001)
8. Gustafsson, R., Orlov, A., Badger, C., Griffiths, P., Cox, R., Lambert, R.: A comprehensive evaluation of water uptake on atmospherically relevant mineral surfaces: drift spectroscopy, thermogravimetric analysis and aerosol growth measurements. Atmos. Chem. Phys. **5**, 3415–3421 (2005)
9. Townsend, I.: Automatic pressure filtration in mining and metallurgy. Miner. Eng. **16**, 165–173 (2003)
10. Semiyaga, S., Okure, M.A.E., Niwagaba, C.B., Nyenje, P.M., Kansiime, F.: Optimisation of centrifuge operating conditions for dewatering physically conditioned faecal sludge from urban slums. Environ. Technol. Inno. **8**, 28–39 (2017)
11. Addai-Mensah, J.: Enhanced flocculation and dewatering of clay mineral dispersions. Powder Technol. **179**, 73–78 (2007)
12. Ruiz-Hernando, M., Martinez-Elorza, G., Labanda, J., Llorens, J.: Dewaterability of sewage sludge by ultrasonic, thermal and chemical treatments. Chem. Eng. J. **230**, 102–110 (2013)
13. Wills, B.A., Napier-Munn, T.: Wills' Mineral Processing Technology: An Introduction to the Practical Aspects of Ore Treatment and Mineral Recovery, 7th edn. Elsevier Science & Technology (2006)
14. Mpofu, P., Addai-Mensah, J., Ralston, J.: Influence of hydrolyzable metal ions on the interfacial chemistry, particle interactions, and dewatering behavior of kaolinite dispersions. J. Colloid Interf. Sci. **261**, 349–359 (2003)
15. Sobeck, D.C., Higgins, M.J.: Examination of three theories for mechanisms of cation-induced bioflocculation. Water Res. **36**, 527–538 (2002)
16. Nielsen, P.H., Keiding, K.: Disintegration of activated sludge flocs in presence of sulfide. Water Res. **32**, 313–320 (1998)
17. Biggs, S., Habgood, M., Jameson, G.J., Yan, Y-d: Aggregate structures formed via a bridging flocculation mechanism. Chem. Eng. J. **80**, 13–22 (2000)
18. Watanabe, Y., Tanaka, K.: Innovative sludge handling through pelletization/thickening1in memorium. Yasuhiko watanabe and dr. Tanaka wish to respectfully dedicate this article in the late dr. Richard s. Engelbrecht's memory, and in honor of his many contributions to the profession. Water Res. **33**, 3245–3252 (1999)
19. Gálvez, E.D., Cruz, R., Robles, P.A., Cisternas, L.A.: Optimization of dewatering systems for mineral processing. Miner. Eng. **63**, 110–117 (2014)
20. Li, W., Lemougna, P.N., Wang, K., He, Y., Tong, Z., Cui, X.: Effect of vacuum dehydration on gel structure and properties of metakaolin-based geopolymers. Ceram. Int. **43**, 14340–14346 (2017)

21. Zhao, Y.Q., Bache, D.H.: Integrated effects of applied pressure, time, and polymer doses on alum sludge dewatering behaviour. Waste Manag. **22**, 813–819 (2002)
22. Wu, R.M., Lee, D.J., Wang, C.H., Chen, J.P., Tan, R.B.H.: Novel cake characteristics of waste-activated sludge. Water Res. **35**, 1358–1362 (2001)
23. Watson, J.L., Li, Z.: Application of magnetic forces to disk vacuum filtration in the laboratory and plant. Miner. Eng. **12**, 1253–1262 (1999)
24. Chu, C.P., Lee, D.J.: Experimental analysis of centrifugal dewatering process of polyelectrolyte flocculated waste activated sludge. Water Res. **35**, 2377–2384 (2001)
25. Raats, M.H.M., van Diemen, A.J.G., Lavèn, J., Stein, H.N.: Full scale electrokinetic dewatering of waste sludge. Colloids Surf. A Physicochem. Eng. Asp. **210**, 231–241 (2002)
26. Kim, T.-H., Kim, S., Kim, S.-W., Kim, J., Park, C., Lee, M., Lee, J.: Effects of various pretreatments for enhanced anaerobic digestion with waste activated sludge. J. Biosci. Bioeng. **95**, 271–275 (2003)
27. Bien, J.B., Kempa, E.S., Bien, J.D.: Influence of ultrasonic field on structure and parameters of sewage sludge for dewatering process. Water Sci. Technol. **36**, 287–291 (1997)
28. Bien, J., Wolny, L.: Changes of some sewage sludge parameters prepared with an ultrasonic field. Water Sci. Technol. **36**, 101–106 (1997)
29. Yin, X., Han, P., Lu, X., Wang, Y.: A review on the dewaterability of bio-sludge and ultrasound pretreatment. Ultrason. Sonochem. **11**, 337–348 (2004)
30. McFarlane, A., Bremmell, K., Addai-Mensah, J.: Improved dewatering behavior of clay minerals dispersions via interfacial chemistry and particle interactions optimization. J. Colloid Interf. Sci. **293**, 116–127 (2006)
31. Bray, H.J., Redfern, S.A.T.: Kinetics of dehydration of ca-montmorillonite. Phys. Chem. Miner. **26**, 591–600 (1999)
32. Landais, P.: Advances in geochemical research for the underground disposal of high-level, long-lived radioactive waste in a clay formation. J. Geochem. Explor. **88**, 32–36 (2006)
33. Ohkubo, T., Saito, K., Kanehashi, K., Ikeda, Y.: A study on hydration behaviors of interlayer cations in montmorillonite by solid state nmr. Sci. Technol. Adv. Mat. **5**, 693–696 (2004)
34. Bartelt-Hunt, S.L., Smith, J.A., Burns, S.E., Rabideau, A.J.: Evaluation of granular activated carbon, shale, and two organoclays for use as sorptive amendments in clay landfill liners. J. Geotech. Geoenviron. **131**, 848–856 (2005)
35. Fonseca, C.G., Vaiss, V.S., Wypych, F., Diniz, R., Leitão, A.A.: Structural and thermodynamic investigation of the hydration-dehydration process of Na^+-montmorillonite using DFT calculations. Appl. Clay Sci. **143**, 212–219 (2017)
36. Tambach, T.J., Hensen, E.J.M.E., Smit, B.: Molecular simulations of swelling clay minerals. J. Phys. Chem. B **108**, 7586–7596 (2004)
37. Zheng, Y., Zaoui, A., Shahrour, I.: A theoretical study of swelling and shrinking of hydrated wyoming montmorillonite. Appl. Clay Sci. **51**, 177–181 (2011)
38. Zheng, Y., Zaoui, A.: How water and counterions diffuse into the hydrated montmorillonite. Solid State Ionics **203**, 80–85 (2011)
39. Marry, V., Turq, P., Cartailler, T., Levesque, D.: Microscopic simulation of structure and dynamics of water and counterions in a monohydrated montmorillonite. J. Chem. Phys. **117**, 3454–3463 (2002)
40. Yoon, R-H, Asmatulu, R., Yildirim, I., Jansen, W., Zhang, J., Atkinson, B., Havens, J.: Development of dewatering aids for minerals and coal fines. Minerals & Coal Technologies (US) (2004)
41. Lampinen, P., Brodin, P.: Novel dewatering aids for mineral concentrates and coal
42. Svarovsky, L.: Solid-Liquid Separation. Butterworth-heinemann (2000)
43. Koster, V.G., Guggenheim, S.: Dehydration of a ca-and a mg-exchanged montmorillonite (swy-1) at elevated pressures. American Mineralogist (USA) 72 (1987)
44. Koster, V.G., Guggenheim, S.: The effect of pressure on the dehydration reaction of interlayer water in na-montmorillonite (swy-1). Am. Miner. **69**, 872–879 (1984)
45. Koster, V.G., Guggenheim, S.: Dehydration of k-exchanged montmorillonite at elevated temperatures and pressures. Clay Clay Miner. **34**, 281–286 (1986)

46. Riera-Franco de Sarabia, E., Gallego-Juárez, J.A., Rodríguez-Corral, G., Elvira-Segura, L., González-Gómez, I.: Application of high-power ultrasound to enhance fluid/solid particle separation processes. Ultrasonics **38**, 642–646 (2000)
47. Tiehm, A., Nickel, K., Zellhorn, M., Neis, U.: Ultrasonic waste activated sludge disintegration for improving anaerobic stabilization. Water Res. **35**, 2003–2009 (2001)

The manufacturer's authorised representative in the EU is Springer Nature Customer Service Centre GmbH, Europaplatz 3, 69115 Heidelberg, Germany. If you have any concerns regarding our products, please contact ProductSafety@springernature.com

Printed and bound by CPI Group (UK) Ltd, Croydon, CR0 4YY

30/09/2026

02225288-0006